# Health Reform Policy to Practice

Oregon's Path to a Sustainable Health System

*A Study in Innovation*

# Health Reform Policy to Practice

## Oregon's Path to a Sustainable Health System
*A Study in Innovation*

Edited by

**Ron Stock, MD**
Oregon Health Authority Transformation Center; Family Medicine, Oregon Health & Science University, Portland, OR, United States

**Bruce Goldberg, MD**
OHSU-PSU School of Public Health; Oregon Rural Practice Based Research Network, Oregon Health & Science University, Portland, OR, United States

Academic Press is an imprint of Elsevier
125 London Wall, London EC2Y 5AS, United Kingdom
525 B Street, Suite 1800, San Diego, CA 92101-4495, United States
50 Hampshire Street, 5th Floor, Cambridge, MA 02139, United States
The Boulevard, Langford Lane, Kidlington, Oxford OX5 1GB, United Kingdom



**Notices**
Knowledge and best practice in this field are constantly changing. As new research and experience broaden our understanding, changes in research methods, professional practices, or medical treatment may become necessary.

Practitioners and researchers must always rely on their own experience and knowledge in evaluating and using any information, methods, compounds, or experiments described herein. In using such information or methods they should be mindful of their own safety and the safety of others, including parties for whom they have a professional responsibility.

To the fullest extent of the law, neither the Publisher nor the authors, contributors, or editors, assume any liability for any injury and/or damage to persons or property as a matter of products liability, negligence or otherwise, or from any use or operation of any methods, products, instructions, or ideas contained in the material herein.

**British Library Cataloguing-in-Publication Data**
A catalogue record for this book is available from the British Library

**Library of Congress Cataloging-in-Publication Data**
A catalog record for this book is available from the Library of Congress

ISBN: 978-0-12-809827-1

For Information on all Academic Press publications
visit our website at https://www.elsevier.com/books-and-journals

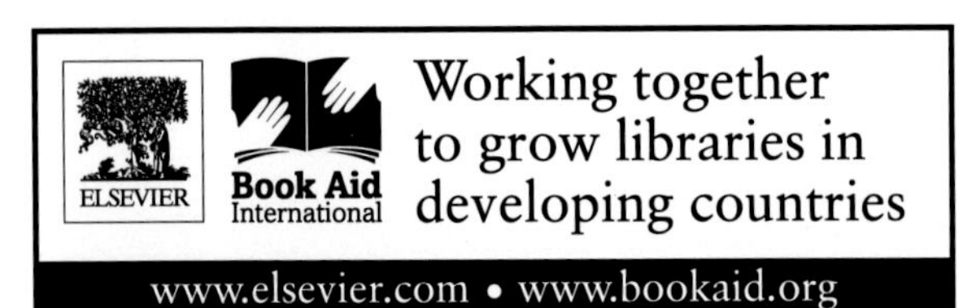


*Publisher:* Mica Haley
*Acquisition Editor:* Erin Hill-Parks
*Editorial Project Manager:* Tracy Tufaga
*Production Project Manager:* Lucía Pérez
*Designer:* Christian Bilbow

Typeset by MPS Limited, Chennai, India

# Contents

## Section III
## **Future Implications for State and National Health Reform**

# List of Contributors

**Kelly Ballas** Independent Financial Consultant, Portland, OR, United States

**Mike Bonetto** Tenfold Health, Bend, OR, United States

**Deborah J. Cohen** Oregon Health & Science University, Portland, OR, United States

**Emilee Coulter-Thompson** Oregon Health Authority Transformation Center, Portland, OR, United States; University of Michigan, Ann Arbor, MI, United States

**Lori Coyner** Oregon Health Authority, Portland, OR, United States

**Bob Dannenhoffer** Douglas Public Health Network, Roseburg, OR, United States

**Senator T. Daschle** The Bipartisan Policy Center, Washington, DC, United States

**Chris DeMars** Oregon Health Authority Transformation Center, Portland, OR, United States

**Kristen Dillon** Family Physician, Hood River, OR, United States; PacificSource Columbia Gorge CCO, Hood River, OR, United States

**Tina Edlund** Health Management Associates, Portland, OR, United States

**Kevin Ewanchyna** Samaritan Health Plans, Corvallis, OR, United States

**Bruce Goldberg** Oregon Health & Science University-Portland State University School of Public Health, Portland, OR, United States

**Merwyn (Mitch) Greenlick** Oregon Health & Science University, Portland, OR, United States

**Jennifer D. Hall** Oregon Health & Science University, Portland, OR, United States

**K. John McConnell** Oregon Health & Science University, Portland, OR, United States

**Leann R. Johnson** Oregon Health Authority, Portland, OR, United States

**Cathy Kaufmann** Health Management Associates, Portland, OR, United States

**Governor John A. Kitzhaber** Independent Consultant, Writer and Speaker on Health Policy, Politics, and Strategy, Portland, OR, United States

**Christopher F. Koller** Milbank Memorial Fund, New York, NY, United States

**Nicole Merrithew** CareOregon, Portland, OR, United States

**Piper Nieters Su** McDermott + Consulting, Washington, DC, United States

**Susan Otter** Oregon Health Authority, Portland, OR, United States

**Liz Powers** Winding Waters Medical Clinic, Enterprise, OR, United States

**Rebecca Ramsey** CareOregon, Portland, OR, United States

**Daniel A. Reece** Oregon Health Authority Transformation Center, Portland, OR, United States

**Stephanie Renfro** Oregon Health & Science University, Portland, OR, United States

**Jennifer Richter** Yamhill Community Care Organization, McMinnville, OR, United States

**Evan Saulino** Oregon Health Authority, Portland, OR, United States; Providence Health and Services, Portland, OR, United States

**Eli Schwarz** Oregon Health & Science University, Portland, OR, United States

**Jeanene Smith** Health Management Associates, Portland, OR, United States

**Ron Stock** Oregon Health Authority Transformation Center, Portland, OR, United States; Family Medicine, Oregon Health & Science University, Portland, OR, United States

**Somava S. Stout** Institute for Healthcare Improvement, Cambridge, MA, United States; Cambridge Health Alliance, Cambridge, MA, United States; Harvard Medical School Center for Primary Care, Boston, MA, United States

**Latricia Tillman** Multnomah County Health Department, Portland, OR, United States; Oregon Health Authority, Portland, OR, United States

**Patty Wentz** Strategies 360, Portland, OR, United States

# Acknowledgments

As detailed throughout this book, Oregon's efforts to improve its health care system were made possible by the work and input from thousands upon thousands of people—policy makers, clinicians, administrators, consumers, community health advocates, and so many more. This was both deliberate and organized, as well as unplanned and fortuitous. And it was the combination of all that work and civic engagement that resulted in what is documented in this book. Oregon's health reform was, and is, the work of a great many and not an influential few. As such, we would like to acknowledge all those that participated in some manner and those that continue to move this important work forward.

We would also like to extend a sincere thanks to all those who contributed chapters to this edition. Their work has helped tell a story we can all learn from.

Finally, we would like to acknowledge Vern Katz, MD. His encouragement and assistance, along with support from the editors and staff at Elsevier, made this project possible.

**Ron Stock and Bruce Goldberg**
*Portland, OR, United States*

# Introduction

*Reforming the healthcare system is hard, REALLY HARD.*

—Oregon Physician

A broken and expensive health care system cannot be fixed by doing more, or less, of the same. For more than two decades, Oregonians have participated in a dialogue to create an innovative approach for providing high-quality, affordable, and effective health care while simultaneously activating their citizens to participate and improve their own health. Historically, state health care programs have pursued financial sustainability by cutting people from care, reducing provider rates, or slashing services. In 2011, Oregon embarked on a "Fourth Path," redesigning its clinical delivery system rather than cutting services in response to financial uncertainties. The intent of the new system is to reduce waste, improve individual health and preventive care outcomes, reduce the utilization of services, create local and statewide fiscal accountability, and align financial incentives.

Oregon is "on board" with the national "triple aim" movement of pursuing better quality and better experience of care at a lower cost. After 5 years, this path has led to remarkable improvements in the care provided to Medicaid recipients—more primary care visits, lower emergency room use, reduced hospital admissions for several chronic conditions because of improved outpatient care, reduction in readmissions to the hospital, more children and adults getting preventive care, and screening for social determinants of health. At the same time, Oregon has reduced Medicaid per capita spending by 2%, saving the state and federal government hundreds of millions of dollars and keeping cost growth to a more sustainable rate. However, this is not only an Oregon story, but a national one, as other states, payers, and purchasers are implementing innovative health care reforms in a political landscape fraught with uncertainty.

The following chapters are a testament that the will, knowledge, and skills of a committed group of clinicians, health system leaders, policymakers, and legislators can make a difference to improve the health of their fellow citizens. This is not a rigorous academic textbook, but rather a collection of essays from those that were actively engaged in leading or implementing changes in care delivery across the state. Many of the authors are former and current content experts that experienced the development of the coordinated care model from the grassroots, many were instrumental in authoring the health and public policies, and creating the operational plans that have led to its implementation. We also included national experts to provide additional context and perspective.

It was our intent to create a manuscript that would tell the Oregon health reform story, demonstrating how a comprehensive statewide health care reform effort is complex, rooted in the culture and experiences of its citizens, requiring multiple implementation "levers" to be pulled simultaneously, and requiring strong leadership to implement and sustain reforms. We describe the key components to developing a state- and community-focused health policy, the influence of the national Affordable Care Act (ACA) legislation, and the implications that may impact other states, purchasers, and payers. We attempt to illustrate how an implementation strategy is challenged by a fluid, rapidly changing health care environment that requires coordination among a diverse group of stakeholders, using tactics, and cross-cutting strategies that constantly require evaluation and improvement. However health reform efforts do not happen spontaneously, it involves a health policy approach at the state and local level that aligns and not only supports the delivery of care in practice but also creates an environment in our communities that addresses those factors that contribute to the holistic health of its population. And maybe most important, it requires individuals with the knowledge and skills to work together, creating those relationships that lead to a common vision. For some, this meant giving up some personal, business, and professional interests for the common "good" of their neighbors and community.

But the real success stories are those of the people, families, and communities impacted by health reform. For example, Malik, an 8-year-old boy, had been going to the emergency room once or twice a month because of asthma attacks. He had great doctors and was on all the right medicines, but he was still wheezing. His mom finally found a coordinated program whereby a community health worker visited their home, identified household triggers for his attacks, and provided a hypoallergenic vacuum cleaner. The worker also provided and taught Malik how to use a home nebulizer to convert his medication into a quicker-acting mist, and provided a portable miniinhaler to help him overcome wheezing attacks at school or on the playground. Malik stopped needing hospitalizations, he was healthier, and the cost of his care was dramatically reduced. This is a great example of how changing care delivery can improve lives and reduce costs, and one that invites the question as to why he did not get this help sooner. The answer is multifactorial and rooted in a fragmented health system that focuses and rewards the quantity of visits and not improvement in health, and that at the time had no mechanism to reimburse for the services of the community health worker that ultimately led to better health.

This story, like the stories in the following chapters, shares lessons that hopefully other states, regions, and health systems can learn from that will lead to an acceleration of broader health reform efforts. We asked our authors to "tell it like it is," giving them the editorial freedom to share with the reader their perspective and to reflect on their successes, challenges, and failures. The book is divided into three sections. The first section, "Origins of Oregon Health Reform," describes the history of Oregon health reform over two decades, how policy was shaped by community grassroots engagement, requiring strong leadership at the highest levels, and designed to create the Coordinated Care Model. In the second section, "Implementation of the Coordinated Care Model: Key Components," chapters discuss key components that were implemented, describing not only what was done but also the lessons that informed future work. Finally, in the third section, "Future Implications for State and National Health Reform," we asked key state and national leaders to reflect on the work done in Oregon and discuss the implications of this work more broadly. This book aims to evaluate where we have come after nearly a decade of reform, how we got there, and what is next. Health delivery reform has been, and will continue be, an iterative process, the work is far from complete.

The path to health reform in Oregon in the past decade was born out of a "perfect storm" of too many Oregonians without access to health care services; an inconsistent, and sometimes unknown, quality of care; and a national and state financial crisis. But Oregon had leadership with a deep knowledge of the issues, and the skills and political will to address them. It is not that we do not know what works or the multitude of solutions that would move to positive changes in the health care system. We have decades of regional and national experimentation and demonstration projects, many of them occurring in Oregon practices and health systems. However, we struggle to scale innovations that succeed in one setting across practices and communities. In this book, you will learn that many of the innovative changes were those that have been known and demonstrated effective for a long time, and that given a supportive environment, these ideas can be scaled to larger populations. Health delivery reform has also been a lesson in listening. In creating a true learning environment that was capable of making the iterative changes along the path, we needed to listen to the national health reform dialog, for example, ACOs, alternative payment demo projects, quality improvement, and dissemination methodology successes. But in order to implement these into the practices and communities in Oregon, we also needed to listen to the needs of our constituents and to help them, and us, understand how to make this relevant in their community context.

We realized that knowledge of what needs to be done, without the skills to apply and sustain change, would be insufficient to accomplish our goals. This book demonstrates the deep understanding of public health care funding, and the relationships with federal partners, needed to develop effective state health policy. It also describes the skills needed to implement clinical delivery change, such as the quality improvement methodology and the tools to bring teams together to plan, do, study, and act; and skills to facilitate dialog within and between organizations to develop a culture of peer-to-peer interaction and disseminate key learnings.

Finally, health care change of this magnitude requires an environment in which there is the *will* to change, to listen, to create, and to innovate despite forces that would otherwise deter. The path that Oregon chose was supported by a long-standing culture of community-oriented approaches to change driven by progressive, pragmatic, independent thinkers. The will to change started at both the top and the bottom, from the Governor's office to the community grassroots. This cannot be underestimated in terms of its influence on the changes that have occurred over the past 5 years. Oregon, as do all states, has its "warts," naysayers, and challenges, yet in this book you will hear from authors and contributors who are passionate, persistently committed, and at "the point of the spear" of health care change.

We invite you to read this book, a tale of one state's journey to reform health care delivery and to improve the "health" of its citizens. As illustrated in the example of Malik, an 8-year-old boy with asthma, there are real human consequences of the reform efforts. Oregon is certainly not at the end of its health reform journey, although it feels as though great progress has been made. This book examines how we got there from the perspectives of those involved in the listening, collaborations, and hard work of the health policy and care delivery changes that have been required.

**Ron Stock** and **Bruce Goldberg**
*Portland, OR, United States*

Section I

# Origins of Oregon Health Reform

Chapter 1

# The Oregon Narrative: History of Health Reform in Oregon

**Mike Bonetto**
*Tenfold Health, Bend, OR, United States*

Oregon has a long history of using bold innovations to reform health care delivery and finance. While these efforts have taken a few forms, certain goals have remained constant. One aim is to provide medical insurance to as many Oregonians as possible. A second is to improve the quality and efficiency of the health care provided. And a third is to involve community members and stakeholders in drawing up reforms. State efforts to develop comprehensive health reform policy began in the 1980s. These reforms first put Oregon on the health care map. The state's newest developments, Coordinated Care Organizations (CCOs), have maintained Oregon's reputation as an important laboratory for fiscally and socially responsible health care.

## 1980s—STATEWIDE DIALOG, MOBILIZATION, AND ACTION

### Hard Choices and the Death of a Boy

In the early 1980s, Oregon began analyzing options to cover the uninsured and medically vulnerable in the state. Oregon Health Decisions, a nonprofit organization, formed the Citizens Health Care Parliament in 1983 to convene community forums across the state. The group issued policy papers focused on possible solutions (Brown, 1991). In 1987, the legislature passed a bill that refused to fund certain transplants under Medicaid: Because the cost of such procedures bought only marginal improvements in health to a small number of people, it was felt that Medicaid money could be better spent elsewhere. For instance, proponents noted that in 1 year, 1500 women on Medicaid could be provided with prenatal and delivery care for the same dollars that would cover projected 34 transplants to Medicaid beneficiaries (Thorne, 1990). The legislature opted to use limited resources to meet the needs of the many as opposed to the needs of the few.

**Health Reform Policy to Practice. DOI: http://dx.doi.org/10.1016/B978-0-12-809827-1.00001-X**

Soon after this decision was adopted, Oregon landed in the national spotlight. A 7-year-old boy named Coby Howard contracted acute lymphocytic leukemia and needed a bone marrow transplant. His mother's request that Medicaid pay for the transplant was denied and he soon died, causing an outpouring of outrage. "By the time he died, Coby Howard had become one of the country's most emotional and visible symbols of the debate over state decisions to limit Medicaid funding for transplants," wrote the *Los Angeles Times* (Japenga, 1987).

## Goal Setting and the Oregon Health Plan

The national attention led Governor Neil Goldschmidt to appoint a workgroup in 1987 comprised of health care providers, insurers, businesses, consumers, labor representatives, and lawmakers that focused on three basic questions: (1) who should be covered, (2) what should be covered, and (3) how should the coverage be financed and delivered. The workgroup agreed that (Oregon Department of Human Services, Office of Medical Assistance Programs, 2006):

- All citizens should have access to a basic level of health care.
- Society is responsible for financing the care of poor people.
- The process must be based on criteria that are publicly debated, reflects a consensus of social values, and considers the good of society as a whole.
- The health care delivery system must encourage use of services and procedures that are effective and appropriate, and discourage overtreatment.
- Health care is one important factor affecting health; funding for health care must be balanced with other programs that also affect health.
- Funding must be explicit and economically sustainable.
- There must be clear accountability for allocating resources and for the human consequences of funding decisions.

Under the leadership of state senate president John Kitzhaber, MD, the legislature passed groundbreaking bipartisan legislation in 1989 that was collectively referred to as the Oregon Health Plan (OHP). Kitzhaber was the chief author of the legislation and is known as the intellectual father of Oregon's reform efforts, and often seen as an once-in-a-lifetime leader in health policy. His early work as an emergency room physician and state legislator provided him great insight into society's policy dilemma. He regularly treated uninsured patients who had been denied Medicaid coverage due to state budget cuts. Yet many of those patients were being seen in the emergency department with costly debilitating illnesses that could have been prevented with appropriate early intervention. This drove him to eliminate the inequities of insurance coverage while using a more transparent and rationale approach to establish the state's Medicaid budget.

A central tenet of the OHP was the responsibility of both the private and public sectors to, in partnership, achieve near-universal coverage of more than 450,000 uninsured Oregonians. The critical features were that: (1) all persons with incomes below the federal poverty level would be eligible for Medicaid, (2) the Medicaid benefit package would consist of a prioritized list of diagnoses and treatments, (3) the legislature would draw a line on the list below which treatments would not be covered, (4) the legislature would not be allowed to reduce reimbursement rates to Medicaid providers, (5) Medicaid services would be provided through managed-care plans, (6) employers would be required to insure their employees, with the prioritized list as a basic benefit package, and (7) a high-risk pool would be created to serve those who were denied health insurance due to preexisting conditions (Bodenheimer, 1997).

## Oregon's Controversial Prioritized List of Services

In 1989, the legislature created an 11-member body to develop a prioritized list of services, an action *The New York Times* later labeled a "brave medical experiment" (*New York Times*, 1990). The Oregon Health Services Commission was tasked to define a systematic method for balancing medical outcomes and the social values attached to those outcomes against costs of the treatments—a process that was intended to result in all Oregon residents below the poverty line being covered, but for fewer services. The process was extremely open with nearly 50 community meetings, 11 public hearings, and a telephone survey to learn the values and preferences of Oregonians about prioritization of medical treatments (Brown, 1991). "The Oregon approach to rationing, which simultaneously drew on public preferences and cost–benefit analyses, thus represented an unusual marriage of health services research and deliberative democracy," noted a review in the *Canadian Medical Association Journal* (Oberlander, Marmor, & Jacobs, 2001). The commission eventually established 17 categories of health conditions—ranging from potentially fatal, but treatable, acute conditions to others such as maternity and newborn services. All diagnoses and treatments were assigned to one of the categories and then ranked based on specific criteria, including life expectancy, quality of life, cost-effectiveness, and the number of individuals who might benefit from each treatment.

The Health Services Commission's first list was completed in 1990 using an ineffective mathematical formula that included a "net benefit value"—it highlighted the expected outcomes of given treatments for hundreds of health conditions. The formula produced a list of 1600 condition/treatment pairs. The list was immediately dismissed as extremely flawed. For instance, crooked teeth received a higher ranking than early treatment for Hodgkin's disease and dealing with thumb sucking scored better than hospitalizing a child for starvation (Fox and Leichter, 1991).

After significant revisions, the commission released a new list in 1991 that had been reduced to over 700 health services and treatment pairs and relied more heavily on public values and clinical judgment. The revised list was seen as more defensible, both medically and politically (Table 1.1).

**TABLE 1.1 Example of 1995 Prioritized List**

| |
|---|
| *The five top items* |
| Line 1. Diagnosis: severe or moderate head injury, hematoma or edema with loss of consciousness. Treatment: medical and surgical treatment. |
| Line 2. Diagnosis: insulin-dependent diabetes mellitus. Treatment: medical therapy. |
| Line 3. Diagnosis: peritonitis. Treatment: medical and surgical treatment. |
| Line 4. Diagnosis: acute glomerulonephritis, with lesion of rapidly progressive glomerulonephritis. Treatment: medical therapy, including dialysis. |
| Line 5. Diagnosis: pneumothorax and hemothorax. Treatment: tube thoracostomy or thoracotomy, medical therapy. |
| *The five bottom items* |
| Line 741. Diagnosis: mental disorders with no effective treatments. Treatment: evaluation. |
| Line 742. Diagnosis: tubal dysfunction and other causes of infertility. Treatment: in vitro fertilization, gamete intrafallopian transfer. |
| Line 743. Diagnosis: hepatorenal syndrome. Treatment: medical therapy. |
| Line 744. Diagnosis: spastic dysphonia. Treatment: medical therapy. |
| Line 745. Diagnosis: disorders of refraction and accommodation. Treatment: radial keratotomy. |
| *Six items near the 1997 cutoff line* |
| Line 576. Diagnosis: internal derangement of the knee and ligamentous disruptions of the knee, grade III or IV. Treatment: repair, medical therapy. |
| Line 577. Diagnosis: keratoconjunctivitis sicca, not specified as Sjögren's syndrome. Treatment: punctal occlusion, tarsorrhaphy. |
| Line 578. Diagnosis: noncervical warts, including condyloma acuminatum and venereal warts. Treatment: medical therapy. |
| Line 579. Diagnosis: anal fistula. Treatment: fistulectomy. |
| Line 580. Diagnosis: relaxed anal sphincter. Treatment: medical and surgical treatment. |
| Line 581. Diagnosis: dental conditions (e.g., broken appliances). Treatment: repairs |

Source: Adapted from Oregon Health Plan Administrative Rules [Bodenheimer, T. 1997. The Oregon Health Plan – Lessons for the Nation. First of Two Parts. *The New England Journal of Medicine* 337(9): 651–655].

## Other Kinds of Insurance Expansion

### *Employer Mandate*

The 1989 legislation also included provisions for extending insurance to working people. A mandate, with an effective date of 1993, was based on a traditional play-or-pay model and would have required employers with 50 or more employees who worked at least 17.5 hours to either provide health insurance or pay into a special state insurance fund that would offer coverage for their employees and dependents. However, as with all employer mandates, a federal exemption from the Employee Retirement Income Security Act (ERISA) was needed for its implementation. With no such approval, the employer mandate never went into effect and sunset in 1996.

### *The Oregon Medical Insurance Pool*

The Oregon Medical Insurance Pool (OMIP) was an attempt to cover another group of uninsured Oregonians. It was established in 1987 and first began offering health insurance policies in June 1990. As the state's high-risk health insurance pool, it provided insurance to Oregonians denied coverage because of preexisting health conditions as well as those who lost group coverage and did not have portability coverage. OMIP was overseen by a nine-member board made up of five representatives of the health insurance and reinsurance communities, a medical provider, two consumer advocates and the Oregon Insurance Commissioner.

To keep OMIP premiums generally affordable for the enrollees, the high-risk pool was structured to distribute claim and administrative costs between the enrolled members and the private insurance market. OMIP covered approximately 15,000 individuals on an annual basis. OMIP member premiums covered a little over 50% of the total program costs with the remaining costs (losses) recovered through assessments of Oregon health insurance and reinsurance companies based on people enrolled in Oregon. The need for OMIP ended with the implementation of the Patient Protection and Affordable Care Act in 2014.

### *Small Employer Health Insurance*

In 1987, the legislature also established the Insurance Pool Governing Board (IPGB) to develop and offer health insurance plans to employers with fewer than 25 employees. To be eligible, an employer could not have provided health insurance for a period of 1 year prior to enrollment of their employees. The plans were not subject to state benefit mandates, initially included a declining state tax credit (nonrefundable and phased out in 1993) and were considered to have a lower cost than the regular market. It was overseen by a seven-member board comprised of representatives from the business and consumer advocate communities.

## 1990s—OHP IMPLEMENTATION—CHALLENGES AND BENEFITS

### Explicit Versus Implicit Rationing

Oregon's plan was to take the prioritized list and draw a line above which services would be covered and those below which would not. Not surprisingly, Oregon's decision to cover only a specific number of diagnoses and treatments within the Medicaid population ignited a national debate on the rationing of health care. The state's attempt to explicitly determine which services should be covered based on value and available resources was contrary to existing state and federal policies. At the time, it was perceived as pursuing contradictory goals—guaranteeing access to health care for all while also containing health care costs. However, Oregon leaders assumed that by having a more limited benefit package than traditional Medicaid, all Oregonians with incomes below 100% of the federal poverty level could be covered.

Senator Al Gore (TN) was one of the most vocal opponents to Oregon's plan, along with the Children's Defense Fund, Families USA, the American Academy of Pediatrics, the National Association of Community Health Centers, and the National Association of Children's Hospitals. These groups believed that Oregon's proposed system of rationing would hurt the most vulnerable groups in the state, mainly its poorest women and children. They argued that drawing a line to indicate acceptable services would force the state to only treat patients whose needs happen to be sufficiently cost-effective (Gore, 1990). Others would be left to pay their own way—or suffer the consequences of their illnesses. This would, in essence, create a two-tiered system with even more inequity (Leichter, 1999). Critics also noted that rationing was unnecessary since there was more to be done by eliminating administrative waste and striking better deals with drug companies and providers. In addition, since Oregon ranked 46th in Medicaid spending as a percentage of total state spending, opponents believed that, before rationing care, the state should first invest more money the way other states did (Gore, 1990).

Proponents of Oregon's plan saw things differently. They observed rationing happening every day—although it was done implicitly. Hundreds of thousands of Oregonians were uninsured and had no access to health care. In fact, every state already limited the range of services provided to Medicaid patients and denied payment for all services to low-income individuals who were ineligible for Medicaid. Proponents argued that in a world of finite resources, including Medicaid budgets, it is morally irresponsible not to ration marginally beneficial services (Strosberg, 1992). Supporters highlighted that there were already three tiers of coverage—with the uninsured in the bottom tier.

## Oregon Health Plan Goes Live

It is important to note that the Medicaid program is a joint state-federal program and any proposal to modify or eliminate services, change eligibility, impact beneficiary choice or change provider reimbursement requires federal approval. Thus, for Oregon to implement its new prioritized list, it would first have to receive approval from the Health Care Finance Administration (HCFA). This approval would be through a waiver under Section 1115 of the Social Security Act, which allows a state to modify its Medicaid program.

The early days of attempting to gain federal approval, Oregon's request was met with barriers and skeptics. Oregon sent its first Medicaid waiver application to HCFA under the George H. W. Bush administration in 1991. Representative Henry Waxman (CA), the then chair of the Health and Environment Subcommittee of the House Energy and Commerce Committee, and Senator Gore were the two most outspoken opponents of Oregon's waiver application. Senator Gore was quoted at the time as saying that Oregon is "trying to play God by playing with spreadsheets" (Brown, 1991). The combined influence of Gore and Waxman led to a denial in 1992 that cited possible violations of the Americans with Disabilities Act.

The Oregon Health Services Commission then further refined the prioritized list to better reflect the probability of death or disability with and without treatment. In addition, Oregon's senators Bob Packwood and Ron Wyden were instrumental in gaining approval from the Clinton administration. In the spring of 1993, HCFA approved Oregon's 1115 waiver to implement its new plan. In 1994, Oregon expanded Medicaid to include individuals under 100% of the federal poverty level, which provided a health care package via the prioritized list to 120,000 Oregonians in its first year. Due to this expansion, Oregon's uninsurance rate fell from 18% in 1994 to 10% in 1998—an impressive achievement.

While the prioritized list provided some level of budget certainty, it did not fully pay for the Medicaid expansion. To help fund the early years of the OHP, the legislature in 1993 also increased state general fund expenditures and passed a ten-cent tobacco tax. The OHP was initially funded through line 606 of 745 possible treatments. As originally planned, the line was to move based on available funding. In 1995, the legislature funded to line 581 and to line 578 in 1997. In 1999, the legislature approved funding to line to 564—however, HCFA never approved the change.

## Managed Care and Cost Containment

Managed care was crucial to keeping costs in line under the OHP. From the outset, the success of the OHP was contingent on enrolling beneficiaries into managed care plans to better manage and coordinate care, thus reducing expenses. Oregon has a long history with managed care dating back to the 1940s with the establishment of Kaiser Permanente. Over 85% of OHP enrollees were initially enrolled in one of 13 managed care plans. The initial plans were nearly all large commercial plans. This was short-lived as many commercial managed care plans did not have expertise or experience managing a more vulnerable and at-risk Medicaid population.

As the commercial plans exited, local independent practice associations (IPAs) mobilized to become the backbone of Oregon's Medicaid delivery system. Most IPAs were formed during the early days of managed care in the late 1980s and early 1990s. IPAs allowed independent physician practices to come together and jointly contract with health insurers looking to partner with provider groups willing to take financial risk and manage care. This led to more local control in many areas and the development of community partnerships essential to managed care. The Oregon legislature eventually carved out a special place for IPA-like entities by allowing them to contract with the state to cover the Medicaid population without having to meet all the financial risk-based capital requirements of traditional insurers. The new organizations were called Fully Capitated Health Plans.

## The Family Health Insurance Assistance Program

The Family Health Insurance Assistance Program (FHIAP) was established in 1997 after Oregon's employer mandate did not receive the necessary federal waivers. FHIAP provided health insurance premium assistance to Oregonians who were not eligible for Medicaid and had incomes below 170% of the federal poverty level. FHIAP offered monthly premium subsidies to adults on a sliding scale, ranging from 50% to 95% of the cost of insurance. All children under the age of 19 were subsidized at 100% regardless of family income. The adult members' monthly subsidies decreased as their income increased.

FHIAP offered members numerous health insurance plan choices offered by a variety of carriers throughout the state. The plans offered comprehensive medical benefits including prescription drug coverage, reasonable out-of-pocket costs, and a wide array of provider choice dependent on the carrier chosen. FHIAP contracted with five large domestic individual private market carriers in Oregon to provide plans, many of which were available statewide.

From its inception, FHIAP enrollment was limited by the amount of state and federal funds available. FHIAP provided subsidies to approximately 15,000 individuals annually and was generally closed to new enrollees and at times there were 40,000 to 50,000 people on a waiting list.

## 2000s—POLICY REFINEMENT AND THE QUEST FOR SUSTAINABILITY

### Creation of OHP2

After John Kitzhaber's tenure as senate president, he was elected governor in 1994 and reelected in 1998 for two terms that ran from 1995 to 2003. During this time, he focused on expanding and preserving the framework of the OHP. In 2001, through the leadership of Governor Kitzhaber, the legislature again looked to expand Medicaid coverage by shrinking the benefit package. This time the goal was to cover an additional 46,000 Oregonians by expanding Oregon's Medicaid and FHIAP populations to 185% of the federal poverty level. The plan separated the OHP into two parts, OHP Plus and OHP Standard. OHP Plus covered categorically eligible Medicaid populations (such as pregnant women and children) and its benefit package would remain tied to the prioritized list. OHP Standard covered single adults, couples, and parents not eligible for Medicaid under federal regulations, but with a reduced package of services estimated at 78% of OHP Plus's value. The OHP Standard benefit eliminated coverage for hearing, vision, some dental services, durable medical equipment, and nonemergency transportation (McConnell & Wallace, 2004). In addition, OHP Standard included copayments and increased monthly premiums.

Since launching the OHP, Oregon had become increasingly frustrated by its inability to move the line on the prioritized list due to federal pushback. The Centers of Medicare and Medicaid Services (CMS) (formerly HCFA) had sharply constrained Oregon's ability to use the list as a cost control instrument (Oberlander, 2006). Thus, the legislature looked to cap OHP Standard enrollment based on available funds. CMS approved Oregon's new plan in 2002 and OHP2 went into effect in February 2003.

### The Surprising Impact of Cost-Sharing and Premium Requirements

The policy changes to OHP Standard at the time were seen as pragmatic and fair. The OHP's premise was for everyone to have something rather than for some to have everything while others received nothing. OHP Standard reflected a bipartisan solution of covering more individuals through increased copays and premiums more closely mirroring the commercial insurance market. In addition, stricter administrative regulations were put in place that eliminated previous premium exemptions for several vulnerable populations (i.e., the homeless and victims of domestic violence) and required a 6-month lockout for people who missed a single monthly payment.

The amount of cost-sharing was determined on a sliding scale based on income. Premiums cost $6–20 per month. Copays ranged from $5 for an outpatient physician visit, $50 for an emergency department visit, to $250 for an inpatient hospital admission (Table 1.2).

It was unknown at the time how such cost-sharing and administrative provisions would impact enrollment. However, it soon became evident that the new requirements drastically reduced the number of people served. After the implementation of OHP2, Oregon's Medicaid expansion population (OHP Standard) plummeted 53% in the first year—from 104,000 to 49,000. Enrollment fell again over the next 18 months by another 50%, leaving only 24,000 covered by OHP Standard in 2005.

The precipitous drop in enrollment was due to two significant factors. First, the increased cost-sharing and decreased benefit package of OHP Standard made it less attractive and affordable for many eligible individuals. Retrospective analysis showed that those who left OHP Standard were disproportionately the most economically vulnerable (Wright et al., 2005). Forty-four percent identified increased cost-sharing as the main reason they

**TABLE 1.2 The 2003 Cost-Sharing Requirements for OHP Standard and Plus**

| Plus | Standard |
|---|---|
| **Premiums** | |
| Unchanged | Increased to $6–20/person depending on income (unchanged for single people) |
| No premiums | Family premium discounts no longer applied |
| | Premium exemption for enrollees with zero income eliminated |
| | Missing payment resulted in loss of eligibility for 6 months |
| **Copays** | |
| Unchanged | Nonvoluntary, enrollee can be denied service for not paying a copay |
| Voluntary only, enrollee cannot be denied service for not paying a copay | $250 inpatient visit |
| $3 outpatient visit | $50 emergency department visit |
| $2 generic, $3 brand-name drugs | $5 outpatient visit |
| | $3 for laboratory test or X-ray |
| | $5 physical/speech/occupational therapy |
| | $2 generic, $15 brand-name drugs |

Note: All copay changes were in effect only from March 2003 to June 2004.
Source: Wright, B.J., et al., 2010. Raising premiums and other costs for Oregon Health Plan enrollees drove many to drop out, *Health Affairs 29*(12): 2311–2316.

left, and those who cited cost-sharing as their main reason were very poor, with incomes between 0% and 25% of poverty (Wright et al., 2005). Not surprisingly, those who left had a more difficult time accessing care and were more likely to have medical debt. These enrollment changes highlighted a real lack of understanding by policymakers of the price sensitivity in low-income populations (Wright et al., 2005).

Second, Oregon's economy struggled during 2001–03 and the legislature responded by passing an over $500 million tax increase that was soon put out to a statewide vote in 2004, known as Ballot Measure 30. Measure 30 ultimately failed and OHP Standard endured $40 million in cuts, resulting in the elimination of outpatient mental health and chemical dependency services.

## Expanded Access to OHP Standard

In 2005 the legislature made some modest changes to protect those with lower incomes. OHP beneficiaries with family income less than 10% of the federal poverty level were now exempt from paying premiums. For those with incomes above 10%, a grace period of up to 6 months was enacted to help maintain their enrollment while they paid any overdue premiums. The legislature also eliminated a 6-month disqualification period for failure to pay premiums.

OHP Standard enrollment continued to hover around roughly 20,000 members due to limited funding. By 2008, Oregon was positioned to increase funding for OHP Standard and expand coverage. There was significant demand for OHP Standard coverage as a waiting list of nearly 90,000 Oregonians grew in preparation for the expansion. Instead of using a first-come, first-serve process, Oregon initiated an innovative lottery drawing for those on the waiting list (Baicker et al., 2013). This process was selected because it gave everyone on the waiting list an equal opportunity to be selected. Those selected had the opportunity to apply for coverage and enroll if they were eligible.

Out of approximately 30,000 selected from the list, 18,000 then submitted applications and 9000 were approved for coverage. This approach also gave researchers a unique opportunity to investigate the impact of coverage through a randomized control trial. It was shown that both participation in the waiting list and take-up of the program were higher for individuals who were somewhat older and sicker than the overall Medicaid population (Allen et al., 2010).

## Covering More Children: The Healthy Kids Initiative

Drawing on the original goals of the OHP, the legislature approved the Healthy Kids Program in 2009, covering over 100,000 children through both

Medicaid and private insurance. The objective was to offer health coverage to all children and fund it through a 1% tax on commercial insurance. The program successfully lowered Oregon's children's uninsurance rate from over 11% to 5%.

Those uninsured children who did not qualify for Medicaid coverage (family income between 201% and 300% of the federal poverty level) could receive a premium subsidy to purchase health insurance from insurance carriers contracted by the state. Uninsured children whose family income was above 300% of the federal poverty level could enroll in the program by paying the full premium cost. Unlike the OHP lottery results, it was found that the children who applied to or enrolled in Healthy Kids were not in particularly poor health (Wright et al., 2011). However, it was also found that nearly a third of enrolled children had not received routine health care in the previous year (Wright et al., 2011).

## 2010–12—THE PERFECT STORM FOR DELIVERY SYSTEM REFORM

### The Evolution of a Burning Platform

The 2007–08 financial crisis, which led to the Great Recession from 2008 to 2010, significantly impacted Oregon with huge job losses and contributed to a spike in Oregon's unemployment rate—it jumped from 5.3% in 2007 to 10.6% in 2010 (Oregon Office of Economic Analysis, 2010). The downturn in employment dramatically decreased state revenues. As a result, state budgets across the nation were drastically cut, prompting Congress to pass the Federal American Recovery and Reinvestment Act (ARRA) of 2009. ARRA provided states with temporary financial support, and for Oregon, resulted in nearly $900 million of *one-time* revenue for Medicaid and other health-related programs during the 2009–11 biennium (Oregon Legislative Fiscal Office, 2011).

### A Gifted Leader Takes Charge

The year 2010 was a gubernatorial election year and former governor John Kitzhaber reemerged as the Democratic nominee and won a third term as governor in the November general election. Due to the economic recession and the elimination of the one-time ARRA funding, the state was facing a $3.5 billion shortfall, with over $1 billion of the shortfall focused on Medicaid and human services. Once Governor Kitzhaber took office and was responsible for creating a balanced budget, the looming budget deficit allowed him to change the narrative of resource allocation and value-based investments across multiple sectors. Health care became a primary focal point.

Kitzhaber had long argued that there was an urgent need to resolve how health care was financed and delivered—to *everyone* in the state. The governor was a gifted leader on health care and used this opportunity to garner widespread support by highlighting how rising health care costs negatively impacted corporate margins, reduced the capacity of firms to grow, compromised the state's competitiveness in the global economy, slowed the rate of job growth, suppressed wage increases for existing workers, fostered labor disputes and lost productivity, and caused companies to send US jobs off shore.

Kitzhaber believed that the health care system should help produce healthy citizens, not simply finance and deliver health care. He saw health care as a means to an end, not an end in itself. This led to Kitzhaber's adoption of the Triple Aim in health care: (1) to improve the health of a defined population; (2) to reduce per capita cost; and (3) to improve the patient experience (in terms of clinical outcomes, patient safety, and patient satisfaction). The Triple Aim was developed by Donald Berwick, MD, and the Institute for Healthcare Improvement.

Kitzhaber was also masterful in reframing the debate on how to fundamentally transform Medicaid. In a profile, *The Nation* wrote that Kitzhaber "has earned a national reputation for thinking holistically and eliminating compartmentalized policy and funding silos that ought to be dealt with as parts of a continuum" (Abramsky, 2013). Using his experience as a legislator and governor, he cited the three traditional approaches that states take to save dollars on Medicaid: reduce provider payments; reduce the number of people covered; and/or reduce covered benefits. However, it was clear that these approaches had proven unsuccessful in reducing the actual cost of care and had squelched investments in health improvement that could lower future costs. Kitzhaber articulated a fourth way: Rather than reducing spending in an inefficient system, change the system for better efficiency, value, and health outcomes. Among other actions, this meant retaining the prioritized list of services.

## Immediate Budget Implications

Governor Kitzhaber came into office after the passage of President Obama's Patient Protection and Affordable Care Act (ACA). While Kitzhaber acknowledged that the ACA would go a long way toward closing the coverage gap, he was concerned that it would do little to stem the escalating cost of care and that states would need to lead by example.

Although the OHP, with its emphasis on managed care, had served as an innovative model, it still had flaws that were inherent to other Medicaid programs. The state contracted with nearly 40 managed care organizations to provide physical, mental, or oral health care. This created duplication and inefficiency. For example, an OHP beneficiary could be enrolled in multiple managed care plans at the same time—yet have very little, if any, care coordination or measureable outcome of health improvement. In addition, there was little accountability across the managed care organizations and no real incentive to control costs. A fixed monthly payment to the managed care plans was largely based on the number of encounters (office visits, hospitalizations, etc.) that the plan's beneficiaries had, so more encounters actually generated a higher monthly payment. If utilization was controlled through aggressive care management, managed care plans would see a *decrease* in their monthly capitated payment.

These perverse incentives prevented much innovation from changing the way health care was financed and delivered. Kitzhaber emphasized that those same perverse incentives were responsible for the 30% of health care spending in the United States that was ineffective, duplicative, or wasteful (Oregon Health Policy Board, 2012). The Governor argued that Oregon should focus on capturing a small percentage of that 30% through better managed care and realigned financial incentives.

The Governor proposed a budget in early 2011 that assumed cost savings of $239 million in Medicaid through *health care transformation*. Those savings were calculated based on best practice interventions by better integrating physical, behavioral, and oral health.

## Early Political Discussions and Negotiations

For nearly 2 months in early 2011, a group of 46 stakeholders from across the state (with bipartisan participation) led by the Governor and the Oregon Health Authority met weekly to review policy options in constructing a straw-model. Their work produced House Bill 3650—which provided a framework for a new delivery system—CCOs. CCOs were intended to create community-based organizations governed by health care providers, community members, and organizations responsible for financial risk to create a more coordinated and affordable, patient-centered health care delivery system (Oregon's 1115 waiver request, http://www.oregon.gov/oha/HPA/HP-Medicaid-1115-Waiver/Pages/index.aspx). The essential elements of this model included:

- integration and coordination of benefits and services;
- local accountability for health and resource allocation;
- standards for safe and effective care; and
- a global Medicaid budget tied to a sustainable rate of growth.

## Oregon-Style Collaboration

Ultimately, House Bill 3650 was met with strong bipartisan support and passed the House chamber, equally divided between the parties, with a vote of 59-1. House Bill 3650 simply established a high-level framework for CCOs; the legislature required a CCO implementation plan during the 2012 legislative session for final approval. As a result, four statewide workgroups comprising 133 people met from August through November 2011 to flush out details on how CCOs would function, specifically around governance, accountability metrics, and a global budget.

In addition, a series of eight community meetings were held around the state that solicited feedback from more than 1200 people. Monthly public comment was also taken at all Oregon Health Policy Board meetings. This activity and input were overseen by the Oregon Health Policy Board, which put together a CCO implementation proposal for the 2012 legislature.

Key highlights of the proposal included (Oregon Health Policy Board, 2012):

- *Governance*: A majority on the CCO governing boards should be comprised of representatives from entities that share financial risk. Also included are representatives from the major components of the health care delivery system. In addition, CCOs should be responsible for convening community advisory councils (CAC) to assure a community perspective. A member of the CAC should also serve on the CCO governing board.
- *Global budgets*: The budgets should include services that are currently provided under managed care in addition to Medicaid programs and services that have been provided outside of the managed care system, with the exception of psychiatric drugs and Medicaid-funded long-term care. This inclusive approach should enable CCOs to fully integrate and coordinate services and achieve economies of scale and scope. A goal is to allow CCOs maximum flexibility to dedicate resources toward the most efficient forms of care.
- *Accountability*: CCOs performances should be transparent via contractual outcomes and quality measures. These function both as an assurance that CCOs are providing quality care for all of their members and as an incentive to encourage CCOs to transform care delivery in alignment with the direction of HB 3650. Accountability measures and performance expectations for CCOs are to be introduced in phases to allow CCOs to develop the necessary measurement infrastructure and enable OHA to incorporate CCO data into performance standards.

The implementation proposal also highlighted the application process for prospective CCOs through a noncompetitive Request for Applications (RFA) much like the process developed by the federal government for Medicare Advantage plans. The RFA would describe the criteria that organizations must meet to be certified as a CCO, including relevant Medicare plan requirements. The request for applications was to be open to all communities in Oregon and not limited to certain geographic areas.

The implementation proposal was delivered to the legislature in January 2012 and the legislature approved it in February 2012 by passing Senate Bill 1580, again with wide bipartisan support.

## CMS Waiver Approval and Funding

Federal approval was required to make the necessary changes to launch the CCOs. A significant component of the approval included an initial upfront investment from CMS that was based on a financial analysis demonstrating that Oregon's total Medicaid spending would be dramatically reduced under the CCO model. These savings would accrue to both Oregon and federal budgets—thus allowing Oregon to negotiate an upfront investment that would pay for itself due to a lower trajectory of cost growth over time. The initial analysis showed that Oregon's model could save approximately 5 billion dollars in state and federal Medicaid funds over 10 years. This would be accomplished by lowering its annual per capita cost growth by 2 percentage points—from 5.4% to 3.4%. Oregon eventually negotiated an upfront investment from CMS of $1.9 billion over 5 years to help CCOs transition to new models of care—with the agreement that Oregon's Medicaid annual per capita costs will never grow no more than 3.4%.

CMS approved Oregon's 5-year waiver request to transition to CCOs, which included upfront funding in May 2012. Eight CCOs went live in August 2012, five in September 2012, two in November 2012, and one in January 2013. By the end of 2012, CCOs enrolled approximately 600,000 Oregon Medicaid members (90% of the Medicaid population).

## CONCLUSION

When states try to get Medicaid costs under control, they often reduce provider payments and/or lower the number of people covered. Oregon, under the inspired leadership of John Kitzhaber, first as senate president and then as governor, followed a different path. In Oregon's health care reform, an overarching goal was to expand insurance coverage to people who would ordinarily not qualify for Medicaid. In 1989, legislation collectively known as the OHP was developed to cover all Oregon residents below the poverty line, but for fewer services. The innovative mechanism for this rationing was a prioritized list of services. Although some congressmen and health associations objected, the strategy was vindicated when the uninsurance rate in the state fell from 18% in 1994 to 10% in 1998. By 2010, budget constraints resulting from the Great Recession, along with rising costs, led to a major modification of the OHP. Kitzhaber, now governor, proposed a new focus. Rather than simply trying to reduce spending in an inefficient system, Oregon, he said, should work for changes that would increase its efficiency and ability to promote the health of its citizens. Out of these discussions, the framework for CCOs was developed in 2011. These organizations went live in 2012 and are community-based organizations governed by health care providers, community members, and organizations responsible for financial risk. CCOs are responsible for (1) integrating and coordinating benefits and services; (2) local health and resource allocation; (3) providing safe and effective care; and (4) managing a global Medicaid budget tied to a sustainable rate of growth.

The following chapters detail how various aspects of CCOs, such as primary care, global budgets, and community engagement, are operating within Oregon. How can health policy analysts gauge the success of this overall program? The simplest tack is to look at the Triple Aim. Have CCOs improved the health of Oregonians? Have they reduced per capita costs? Have they bettered the patient experience, in terms of quality and satisfaction? These are questions this book will consider.

## REFERENCES

Abramsky, S. (2013, May 22). John Kitzhaber's Oregon dream. *The Nation*, June 10–17. Available from: https://www.thenation.com/article/john-kitzhabers-oregon-dream/

Allen, H., et al. (2010). What the Oregon Health Study can tell us about expanding Medicaid. *Health Affairs (Millwood)*, *29*(8), 1498–1506.

Baicker, K., et al. (2013). The Oregon experiment—Effects of medicaid on clinical outcomes. *The New England Journal of Medicine*, *368*(18), 1713–1722.

Bodenheimer, T. (1997). The Oregon Health Plan—Lessons for the nation. First of two parts. *The New England Journal of Medicine*, *337*(9), 651–655.

Brown, L. D. (1991). The national politics of Oregon's rationing plan. *Health Affairs*, *10*(2), 28–51.

Fox, D., & Leichter, H. (1991). Rationing care in Oregon: The new accountability. *Health Affairs*, *10*(2), 7–27.

Gore, A., Jr (1990). National Policy Perspectives. Oregon's bold mistake. *Academic Medicine*, *65(II)*, 634–635. Available from: http://journals.lww.com/academicmedicine/Abstract/1990/10000/National_policy_prospectives__Oregon_s_bold.7.aspx

Japenga, A. (1987, Dec 28). A transplant for Coby: Oregon boy's death stirs debate over state decision not to pay for high-risk treatments. *Los Angeles Times*. Available from: http://articles.latimes.com/1987-12-28/news/vw-21384_1_marrow-transplant

Leichter, H. M. (1999). Oregon's bold experiment: Whatever happened to rationing. *Journal of Health Politics, Policy and Law*, *24*(1), 147–160.

McConnell, J., & Wallace, N. (2004). Impact of premium changes in the Oregon Health Plan. The Office for Oregon Health Policy and Research.

Oberlander, J. (2006). Health reform interrupted: The unraveling of the Oregon Health Plan. *Health Affairs*, *26*(1), w96–w105.

Oberlander, J., Marmor, T., & Jacobs, L. (2001). Rationing medical care: Rhetoric and reality in the Oregon Health Plan. *Canadian Medical Association Journal*, *164*(11), 1583–1587. Available from: https://www.ncbi.nlm.nih.gov/pmc/articles/PMC81116/

Oregon Department of Human Services, Office of Medical Assistance Programs. (2006). Oregon Health Plan: An historical perspective.

Oregon Health Policy Board. (2012). *Coordinated care organizations implementation proposal*. Available from: http://www.health.oregon.gov/OHA/OHPB/health-reform/docs/cco-implementation-proposal.pdf

Oregon Legislative Fiscal Office. (2011, August). Budget highlights: 2011-13 legislatively adopted budget. Available from: https://www.oregonlegislature.gov/lfo/Documents/2011-13%20Budget%20Highlights.pdf

Oregon's Brave Medical Experiment, *The New York Times*. May 12, 1990. Available from: http://www.nytimes.com/1990/05/12/opinion/oregon-s-brave-medical-experiment.html

Strosberg, M. A. (1992). Introduction. In M. Strosberg, J. Wiener, & R. Baker, with I. Alan Fein. *Rationing America's medical care: The Oregon Plan and beyond*. Washington, DC: Brookings Institution.

Thorne, J. I. (1990). The Oregon Plan: Rejecting invisible rationing. *The Internist*, July/August, 9–11.

Wright, B. J., et al. (2005). The impact of increased cost sharing on medicaid enrollees. *Health Affairs*, *24*(4), 1106–1116.

Wright, B. J., et al. (2010). Raising premiums and other costs for Oregon Health Plan enrollees drove many to drop out. *Health Affairs*, *29*(12), 2311–2316.

Wright B., Allen H., Carlson M.J. 2011. The Healthy Kids evaluation survey: early results from a baseline survey of program applicants. Prepared for the Office for Oregon Health Care Policy and Research, Salem, OR.

Chapter 2

# State-Level Design: The Coordinated Care Model

**Bruce Goldberg**[1] **and Ron Stock**[1,2]
[1]*Oregon Health & Science University, Portland, OR, United States,* [2]*Oregon Health Authority Transformation Center, Portland, OR, United States*

## BACKGROUND

The circumstances in Oregon that set the stage for its health system transformation were no different than what was occurring around the nation. Oregon, like everyone else, was experiencing health care costs that were rising at a rate that far outpaced all other economic indicators. For states, when health care costs outpace the rise in state revenues, it has the effect of confiscating funds from other important state funded functions such as education, transportation, social services, public safety, and the arts. Compounding this, Oregon was coming out of the "Great Recession." State revenues had dropped significantly resulting in increased pressures on the state budget to fund the important functions of government.

At the same time, there was a belief in Oregon that the state could do better. In addition to cost issues, like elsewhere, health care quality was uneven at best and the health care system was difficult to navigate and increasingly complex. Oregon had a long history of progressive health reform and the election of a Governor with a track record of effective health reform set the stage for change.

## THE FOURTH PATH—CHANGING HOW HEALTH CARE IS DELIVERED

When health care costs outpace revenue, traditional budget balancing has usually resorted to one or more of the following: cutting people from care, reducing benefits or cutting provider reimbursements, and shifting costs to consumers. States cut people from care very bluntly by regulating their Medicaid roles— reducing eligibility when times are tough and covering more people when budgets allow. Employers have done this as well, either by reducing the number of employees they offer insurance to or through

**Health Reform Policy to Practice. DOI: http://dx.doi.org/10.1016/B978-0-12-809827-1.00002-1**

pricing. For example, covering the employee but making the costs of family coverage unaffordable to low-wage workers.

Reducing covered services has been another mechanism to cut costs. This manifests itself in a variety of ways such as limiting the number of services or visits covered, cutting some entirely, or establishing financial limits.

And lastly, costs are contained by reducing payments to providers. This has been in the form of direct reductions in provider payments or by cost shifting onto consumers through increased copays, coinsurance, and deductibles. Payments are also indirectly "reduced" if they are held steady over the course of years and do not keep up with general inflation.

These methods have been used for decades in one form or another and yet have been ineffective in controlling the overall growth of health care costs or in greatly improving the health of our populations. As such Oregon, having done all the above, looked for a better way—a fourth path so to speak, one that would help the state better reach the "triple aim." That path was to fundamentally change how health care is delivered.

The concept here was threefold. First, was the tremendous amount of waste in the delivery of health care and the need to create an environment that actively worked to reduce it. Dr. Don Berwick and his colleagues had outlined a path for incrementally reducing waste in six categories: overtreatment, failures of care coordination, failures in execution of care processes, administrative complexity, pricing failures, and fraud and abuse (Berwick & Hackbarth, 2012). Leaders within Oregon's health policy and health care delivery institutions understood that while there might be different opinions about the extent of that waste, there was agreement on the fact that there was a substantial amount of resources that could be used more effectively. In addition, there was widespread agreement that most previous attempts to reduce that waste were focused on regulatory reforms that had limited success. As such, there was an effort to focus on creating a climate and incentives to changing how care was delivered to reduce waste.

Second, was the belief that social, environmental, and lifestyle factors had as much of an influence if not more on our health and health status than medical care itself (McGinnis, Williams-Russo, & Knickman, 2002). Yet, there were many structural and cultural barriers that prevented the health care system from effectively addressing them. This was best illustrated by an often-repeated story Governor Kitzhaber told about an elderly woman with congestive heart failure who needed to be hospitalized because the effects of a heat wave exacerbated her tenuous cardiac status. He would go on to say that our health care system would easily pay tens of thousands of dollars for her cardiac intensive care unit stay, but not for the $300 air conditioner that would have kept her from having the exacerbation in the first place. As such there was a need for a system that could and would make rationale and appropriate decisions about how to best use health dollars to better improve health and reduce cost.

Finally, there was the belief that "all health care is local" and that to change how health care is delivered would require a system that brings together local health care providers, public health agencies, behavioral health organization, and other community human and social service agencies in working toward a common purpose. This would require a new local organization accountable for both the health of the community as well as for the financing of health care services.

## ELEMENTS OF THE COORDINATED CARE MODEL

The core elements of Oregon's health reforms were developed over a series of weekly meetings held with approximately 50 stakeholders from across the state. The group included representatives from health care and consumer groups as well as a bipartisan group of legislators. Their work produced a framework for a new health care delivery system, Oregon's Coordinated Care Organizations (CCOs).

There are five critical elements upon which CCOs were founded: local accountability, a global budget, flexibility in the use of services and health care dollars, coordinated care, and metrics. These elements formed the basis of the coordinated care model and were the foundations upon which CCOs were based.

### Local Accountability and Governance

As mentioned previously, one of the guiding principles of Oregon's health system transformation was that "all health care is local." There needed to be strong local accountability for both the finances of health care as well as for the health of the population. Local health care delivery systems were not truly integrated "systems," but rather a collection of providers such as hospitals, pharmacies, doctors, nurses, physical therapists who were all operating in a geographic locale but were not working together for a common goal. Each entity was doing a good job in caring for its patients, but often devoid of substantive collaboration with others. That is the system at its best. At its worst, individual providers compete with each to obtain a greater share of profitable patients and to minimize exposure to nonprofitable patients. In essence, they compete around those payers and diagnoses that provide the greatest financial reward.

The goal was to create local accountability for health and the financing of health care that would bring together the various parts of the health care sector and consumers into a joint governing body where decisions could be made collectively. With an understanding that if operating under a budget, increased payments to one sector, e.g., hospital services would mean a reduction somewhere else. Having all stakeholders sharing in governance would help assure that dollars were best utilized across the community for the common good.

This of course was easier said than done. How do you set up a functional governance model that does not become a so large and inclusive of every stakeholder and still at the same time remains somewhat representative of the health interests at large? Oregon's legislature grappled with this and vacillated between being prescriptive about numbers and seats on the boards of CCOs and giving broad leeway to these new organizations. In the end, they rested on a governance structure that included: all entities within the CCO taking financial risk, the major components of the health care delivery system, at least two health care providers in active practice (representing primary care and mental health/chemical dependency), at least two community members and at least one member of the CCO's Community Advisory Council (CAC). Each CCO is also required to have a CAC of which more than 50% of its members must be consumers and must also include a representative from each county government in the service area. The duties of the CAC include developing and reporting on progress around a Community Health Improvement Plan. Likewise, each CCO is required to have a Memorandum of Understanding (MOU) with their local public health authority, tribes, and area agency on aging.

Two pieces of legislation created CCOs, SB 1580, and HB 3650. They can be found at: https://olis.leg.state.or.us/liz/2012R1/Downloads/MeasureDocument/SB1580 and https://olis.leg.state.or.us/liz/2011R1/Downloads/MeasureDocument/HB3650.

The politics surrounding the final decisions around governance structure is interesting and discussed in more detail elsewhere in the book. The initial vision was a governing body equally split among three constituencies: consumers, providers, and those taking financial risk. In the end, there was a requirement that at least 50% of board needed to be financial risk takers. This was due to the political influence of Oregon's Medicaid Managed Care Organizations that were being replaced and reconstituted by the new CCO organizations.

In addition, the legislature stayed silent on the type of corporate structure for a CCO, namely whether they needed to be not-for-profit entities. A number of policy makers had a strong belief that CCO's should be nonprofit entities with boards that had at least 51% community members. Chapter 18, What Is Next for Oregon: Refining the CCO Model, by Representative Mitch Greenlick discusses the rationale behind this at length.

Certainly, managing finances and outcomes has been the focus of managed care organizations for quite some time. What made CCOs different was that they would all be locally controlled and governed by structures that brought together consumers and health care providers for joint decision making.

## Global Budget With Per Capita Growth

A global budget with a 3.4% per capita fixed rate of growth is probably the most important of the five elements. The global budget was envisioned to be a way to foster innovation, reduce waste, and better moderate health care costs.

Prior to Oregon's reforms, there were multiple streams of Medicaid funding. The largest were funding for three separate managed care entities: physical, mental, and dental health care organizations. However, Oregon like other states also had a variety of other smaller funding streams that relied on Medicaid dollars—such as family planning, transportation, and home visiting to name a few. Each of these funding streams serving overlapping populations was managed separately and often without coordination. The global budget aggregated Medicaid funding streams into a single integrated budget for which CCOs accepted full risk. The budget grows at 3.4% per capita each year, a rate commensurate with predictions of growth for state revenue and other economic indicators over the next decade. The rate of growth is per capita and is statewide, not for each individual CCO. In addition, and most importantly, if CCO enrollment grows, the budget grows to reflect the increase in enrollment. In essence, the budget is based on an amount attributable to per person enrollment and the per person costs increase at 3.4% a year.

Of note is that long-term care services are statutorily excluded from the global budget. While framers of the CCO model sought to include these, there was strong opposition by long-term care advocacy groups. Oregon has a long tradition of innovation in the funding of long-term care services and has developed a social model with one of the highest rates of community and home rather than institutional care. There were concerns that CCOs could disrupt what had been by all accounts a very successful model that was effective at controlling costs and meeting recipient needs.

The per capita growth of 3.4% a year was 2% less than what the Federal Office of Management and Budget had predicted that Medicaid would grow in the upcoming years. By reducing this trend by 2% points, it was estimated that Oregon was situated to save approximately 5 billion dollars in state and federal Medicaid funds over the next 10 years.

The global budget was also a means for moving away from yearly actuarial predictions and calculations of rates. These methodologies tend to focus on trending forward expected costs, the costs of new technologies and medical advances, often without the same emphasis on trending forward the cost reductions that could come with implementing systems for delivering care with greater efficiency and eliminating waste. This is understandable as such undertakings are voluntary and cannot be counted on, while the expense of a new drug is real. As such they continue to imbed last year's inefficiencies in rate calculations and continue to trend that wasteful expense forward year to year. A global budget was a means to move away from that system and to set a budget that grew at a predictable rate and to provide some incentive to innovate. Given the estimates of such a large percentage of waste in the health care system, it was theorized that setting a rate that grew at general inflation, was realistic, and could be achieved as providers and systems moved to reduce the imbedded waste.

In addition, such a budgetary system rewards efficiency. Under such a system, organizations that are efficient in lowering costs do not see their following year's payment rate based on the previous year's performance, and hence decreased because of their improvements. Rather the per capita payment continues to grow and any savings and efficiencies can be reinvested in communities to help improve health and care.

The idea was to both incent systems and create some community-wide accountability for improving health and moderating the inflationary trend in health care costs. CCOs would then be expected to innovate both in care delivery and in the type of payment models that would incent the appropriate care. By having all providers and community members at the table, making these decisions would allow for decision making that would balance the business needs of the varying health care providers with the community needs for health improvement, access to care, and improved quality.

The global budget is not a block grant. It is a per capita budget that fixes the rate of per capita expense growth at 3.4% a year. This is combined with protections for beneficiaries as the state agreed to not reduce the Medicaid benefits provided to recipients or to reduce statewide enrollment. Likewise, it protects beneficiaries from changes in state enrollment as the budget is comprised of a per capita payment for each beneficiary enrolled in Medicaid. For example, if per capita Medicaid costs were $5000 per year and there were 100 beneficiaries enrolled, the global budget would be $500,000 ($5000 × 100). If the next year enrollment increased to 200, the budget would be $1,003,400 (the new per capita cost now inflated by 3.4% to $5017 × 200 beneficiaries). Budgets to individual CCOs are risk adjusted to account for any differences in health status and age of the populations served. In Oregon's waiver, the state and CMS also recognized that there might be unforeseen and unusual circumstances that warranted an adjustment to the inflationary increase, for example, a catastrophic influenza pandemic causing extreme increases in intensive care unit stays across the state. Block grants on the other hand, generally fix expenditures and as such do not have the beneficiary protections outlined above.

## Integrated and Coordinated Care

The goal of transformation was to promote integrated and coordinated care to reduce waste and promote efficiency. While the Global budget was envisioned as way to combine fractured funding streams to drive integration and coordination, there was also an explicit emphasis placed on better coordination of care, team-based patient-centered primary care, and promoting the use of nonphysicians.

As our health care delivery system has become increasingly complex and specialized, it has become evident that better care coordination is necessary to improve outcomes and reduce duplicative and unnecessary spending. It was felt that placing a greater emphasis on coordinating care for complex patients with high health care needs was needed. Likewise, there was clear emphasis placed on the benefits of integrating physical, behavioral, and oral health care. It was also evident that to do this would require not simply blending funding streams, but also creating new care structures and improved mechanisms to share electronic health information.

Finally, there was great interest in promoting the provision of more care outside the clinic walls, and in home and community settings. Community health workers were envisioned as a mechanism to help coordinate care and plans were made to put systems in place to help facilitate such care. This included establishing training programs for community health workers as well as allowing for payment of such services.

## At Risk for Quality (Metrics)

The goal of CCOs was to provide a means for communities to be accountable for not only the costs of health care but also, and most importantly, improving the health of their population. As such a key component of CCOs is being held to meeting standards for a number of quality metrics and being at financial risk for that quality. There was little argument to CCOs being held accountable for improving health. The difficulty was in operationalizing this concept. What metrics would be chosen? Who would decide them? How much CCOs would be placed at financial risk? Who would be responsible for the accuracy of the data?

In its 1115 waiver, Oregon is held accountable for meeting standards around 33 quality and access metrics. The state in turn holds CCOs accountable for performance on those 33 metrics and place CCOs at financial risk for 17 of them. There are established benchmarks with improvement goals that are tied to a percentage of the CCO global budget. That began at 2% of the global budget and increases annually to 5%.

A statutorily mandated committee comprised of both experts and consumers is responsible for establishing the metrics and the improvement goals. Their work is done in open meetings with the opportunity for public input. Chapter 11, Measuring Success: Metrics and Incentive Payments, by Tina Edlund and Lori Coyner details the process.

One of the biggest initial issues for the state and the committee was to develop "transformational" metrics. That is, metrics that would incent the transformational changes in health care delivery that were envisioned. Many quality metrics exist, but most were felt to be the type of quality metric that continues to reinforce a "quantity" driven health care system, e.g., measure surrounding blood sugar or blood pressure testing. Establishing truly "transformational" metrics that help to catalyze health care delivery system change has proven to be challenging.

In addition, one of the goals of Oregon's health system transformation has been to move payment away from paying for quality and toward paying for health care to outcomes. Ultimately, the goal was to move a large percent of the global budget to payment for outcomes, 10%–20% or more. However, there have been many political and operational challenges to achieving such a goal.

## Flexibility

Along with accountability for the dollars and outcomes of care, it is important that there be flexibility in using those dollars to affect outcomes in ways local communities think best. This means allowing CCO's flexibility in how they pay for services as well as in what they pay for. There were multiple examples of the kinds of services, supplies, and personnel that were either prohibited from using Medicaid dollars for payment or limited and regulated in some manner. At the same time, there was a growing understanding both locally and nationally of the importance of social determinants of health and that many of these "health-related" services could have important impacts on health outcomes.

Allowing local communities, the flexibility to innovate, experiment, and use their professional judgments in how and what to pay for was an important piece of Oregon's health reforms. While simple in concept, it was challenging to implement and account for. Given a system that is driven by the accounting of services with CPT codes, how to account for things that had no such code was problematic. Likewise, there were concerns by regulators, particularly at the federal level that Medicaid dollars could be used inappropriately.

However, it was not just services that were at issue, there were also structural barriers surrounding paying certain providers. Medicaid regulations prohibited payment of nonlicensed health care providers. Likewise, some electronic billing and payment systems did not have mechanism for paying such providers. Therefore, as the state looked to build a new health care workforce that increased the use of community health workers and peer counselors, there was a need for greater flexibility in both the regulatory and operational arenas.

## IMPLEMENTATION: OREGON'S 1115 WAIVER AND SUPPORTS

Implementing the CCO model required several ingredients, all of which are extensively discussed later in this book. It required legislation and political support, federal support through an 1115 waiver and technical support from local communities.

Oregon's 1115 Medicaid demonstration waiver (http://www.oregon.gov/oha/HPA/HP-Medicaid-1115-Waiver/Pages/2012-2017-Demonstration.aspx) though detailed and lengthy, at its core is quite simple. It established CCOs as Oregon's Medicaid delivery system, allowed greater flexibility to use federal funds for improving health and provided the state with additional funding to implement. However, unlike many other federal waivers, Oregon agreed to be held accountable for reducing per capita Medicaid trend by 2 percentage points, meeting specific outcomes as measured by quality metrics and to not reducing Medicaid benefits or eligibility. In addition, Oregon agreed to financial penalties for not meeting those cost or quality goals.

The financial penalties established the imperative for change and confirmed a belief by state and local leaders that by working together, Oregon could achieve a better health delivery system for its Medicaid recipients.

### Supports for Change

From the outset, there was also an understanding that in order to change how health care is delivered, there would need to be some degree of technical assistance provided to newly formed CCOs and their local health care delivery systems. In addition, Oregon's health system transformation was predicated on fostering innovation at the local level and then sharing best practices and innovations statewide. As such, a "Transformation Center" was established with the explicit goals of helping to share local innovations—"making good ideas travel faster"—and providing technical support as needed around such issues as integrating physical and behavioral health, payment reforms, quality improvement methodologies, team-based care, use of community health workers, and more. To do so the Transformation Center engaged in such activities as organizing learning collaboratives, peer-to-peer rapid cycle learning, and providing technical assistance in establishing primary care homes, behavioral health integration, and addressing issues surrounding equity and health disparities.

In addition, the position of "Innovator Agent" was established statutorily. Innovator agents were modeled on the concept of the old agricultural extension agent. Each CCO was assigned an Innovator Agent. The Innovator Agents job was twofold—to bring innovative ideas and learnings to CCOs, but also to be a link back to the state agency (Oregon Health Authority) to help address and reduce bureaucratic barriers to innovation and improvements in local care delivery.

## PROGRESS TO DATE

Progress is monitored closely and posted publicly on the Oregon Health Authority website (http://www.oregon.gov/oha/hpa/analytics-mtx/pages/index.aspx). In summary, 4 years into this, all CCOs are meeting their financial

**TABLE 2.1 Factors Contributing to Better Health and Value**

- Incent and support local innovation
- Focus on chronic disease management with a particular emphasis on individuals with complex health conditions
- Focus on comprehensive primary care and prevention
- Integration of physical, behavioral, and oral health
- Alternative payment for quality and outcomes
- Flexible use of health care expenditures
- More home- and community-based care
- New workforce: community health workers/nontraditional health workers
- Electronic health records and better information sharing
- Tele-health
- Establishment of new care teams
- Use of best practices and centers of excellence
- Development of a health care leadership infrastructure that understands how to better coordinate care

milestones and living within their budgets and the state is meeting its commitment to reduce Medicaid spending trend on a per person basis by 2 percentage points. On the quality side, there has been substantial progress in meeting most metrics. Those that are financially incented, more than the others.

For example, at the time of this writing, emergency department utilization is down 29% from the baseline year prior to CCOs. Likewise, hospital readmissions for adults within 30 days are down 33%, adult admissions for heart failure have decreased 32%, patient centered primary care home enrollment is up to 69%, and the percent of children receiving dental sealants increased 65% and member satisfaction with care increased 10%.

What deserves mention is that Oregon's Medicaid delivery system prior to CCOs was highly managed with the vast majority of Medicaid recipients receiving their care in Medicaid managed care organizations. As such this progress does not represent a movement from an unmanaged to a managed system. But rather, shows what can happen when appropriate structures and incentives are established.

What has caused this improvement is certainly open for further speculation and study; however, it appears multifactorial and evolving over time. In addition, it varies geographically. Table 2.1 summarizes some of the factors that have contributed to outcome improvements.

## LESSONS LEARNED

There have been a lot of lessons learned and there will continue to be more. Among the most salient are:

*Governance*: Creating a governance structure for health care organizations that is accountable for finances and quality, is transparent, and that

brings together all the critical components of a local health care ecosystem—providers and consumers—is critical. Making substantive change is hard, and it is critically important that stakeholders feel they have a role in decision making. The health care system is full of examples of change being driven by one sector and leading to lack of participation by others. As such, paying careful attention how projects are governed and assuring that the right mix of people is governing is essential.
*Leadership*: Leadership at all levels is important. Having strong committed leaders from the executive branch, legislative branch, federal partners and the health system stakeholders (providers, health systems, consumers) was crucial. Critical to Oregon's efforts was the ability of health system leaders to look beyond their own individual interests.
*Common vision*: Change cannot occur unless there is a shared common vision for reforms/changes/interventions among all stakeholders. Oregon spent a lot of time bringing people together and communicating to create a common vision for change. But having a common vision is not enough. There must also be a commitment to the goals and deliverables of health reform.
*Changing payment and workforce*: Our health care system is populated by a workforce trained and organized around an old and established system of care. Likewise, that workforce has generally been paid to deliver quantity not outcomes. As such its naïve to expect new ways of doing things without changing both payment and workforce. Oregon spent a lot of effort on creating systems to move payment toward outcomes. Despite there initially being such a small percent of the global budget placed at risk for quality, CCOs have placed a tremendous amount of focus and effort on the metrics. Likewise, there has been a great deal of investment in a new, more community focused workforce that has included substantial numbers of community health workers, peer counselors, and care coordinators.
*Aligning health care payment systems*: Our major health care payment systems—Medicare, Medicaid, and commercial insurance—are very much connected and seriously misaligned. All generally use the same health care delivery system for care, but employ a variety of divergent payment and measurement modalities. Oregon's health care transformation began within the Medicaid system. While Medicaid is an important piece of the state's system, it is by no means a majority. It is difficult for such a single component of the larger system to drive large-scale change without aligning goals, outcomes, payment systems, and outcomes measurements. Such large-scale change will require multipayer initiatives.

In Oregon, as in any state, if you walk into any hospital or health care office, you will likely see individuals who get their health care coverage paid for by a variety of systems. While the percentages in each facility certainly differ, health care providers and consumers face a bewildering array of different operational issues, requirements, standards, and goals. Creating a common means by which everyone works in concert to improve individual and community health and to assure all have access to the health care they need is urgently needed.

Oregon's health care transformation is still in its early developmental stage. To date, CCOs have been successful in improving the quality of care provided to Medicaid recipients and in keeping Medicaid per capita health care costs growing at a sustainable rate. Consumers satisfaction with care appears to be better than it was at baseline. There has been a proliferation of new community-based health care workers and local health and social service institutions are working more cooperatively. A lot of work and learning lie ahead particularly as it relates to addressing the social determinants of health, integrating physical and behavioral health systems, creating stronger more effective working relationships between individual health, population health, and social service institutions, "upstream" investments in community prevention, and much more.

Establishing a CCO in every community may not be feasible and is not necessarily the only mechanism to meeting the triple aim. However, creating a system that is built upon a similar set of principles and foundational ideas might be. Creating a system that relies on local accountability for health and health care spending, operates within budget that both grows at an affordable rate and that sustains and improves community and individual health, and finally, that brings together health and social service institutions to work toward a common set of goals and metrics could be a helpful prescription for change.

## REFERENCES

Berwick, D., & Hackbarth, A. (2012). Eliminating waste in US health care. *JAMA*, *307*(14).

McGinnis, J. M., Williams-Russo, P., & Knickman, J. R. (2002). The case for more active policy attention to health promotion. *Health Affairs*, *21*(2), 78–93.

Chapter 3

# The Politics of Creating the Coordinated Care Organizations

**Governor John A. Kitzhaber**
*Independent Consultant, Writer and Speaker on Health Policy, Politics, and Strategy, Portland, OR, United States*

## INTRODUCTION

The creation of Oregon's Coordinated Care Organizations (CCOs) in 2012 was, in many ways, an unlikely event given the fiscal and political challenges facing the state at the time. When the Oregon legislature convened on January 10, 2011, the state was in the depths of the Great Recession; unemployment was high and the state faced a $3 billion revenue shortfall, one of the largest per capita budget deficits in the nation. The previous year, the state had been deeply polarized by two controversial ballot measures to raise taxes on corporations and higher-income households. Furthermore, while the Democrats held a slim majority in the State Senate, the House of Representatives was evenly split 30:30 between Democrats and Republicans. A deeply polarized electorate, a divided legislature, and an enormous budget deficit: this was the formula for disaster in anyone's book. The fact that things did not turn out that way is a remarkable and inspiring story, and one well worth telling.

Good health care policy does not spontaneously come about. It requires knowledge, expertise, and an ability to successfully navigate across the political landscape. The creation of the CCOs required navigating two discrete sets of politics: the politics of the Oregon state legislature and the politics of Washington, DC. The first set of politics involved the passage of the legislation that established the CCOs; while the second involved gaining federal approval of the Medicaid Section 1115 waivers necessary to implement the care model.

This chapter tells a very personal and narrative story of the politics of Oregon's recent health reforms with reflections on the lessons I learned from decades in public service.

**Health Reform Policy to Practice. DOI: http://dx.doi.org/10.1016/B978-0-12-809827-1.00003-3**

## BACKGROUND

The creation of the CCOs was not Oregon's first foray into health policy. In 1987 the legislature created the Oregon Health Plan (OHP), which laid the groundwork for the health care transformation of 2012. But it was the lessons I learned during my first campaign for public office that would help shape my ability to enact sound public policy.

Mine was an improbable candidacy when, on January 18, 1978, I announced my intention to run for the legislature in House District 45, which encompassed the southern Oregon timber town of Roseburg in Douglas County. I had moved to Roseburg in 1974, and began my career as an emergency room doctor there.

Nobody had any idea that I had been planning a career in public service ever since Bobby Kennedy was assassinated in 1968 while running for president. But I had been thinking about this for years—about how to put it all together—and I had maps of all the precincts in the district with voter registration and turnout information. Having no grassroots support from the party—and never having run before—I wrote a detailed campaign plan, broken down by the day from Election Day back.

On election night, after falling behind early in the evening, I was declared the winner shortly after midnight with a slim margin of 653 votes out of 14,279 cast. My opponent was the only incumbent Republican to lose in Oregon that year and I think I won for a couple of reasons. For one thing, nobody saw me coming, my opponent did not take me seriously, and we worked hard to get the voter turnout we needed, when we needed it.

More importantly, I knew a lot of these people. I had been practicing medicine in the ER for 4 years. These loggers and millworkers were generally pretty conservative guys, but not only had I treated them in the ER, I had fished with them and drank beer with them. They called me "Dr. John," a name that has followed me through union halls to this day. I know they did not always agree with what I was doing, but they believed I was doing it for the right reason. And one of the lessons I learned from this experience, is that once you break bread with someone, once you raise a glass together, once you get to know them—not for their views and opinions but as the *people* behind those views and opinions—people who care and worry about the same things that all of us care and worry about, our families, our jobs, our future—once you do that, something profound happens. You can still disagree, and undoubtedly will, but it is difficult to demonize someone you know simply because their views or politics differ from yours. And that speaks to the magic of personal relationships, which, in politics, is about getting to know the people behind the positions.

Ten years later, this lesson was a major reason we were able to pass the OHP—the first effort in the nation to honestly confront the reality of fiscal limits and reject the complex federal system of categorical eligibility that provided some people with everything and other people with nothing. Instead, the OHP expanded Medicaid eligibility to *all Oregonians* with incomes below the federal poverty level and created the covered benefit by *prioritizing* health services from the most important to least important, based on the health benefit of each service to the entire population being served. This involved the explicit rationing of health services and was extremely controversial at the time. Nonetheless, not only did it pass, it passed with huge majorities: 30-0 in the Senate and 57-3 in the House.

To develop the legislation necessary to create the OHP, I used some of the things I learned back in Roseburg, specifically the importance of getting to know people on a personal basis. I knew this was not going to be an easy lift. Its success depended on giving the stakeholders who would be affected by the legislation—as well as the legislators who had voted for it—some common understanding of the problem we were trying to address, and a sense of ownership in the proposal we were developing to address it.

To accomplish that I convened a "working group" of key stakeholders that met regularly in my office throughout the summer and fall of 1988 and ran through the 1989 legislative session. We met almost every week and developed a set of eight operating principles that we wanted the health care system to reflect, things like: society is responsible for financing care for poor people; there must be a process to determine what constitutes a basic level of care; funding must be explicit and economically sustainable; and there must be a mechanism to establish clear accountability for allocating resources and the human consequences of those decisions.

We then set ground rules for those who were at the table. In essence, these stated that if a proposal was put on the table and it did not work for a given stakeholder, that stakeholder had to offer an alternative that still met the principles. In other words, to be at the table, you could not simply say "no." As the weeks and months passed—as the people around the table spent time together and got to know each other—they gradually began to feel some ownership in the process. I have come to believe that this is a fundamental truth, not just of politics, but also of life itself. It is all about relationships.

With that background, let us fast forward to September 2008, the collapse of Lehman Brothers and the beginning of the global financial crisis. The following year, in a guest editorial in *The New York Times,* Thomas Freidman wrote: "What if the crisis of 2008 represents something much more fundamental than a deep recession? What if it's telling us that the whole growth model we created over the last 50 years is simply unsustainable economically and ecologically and that 2008 was when we hit the wall?" I believe he was right. The Great Recession had created an enormous budget deficit, and it was clear to me that there was an opportunity for *transformational* change embedded in the fiscal crisis. We had no latitude to deal with the budget deficit by raising taxes, because unemployment was desperately high in Oregon.

And we could not do it by simply cutting programs, by ceasing to fund education and public safety. So, the opportunity lay in *redesigning* those programs, updating them to make them relevant to the challenges and environment of the 21st century, rather than the 20th century when they were put into place.

It was the global financial crisis—and the election of Barak Obama and his commitment to tackle health care reform—that inspired me to run for a third term as governor in 2010. On March 23, 2010, President Obama signed the Affordable Care Act (ACA) into law, which, among other things, set up a huge expansion of Medicaid starting in 2014 when all Americans with incomes up to 138% of the federal poverty level could become eligible. Implementing the ACA and transforming Oregon's Medicaid care model became a central theme of my campaign. I was elected in November 2010 and took the oath office on January 10, the first day of 2011 legislative session.

## OREGON LEGISLATIVE POLITICS

When I took office in January 2011, we were facing a dramatically altered Medicaid landscape. We were sitting at the crossroads of an unprecedented expansion of health care coverage through the ACA; at a time when state and federal budgets were more strained than ever, and states were struggling to keep up with the escalating cost of health care; combined with the recession-driven need for increased services. While many governors believed that this crossroad would be the scene of an inevitable disaster, I saw only opportunity. That was why I ran again in 2010. But to capitalize on that opportunity would take innovation at the state level to completely rethink how health care was delivered.

Of all the budgets impacted by the Great Recession, the Medicaid budget faced the deepest cuts. It was the perfect storm. First, as unemployment rose, the amount of revenue available to the general fund through our income tax system declined. Second, many of those who became unemployed now sought their health coverage through the OHP, increasing enrollment. Finally, Oregon lost $675 million in Medicaid funds, which had been coming from the American Recovery and Reinvestment Act (ARRA) since Congress passed it in February 2009. The ARRA expired in June 2011, so those resources were not available for the 2011–13 biennium. The result was a $1.234 billion revenue shortfall in the Medical budget which—if we continued to provide care to all those who were eligible, and with no replacement revenue—would amount to a 39% cut in provider reimbursement.

I had been in public office during several recessionary periods—as a senator during the recession of the early 1980s; and as governor during the recession of 2002. The historic response of state government to a revenue shortfall is to cut from the existing program structure in lean times and add those cuts back in good times—but leaving the structure of *how* services are provided basically intact. This amounts to doing *less of the same* in hopes that later we would be able to do *more of the same.*

Everyone recognized that the current health care system was unsustainable—expensive, fragmented, inefficient, and ineffective in terms of improving population health. Yet no one was motivated to change the system because we continued to fund it. The ARRA funds, for example, simply propped up the *existing* Medicaid program, but delayed the time we would be forced to look not just at the *cost* of services but also at the *way* in which they were being delivered. Indeed, most of the debate over health care reform over the last few decades had been focused on coverage—on how we could pay for more people to have access to the existing system (doing more of the same). In 2011, however, with the Medicaid budget short by over $1.2 billion, it was clear that our choice going into the future was to do a lot less of the same *or* to change what we are doing. Doing more the same was simply no longer an option.

In many ways, this convergence of factors put providers in a serious financial bind: they were facing a potentially huge cut in reimbursement which, in turn, motivated them with a willingness to look at *any* innovation that might reduce the magnitude of that cut. It was, indeed, the perfect storm. We had a significant Medicaid expansion coming down the road at us, a huge hole in our existing Medicaid program, providers willing to look at change and innovation to avoid a huge cut in reimbursement, and a governor with a medical degree and a long history in health policy who wanted to fundamentally transform the way care was delivered.

Through a combination of benefit changes, administrative efficiencies and front-end loading the resources we did have into the first year of Oregon's 2-year biennial budget, we were able to reduce the size of the cut from 39% to around 11%. However, that still left us with a $240 million general fund hole the second year of the biennium—$600 million if we counted the $360 million in federal Medicaid matching dollars. When I was putting together my budget for the 2011–13 biennium, I intentionally chose to not completely fill that funding deficit for two reasons. First, the state had other priorities for the general fund besides health care; and second, I *wanted* a daunting revenue shortfall in the Medicaid budget in the second year of the biennium. This created the imperative to change, because it was clear to me that as long as we continued to fund the current Medicaid care model, there would be no incentive to change it.

Our plan was to make up the $240 million shortfall through cost savings by *transforming* the care model for the Medicaid program to get more value in terms of health outcomes for each dollar spent. So, it was the way we structured the Medicaid budget cliff into the second year of the biennium that provided the political motivation necessary to unify providers and consumers around our reform efforts. Furthermore, by building the $240 million general fund hole into the budget I presented to the legislature, the legislators themselves were motivated to support the health care transformation agenda, to avoid having to make additional cuts if we could not find the cost savings we sought.

To make this idea a reality, a Joint Legislative Committee on Health Care (JLCHC) was created and cochaired by Representative Tim Freeman and Senator Alan Bates. Tim was a conservative Republican from Roseburg, who had served on the Roseburg city council before being elected to the legislature in 2008. Although he had no health care experience, his even temperament and layman's perspective made him the perfect counterpoint to Senator Alan Bates. Doctor Bates, a Democrat, practiced primary care in Medford, had served on the local school board and had been elected to the Oregon House of Representatives in 2000 and to the State Senate in 2004 and again in 2008. He had been a member of the first Health Services Commission in 1989 and brought a provider's perspective and deep knowledge of the OHP to the committee. The JLCHC was a very large committee and met weekly, the purpose of which was to provide an opportunity for the engagement of a broad group of stakeholders, ensuring that their concerns were reflected in the proposed new care model.

As the legislation was beginning to take shape, I convened a smaller group of key stakeholders, which met three times at Mahonia Hall, the governor's residence. The purpose of this ad hoc group was to work though the politics around the proposed legislation as it moved through the legislative process. The idea was to ensure that the major stakeholders who could affect the outcome of this legislation were all talking with each other and that we could anticipate problems before they developed. This was not dissimilar to the workgroup I established to create SB 27 and the OHP in 1988–89.

These efforts converged into draft legislation that created the concept for the new care model, which we called the CCO. This draft became HB 3650 and established the "Oregon Integrated and Coordinated Healthcare Delivery System," which would replace the existing managed care system for the Medicaid program by January 1, 2014—the same day that the Medicaid expansion required by the ACA also went into effect. HB 3650 required the Oregon Health Authority (OHA) to develop a proposal—to be considered by the 2012 legislative session—that would establish the criteria required to become a CCO; set up a global budgeting process, as well as outcome and quality measures. The result was a piece of legislation that was remarkable, both in terms of its content and also by the fact that it passed with significant bipartisan majorities—something that offered a stark contrast to the bitter partisanship that swirled around health care reform in our nation's capital. HB 3650 passed in late June 2011 with a vote of 57-1 in the House and 22-7 in the Senate.

The bill establishing criteria for becoming a CCO, and the process by which to apply, was introduced as SB 1580 in the 2012 session. The content of this legislation was, in many ways, revolutionary and yet I had little doubt that it would pass. With the passage of House Bill 3650, the previous year there was no turning back. The entire state budget had been predicated on the assumption that we would find $240 million of general fund savings by transforming the Medicaid care model, a fact that motivated providers, consumers, and employers to work together. Success also meant that the legislature would have an additional $240 million to spend on other priorities like education. From a political sense, everyone was boxed in; or, in the parlance of Douglas County, we had them in the "squeeze chute."

Ironically, the sticking point turned out to be over something else: medical malpractice. Members, particularly in the Senate, felt strongly that we needed to directly address medical malpractice in SB 1580. This was a very contentious issue as it pitted many Republicans, members of the business community, doctors and hospitals against many Democrats, consumer advocates, organized labor, and trial lawyers. This issue, rather than the far-reaching policy embedded in the bill itself, was the rock on which our health care reform almost foundered.

I proposed to add a section to SB 1580 that would create a process—which I would personally oversee—to bring to the 2013 session a recommendation to address the medical malpractice issue. I made sure that the language in the bill was not prescriptive but that it set some sideboards on the process. Specifically, the language stated that the medical malpractice problem should be addressed in a way that "improved patient safety; more effectively compensated individuals who were injured as a result of medical errors; and reduced the collateral costs associated with the medical liability system, including the cost associated with insurance administration, litigation and defensive medicine." What I attempted to do with this language was to lay out a set of broad parameters, which would incorporate the basic concerns that *both sides* had about the medical malpractice system. With this language, we were able to separate this contentious issue from our larger health care transformation agenda, and SB 1580 went on to pass the Senate by a vote of 18-12 and the House 53-7.

## FEDERAL POLITICS

While the legislature was designing and approving the formation of the CCOs, we were also working with our federal partners in Washington, DC, to lay the groundwork for the waivers we knew would be needed to implement the new care model. Here, we were fortunate on two fronts. First, President Obama had appointed Kathleen Sibelius as Secretary of the Department of Health and Human Services (HHS). Sibelius served as the Kansas State Insurance Commissioner before being elected Governor of Kansas in 2002. Having a former governor, who understood Medicaid as well as insurance issues, as secretary of HHS gave Oregon an ally in

Washington, DC. Furthermore, the president had also appointed my friend Dr. Don Berwick as the acting director of the Centers for Medicare and Medicaid Services (CMS), the agency that would have to approve waivers for Oregon to implement its new care model. Dr. Berwick, trained as a pediatrician, was the president and CEO of the Institute for Healthcare Improvement and we had worked together on several occasions.

I traveled to Washington, DC several times to advocate for Oregon's approach and on February 28, 2011—long before HB 3560 became law—I had lunch with Dr. Berwick to let him know that we were planning a major overhaul of Medicaid, which would undoubtedly need waivers. By late 2011, however, it was becoming clear that we would need more than waivers; we would also need a significant investment of federal funds. Even if we successfully transformed the way we delivered health care within the OHP—and even if that transformation resulted in significant cost savings—this system change could not possibly happen fast enough to realize those savings in the 2011–13 budget, especially given the fact that the first CCOs were not scheduled to be operational until July 2012 at the earliest—and that was an optimistic projection.

I needed to get this problem on the radar screen at the highest levels of the Obama Administration as soon as possible, and the fact that we were able to do so was made possible largely by Scott Nelson and Dan Carol, two of those genuinely brilliant individuals who never get the credit they deserve for the amazing things they do. Both had deep connections in Washington, DC, including their mutual friend Greg Nelson who was Chief of Staff of the White House Office of Public Engagement and Intergovernmental Affairs, directed by Valerie Jarret, one of President Obama's closest advisors.

It was through these connections that on October 3, 2011, I found myself in Valerie Jarrett's private office in the West Wing of the White House. Also at the meeting was Nancy-Ann DeParle who had directed the White House Office of Healthcare Reform and oversaw the passage of the ACA, before becoming the president's Deputy Chief of Staff. It was a good meeting and everyone expressed support for what we were trying to do in Oregon—essentially to put on the ground a reformed health care delivery system that mirrored the ACA, 2 years before the ACA was to take effect. The only push back came around the need for an additional investment of federal funds. Valerie Jarrett pointed out that such an investment would require approval not only from CMS but also from the Office of Budget Management (OMB), the largest agency in the executive branch, which oversees the development of the president's budget, and she said we would have to find a strong justification for what we were asking.

We knew we were not going to realize our projected $240 million in general fund savings from transforming the system in the second year in the biennium, but we had a plan. In addition to Medicaid, there was something called "Designated State Health Programs" or DSHP. These were programs outside of the Medicaid program but which provide *Medicaid-like* services and were thus eligible for federal match. For example, the general fund dollars Oregon sent to the counties to provide mental health services. As such, we set out to identify around $400 million in DSHP dollars to fill our Medicaid shortfall.

In January 2012—three months after my initial meeting with Valerie Jarrett—I traveled again to Washington, DC, to explain the plan we had come up with. This time the meeting included Nancy-Ann DeParle and Marilyn Tavener, who had succeeded Dr. Berwick as the Director of CMS. Dr. Berwick had resigned in December 2011 in the face of heavy Republican opposition and the likelihood that his appointment would not be confirmed. I laid out our plan to cover the shortfall in our Medicaid budget, with the federal investment gradually phasing out over the next 5 years as cost savings from the new care model began to accrue. I told them that the OHA, the agency that oversaw the Medicaid program, would send a formal request for the DSHP match by March 1, 2012. I also asked that the process be expedited because we expected the CCOs to be operational by July 1, 2012.

To say that there was some discomfort with what I put on the table would be an understatement. We were asking for was a 5-year federal investment of over 2 billion dollars. In addition, some of the DSHP match we were requesting was new territory for CMS officials. Nevertheless, I pointed out that Oregon had a positive track record in health care reform with the implementation of the OHP in 1994, which had saved nearly $15 billion in total funds over the past 20 years. I argued that what we were *attempting to do* was an extension of what we had already demonstrated our *ability to do*. Being a physician, having a relationship with Dr. Berwick and having successfully implemented major Medicaid reform in Oregon in the previous decade, gave us some much-needed credibility with the administration.

I said that Oregon was not looking for a handout. We were willing to agree that the federal investment *be contingent upon* achieving a 2% reduction in the per capita Medicaid inflation trend rate by the end of the second year, with no reduction in enrollment or benefit. In early 2012, Oregon's Medicaid inflation rate was 5.4%, so we would have to bring it down to 3.4%. And because CMS was required to provide over 60% match on each state general fund dollar spent on Medicaid, the federal government would *save* the match on every dollar Oregon *did not* spend if we implemented our new care model. However, if—with the initial federal investment—the new care model reduced the Medicaid inflation rate to 3.4%, the investment would be paid back, with CMS and the state realizing a savings of $4.8 billion over 10 years.

After this meeting, things began to move rapidly. A month later, on February 23, the Oregon legislature approved the establishment of CCOs as the state's new Medicaid care model. On March 1, the OHA submitted its request for the 1115 waivers needed to implement the model; and the formal request for DSHP matching funds. In addition, we requested a decision on the DSHP funds by April 1, 2012 and a decision on 1115 waivers by June 1, 2012. Also on March 1, the OHA requested formal applications for CCOs to be submitted by April 27, 2012.

CMS and OMB readily found $250 million in DSHP dollars—about $150 million short of what we needed to fill our budget hole—but in late March indicated that they would be unable to identify additional funds by our April deadline. Over the next 2 weeks, our team worked with staff at OMB and CMS to reach agreement. We would send off a list of proposed DSHP programs we believed should be eligible for the federal match and staff in Washington, DC, would send back a long list of questions and concerns. We would respond and another set of questions would arrive. Finally, I scheduled a trip to Washington, DC, for Monday, April 16th, for meetings with HHS and the White House to see if we could resolve the impasse. Three days before I was to leave, I received a call from Marilyn Tavener who told me that there are two outstanding issues—both of which could be resolved—but *not* by Monday, April 16. On that assurance, I canceled my trip.

Yet, on April 24, just 3 days before the formal applications for CCOs were due, we were still no closer to an answer, with CMS continuing to request additional information. Once again I scheduled a trip to Washington, DC. Two days later, I received a call from HHS Secretary Sibelius assuring me that an answer would be forthcoming on Friday, April 27. Again, on this assurance I canceled my trip. But on Friday, Dr. Bruce Goldberg—the Director of the OHA, the state agency that was responsible for Medicaid—was informed that CMS would only approve $250 million of the $400 million of DSHP funds Oregon had identified for a federal match, which was about the same level they offered at the beginning of April, when our request was first made. No real progress.

I rescheduled my trip to Washington, DC, for May 1 and then I set about to become the smartest person in the room on every detail of the waivers we were seeking; and on the technical intricacies of what could and could not draw down federal match as DSHP funds. In addition, the DC staff had expressed concern over what kind of precedent they would be setting with the waivers and the significant federal investment if they approved our plan. This was not a technical issue, but, in my view, a matter of political will. I called David Agnew, who was President Obama's Director of Intergovernmental Affairs, to see if he could put me in touch with Jack Lew the presidents' Chief of Staff. Jack had been Deputy Director of OMB in the Clinton Administration. I wanted to get this on his radar screen both because of his relationship with the President and because I knew he would understand the budget issues involved. By the time the call was scheduled, I could talk the language of the OMB with the best of them.

I explained why I believed our request was doable, and I was able to respond to every technical issue he raised—and he raised all of them. I told him about the dance we had been doing with CMS and OMB for the last few weeks, that we were running out of time and that I did not think it was going to be resolved at the staff level. I said I was flying out to Washington, DC, on May 1 and wanted a meeting with the President. Lew clearly understood what we were trying to do from a budgetary standpoint and seemed sympathetic. He was vague, however, about meeting the President. What I did not know at the time was that he was in the midst of planning the President Obama's May 1 surprise visit to Afghanistan.

On the morning of May 1, I flew to Washington, DC, with Dr. Bruce Goldberg. I had a meeting scheduled with Secretary Sibelius the morning of May 2 and another meeting later that afternoon with people from CMS, OMB, and the White House. In the weeks preceding our trip, Scott Nelson and Dan Carol had used their amazing connections to gain access to the highest levels of the administration and lobby in support of the proposal I would be presenting on May 2. Among those called for support were Mary Kay Henry, president of Service Employees International Union; Randy Weingarten, the national president of American Federation of Teachers; John Stocks, the president of the National Education Association; Richard Trumpka, president of the AFL-CIO; and former US Senator and Majority Leader Tom Daschle.

I met with Secretary Sibelius at 10 o'clock on Tuesday, May 2, in her office at the Hubert Humphrey Health and Human Services building. The Secretary said that while she was very supportive, she remained concerned about the precedent that HHS would be setting and that we were still far apart on amount of DSHP funds that could be matched. I was incredibly frustrated. I left the meeting feeling like the Secretary had been preparing me for disappointment at the afternoon meeting.

From the Secretary's office, I went to a working lunch with Bruce and Scott Nelson. Charles Miller joined us, an attorney Bruce had retained to look at the legality of whether the additional DSHP funds we had identified could be matched. Miller was a graduate of the University of California Berkeley Law School and served as a clerk for Supreme Court Justice William O. Douglas. He was with a prestigious Washington, DC, law firm and was an expert on Medicaid law. Scott's assessment was that we were being set up for a "no" at the afternoon meeting—that neither CMS nor OMB really wanted to approve what we were requesting. Their main arguments were going to be that portions of the DSHP funds could not be legally matched and that granting Oregon waivers of this nature would set a dangerous precedent.

Bruce had also concluded that CMS and OMB were reluctant to give us the green light, and Scott's comments further reinforced that view. Bruce argued that at the afternoon meeting we should simply accept the $250 million in DSHP funds that CMS had offered to match and not risk walking away with nothing. I told him that there was a point at which you needed to be willing to walk away. I did not think much of the augment about the precedent they might be setting and I thought I could convince the White House—if not the staff from the agencies—to support us. But if we caved in right now, we would never know. I also felt that we had a strong case to make to the public if the administration refused to help us do exactly what the ACA sought to do. Bruce and I got into a heated argument over strategy, but I made it clear we were not going to back down.

Around 3 o'clock, Bruce, Charles Miller, Scott, and I walked over to the Old Executive Office Building through a glorious spring afternoon. At the security checkpoint outside the Old Executive Office Building, we were met by Scott's longtime friend Greg Nelson. Greg said, "You can't imagine how much trouble you have been causing us. For the past week we've been getting calls from labor leaders, former members of congress and just about everyone else from all over the country. Who are you guys?"

Greg accompanied us up to a large ornate room on the second floor of the Old Executive Office Building where we were met by David Agnew and his deputy, Jewell James. No one else was there. Not a good sign. About 5 minutes later, Nancy-Ann DeParle came in to say she was unable to attend the meeting—another bad sign. She pulled me into the hall and went on to say something to the effect that the administration was spending 4 hours a day on Afghanistan, 3 hours a day on the budget, and an hour a day on the OHP. This was a *very* bad sign. I thanked her for her past support; told her there was a quick way to get us out of her hair... say yes.

I went back into the room and, after some further delay, in they all came: CMS Administrator Marilyn Tavener, Medicaid Director Cindy Mann, Paul Dioguardi, the Director of Intergovernmental Affairs for Secretary Sibelius, Jean Lambrew, the Director of the White House Office of Healthcare Reform, and Heather Higginbottom, the Deputy Director of OMB. Both Marilyn and Paul apologized for being late and I felt certain they had been meeting together to prepare themselves for this meeting. We seated ourselves around the large mahogany table and did a round of introductions.

I opened the meeting by thanking them for all the hard work that had been done on Oregon's behalf by both CMS and OMB. I pointed out that Oregon's health care reform effort shared many similarities with the ACA. I also point out that under the ACA, millions of people would become eligible for Medicaid on January 1, 2014. Adding that many people to the current inefficient, hyperinflationary Medicaid delivery system could cause it to collapse in many states; exacerbating access problems for millions of Americans and undermining the goals of the ACA. Oregon's new transformed Medicaid delivery system, however, could give other states a model to replicate in order to meet the logistic and capacity challenges inherent in this huge expansion of the Medicaid program.

I went over the need for the investment of federal funds, the time constraints under which we were operating, our commitment to meet quality and outcome measures and our willingness to accept penalties if we failed to do so. I discussed our commitment to reduce the Medicaid cost trend rate by 2 percentage points with *no reduction* in benefit or eligibility, and the fact that this would provide Oregon and the federal government a combined $4.8 billion in cost savings over 10 years.

So far, so good. Everyone seemed to be paying attention, but it seemed odd that nobody was taking notes and there had been no questions. I plowed on, going over how we had identified the additional $150 million in DSHP funds and had addressed all the technical issues that had been raised. I pointed out that Mr. Miller had done a legal analysis and was prepared to discuss why all of these funds could be matched. Still no questions. The energy in the room was tense. I felt like I was giving a lecture but there was no interaction, no response. So, I moved on to address the concern about setting a new precedent.

"We do not believe that there are any risks, either legal or fiscal," I said. "The only real risk lies in *not* setting a new precedent. To transform the US health care system we must stop clinging to the precedents of the past—which keep us anchored to the status quo—and set *new precedents* for the future, aimed at creating a pathway to a new transformed health care delivery model. The question we should be asking about the Oregon waiver requests is not whether other states can apply for the same DSHP federal matching funds; but rather, whether the Oregon effort is *consistent with the policies* being championed by the Obama Administration. And, in fact, they are."

"Therefore, if other states should enact legislation to transform their Medicaid programs in the same way—*and are willing to be held accountable* for meeting the same high quality standards around access, clinical outcomes and improvement in population health; to reduce the per capita Medicaid inflation trend line by at least two percentage points; and to *fully repay* the initial investment within five years—they should be *encouraged* by CMS to move forward. This would constitute setting a *new precedent* intended to move us toward the objectives of the ACA rather than shackling us to the dysfunctional and unsustainable health care system of the past."

Still no reaction—it was as if I was alone in the room. Nonetheless, I forged ahead and as I neared the end of my presentation, Mark Childress, Deputy Chief of Staff for Planning, joined us carrying what seemed to be a spreadsheet. Childress had worked for Senator Daschle in HHS and was currently in charge of developing the messaging around the ACA, which was under intense attack by the Republicans. He was a "fixer" and a "closer" and has been referred to as "the most powerful man in the White House you've never heard of." While I was still talking, he began whispering to Marylyn Tavener and going over the spreadsheet, which I found distracting—even a bit rude. But when I finished he said, "I think I have some good news, it appears that we are very close." And he handed around copies of the spreadsheet.

There it was—an additional $130 million is DSHP funds that the federal government was willing to match and it was right off our list—only 20 million short of what we had set out to get. I felt a wave of relief and elation as I sat there looking at the spreadsheet. It meant we could close our budget hole in the current biennium and would get a 5-year federal investment of $1.9 billion. The odd body language that had dominated the room now made sense. The group had met earlier with Mark Childress and had been waiting for him to come in and make the political call—which is exactly what it came down to. I was brought back to the moment when I heard Mark saying, "Okay, if we are all in agreement, we need an action plan." He directed staff from OMB and CMS to work with Bruce to develop an MOU, which they would run by Secretary Sibelius later that afternoon and finalize the next day. That was it. I wrote in my notebook: "The deal is done!! 4:43 pm May 2, 2012."

## LESSONS LEARNED

There are many lessons to be learned from the creation and implementation of the Oregon health care transformation, but four in particular stand out. I hope they will be useful to others who seek to change the status quo and build a more just and equitable society.

1. *Leadership is critical.*

   In the political world, leadership involves three things. First is the ability to see a problem and a solution. Second, leadership involves taking the time to do your homework—to develop a deep personal understanding of both the problem and the solution. Third involves the ability to bring together a large and diverse group of people to see that they have a stake in the problem and to give them a sense of ownership in the solution.
2. *Creating the imperative for change.*

   No one will be motivated to change a system if we simply continue to fund it. It is human nature to cling to the status quo rather than reaching for the possible. Change requires creating an imperative that shows people the status quo is not sustainable and that continuing to do more of the same is simply no longer an option.
3. *All productive politics are personal.*

   Politics is about relationships. Making good policy is not simply about being smart or giving inspiring speeches. It is about taking the time to get to know people personally outside of their politics. Building and nurturing relationships is what sustains the development of good public policy when disagreements inevitably occur. It is what keeps people at the table even when they disagree and what gives them the capacity to find common ground.

**4.** *Put together a good team.*

Politics is a team sport and requires a skilled, coordinated, and dedicated team. There were many people responsible for the successful creation and implementation of Oregon's health care transformation and I do not have space here to recognize them all. However, there are a few individuals in particular that I would like to credit. The first is Dr. Bruce Goldberg, who I mentioned earlier. Bruce and his team managed the budget during a difficult time, provided technical support, and were the architects of strategy that resulted in a $1.9 billion investment in Oregon's new care mode. The second person is Mike Bonetto, who served as my Senior Policy Advisor for Health Care. Mike was the mastermind behind the legislation that created the concept as well as the business plan for Oregon's CCOs, and was instrumental in the passage of that legislation. Dan Carol and Scott Nelson who I also mentioned previously were key to lining up political support and Tina Edlund, the Deputy Director of the OHA, provided necessary technical expertise.

Chapter 4

# Strategic Communication and Engaging the Public

**Patty Wentz**
*Strategies 360, Portland, OR, United States*

## INTRODUCTION

Real stories about real Oregonians were the foundation of the Oregon Health Authority's strategy to engage the public and policymakers about health reform.

First, Ted, an 87-year old, had been feeling of the effects of diabetes, kidney failure, and multiple episodes of congestive heart failure. He had four hospitalizations over the course of a few months. After a heart surgery that left him, he joked, "half-hearted," managing the 13 medications, specialists, and therapies necessary for his multiple conditions would have been overwhelming for him and his family if they had been on their own. But they were not on their own. The family was served by a Congestive Heart Failure Home Team, a coordinated care pilot project through his health plan and primary care physician. About a dozen people met every Monday to talk about his case and share information. They helped Ted manage his care and appointments and flag for each other if anything is amiss. After joining the team, Ted had fewer medical appointments because all of his caregivers are sharing information and he stayed out of the hospital. His health improved and the cost of his care decreased.

> *If there are any problems, they talk about it. It doesn't matter which kind of specialist I go to; they can pull it up on the screen. They keep track. It's a team effort and boy, I think it's working.*
>
> **Ted Hanberg, patient with a coordinate care team**

Then there was St. Clair, who had schizophrenia and cognitive challenges and a history of alcohol abuse, but it was his asthma that was driving his

**Health Reform Policy to Practice. DOI: http://dx.doi.org/10.1016/B978-0-12-809827-1.00004-5**

need for medical care. Before he began receiving coordinated care, he had 40 emergency room visits in one year. And Becky, a community health worker, stationed at a local emergency room whose sole purpose was to keep people from coming back unnecessarily.

And, of course, Malik, who was introduced at the beginning of the book.

The stories of these Oregonians were the foundation on which the Oregon Health Authority built and implemented a public education and outreach campaign about the coordinated care model and the state's 1115 Medicaid Waiver. It was the kind of campaign the state's health and human services agencies had never embarked on before. While it used standard best practices such as identifying target audiences, leveraging communications channels, having clear and accessible materials, the campaign was built on three principles:

1. Meet people where they are.
2. Be disciplined and tenacious.
3. Show what is possible.

This chapter describes the communications efforts in the context of those principles, discusses what worked, what didn't, and what can be replicated in other states.

## #1: MEET PEOPLE WHERE THEY ARE

### Messaging for Where People Are

There were obvious challenges in engaging the public about government-led health reform in the middle of a budget crisis. The health care system is complicated enough to try to explain. Combine it with Medicaid policy and budget discussions, and it was daunting.

There were reams of documents that went into great detail about every aspect of coordinated care available on special websites created for the 1115 Waiver and our health system transformation. Work group agendas and minutes, straw man proposals, all waiver documents and public comment, Request for Proposals (RFPs) for the Coordinated Care Organizations (CCOs), everything was posted in as close to real time as possible. Those were helpful for many stakeholders but their density and volume made it difficult for laypeople to grasp what we were trying to do and engage in a meaningful way.

For policy wonks, CCOs would minimize decades-old barriers that came from fee-for-service, ineffective payment structure, and lack of accountability for outcomes. But there was risk that to everyone else, they would look like just another layer of bureaucracy.

To solve this, very early on we developed the strategy of showing how coordinated care would affect real people. Because really, when we talk about health policy, for many in our audience, it is not about policy. It is about their lives. For them, it is about whether the care they need will be available to them and if it is, whether they can afford it? Will their loved ones receive the treatments and medications they need? Health care is personal.

## *Message Strategies*

### Keep Language Simple and Relevant

We worked hard to keep information at an accessible reading level, even when talking about complex issues. That meant being creative and thinking about the essential information that people needed to understand. Like carving a sculpture we worked to cut away everything that was not necessary. This is most clearly exemplified in our public presentations. PowerPoint presentations did not mention the 1115 waiver, even in passing. When we talked about our coordinated care model, and went right to what our vision was for how real people's care would be transformed and what that would mean for them.

### Focus on People, Not Systems

Most of the communication materials focused on what would happen in clinics and homes under the coordinated care model, not in government or health plan offices. We focused our communications campaign on the human element of health care from the perspective of patients, providers, and community leaders. This also helped us show the reach of Medicaid in a way that people can relate to, such as the number of children and families covered by Medicaid locally, or the percentage of babies born in the state on Medicaid.

Our message strategy was greatly assisted by a research partnership with a local health foundation. With their help, we were able to do public opinion research on both the coordinated care model and the reasons the state was moving that direction. This research helped us develop clear and concise talking points and also showed lawmakers what their constituents wanted for their health system. The research also gave us a foundation for message discipline among spokespeople, who were able to feel confident that their messages were resonating and did not second guess the talking points. And frankly, it gave the communications staff an important tool when translating wonky health policy documents and white papers.

The results of the research were illuminating. We learned, for example, that we had been talking about the budget issue all wrong.

It was true that Oregon was facing a $2 billion Medicaid budget hole. In the early days of our campaign, we believed that the burning platform of that reality would show the importance of health reform to bring lower costs. During a work group, one spokesperson even drew a cliff on a white board and showed the state falling off it if we were not successful in landing the waiver. People would be cut from Medicaid, provider rates and services would be reduced. It would be a free fall, our state Medicaid program would have a hard time pulling out of.

The problem with that messaging, we learned, was that the public did not believe it and they certainly did not see it as a reason for making changes to the health care system. From the research, we learned that the financial messages had to be more nuanced. People did not want to lose health care services but they did not believe there was a budget crisis.

**Messaging in Action**

*At the beginning of our remarks today, we promised to help everyone listening impress your friends and family with your ability to explain health reform in Oregon. So here's what you say. It's three things and it's very simple.*

*One: Health care cost too much and doesn't give us the outcomes we need because we pay for the wrong things at the wrong time. Like unnecessary emergency department visits that could have been prevented. Or asthma attacks that didn't have to happen.*

*Two: The reason we pay for the wrong things at the wrong time is because the system is rigged that way.*

*Three: In Oregon, we came together and changed the rules of the game so we can pay for the right care, at the right time, in the right way. That will bring us better health, better care and lower costs.*

*That's it. Three things. The right care at the right time in the right place will bring better health, better care and lower costs. A healthy community at a cost we can afford. That's a win for all of us.*

***Bruce Goldberg, Portland City Club Speech, April 6, 2012***

What did make sense to people, was for state leaders to use the crisis as a leadership moment to make long-term changes, which is what we were actually doing. "Oregon's budget shortfall will force us to change our health system to reduce costs—whether we like it or not. We should use this opportunity to make smart, long-term changes that will reduce waste and inefficiency, cut costs and improve quality for everyone."

We learned that in Oregon, as in other states, most people believe that health care is too expensive and the rising costs are a problem for everyone because they eat up money that Oregon businesses could otherwise spend on creating jobs or increasing wages, and they cost the government money that could be spent on vital services like schools or public safety. Oregonians believed that fixing the health system will help all of us.

Research also told us they were not interested in the details or structure of a CCO or Medicaid waivers. They just wanted the health care system improved and they had some ideas on how it could be better. They believed that there is about 40% waste and inefficiency in the health care system and that by making the system more efficient to serve people better, we can get better health at a lower cost. They also believed that as much as 70% of health care costs are for chronic conditions that could have been prevented in the first place. Expanding preventive care that will avoid these serious health problems and the costs that go with them.

One of the best examples of a story that used all of our messages was about an air conditioner. This was an anecdote Governor Kitzhaber would often tell as an example of how the system was not set up to provide inexpensive preventative care. Using a composite character, he described how an elderly woman with congestive heart failure that was made worse by the summer heat to the point she ended up in the emergency department—but for the cost of an air conditioner, she would be able to stay safe and healthy in her home and waste and inefficiency would be reduced. This was such a powerful image of prevention and common sense that it resonated through all of the work to the point it became shorthand for health system transformation.

## #2: SHOW WHAT'S POSSIBLE—THE STORY BANK

It became clear early in the communications campaign that we needed real life examples to show what a transformed health system would look like. We were fortunate in Oregon in that several of our managed care systems had embarked on efforts and pilot projects that pointed to the future we were trying to create. OHA communications staff talked to patients, community health workers, primary care physicians, dentists, emergency room providers, nurses, administrators, local elected officials, and anyone else who had a story to share about how person-centered care had worked to bring better health, better care, and lower costs to their community. The stories were gathered into a "Story Bank," which was published online and still exists today.

The stories were used in presentations, speeches, and published materials. They were also foundational to helping news media craft stories about Oregon's health reform proposals. Every reporter needed a local example of how it would work and we were able to provide vetted stories to a journalism corps with eroding resources. As a result, we received several front-page stories around the state that broke complex health policy down to the human element. An additional important benefit to this body of work was that it gave the OHA communications staff a much deeper understanding of what leadership was trying to accomplish. As much as possible, staff was sent to meet one-on-one with people, to visit clinics and go to homes. After those experiences, they were able to explain Oregon's health reform vision with passion, enthusiasm, and insight to the members of the media they were working with.

The guidelines for the stories were simple: they had to be short, they had to focus on one person or health care setting, and they had to be aspirational. We felt that if our stories focused too much on the problems with the health care system and showed how many people were being hurt and how it would be overwhelming. The challenges would be too great to overcome and people would feel disempowered to make change. Instead, we wanted to show that, in fact, Oregonians across the state were already improving the health system and improving things for their local community. All we needed to do was to take it to scale. The stories helped us show not just what was wrong, but what could be possible.

The story bank project was both time and effort intensive with increasing demand for new stories. Since one of the goals of the project was to provide content to stakeholders we were able to partner with a local foundation and received a grant to fund a full-time "Story Banker." We hired a former health care reporter from the largest newspaper in the state, who brought his knowledge and journalistic sensibilities to ensuring we had accurate and compelling examples to share (see Exhibits 4.1 and 4.2).

## #3: BE DISCIPLINED AND TENACIOUS

We had our messages. We had our stories. To disseminate them broadly and build support for the health reform vision, OHA implemented a multiyear communications and outreach plan about health system transformation plan with clear objectives, timelines, audiences, spokespeople, communications channels, public input opportunities, materials, and accountabilities. This was our means to get the information out to people so that they could engage in the development of health reform, to understand what we were trying to accomplish, and give their input to make it better (see Exhibit 4.3 for a two-pager with the plan).

Here are the highlights of the plan:

*Objective*: Engage policymakers, the legislature, multiple agencies, the physical health care system, behavioral health care system, local governments, advocates, media, and general public around a unified vision of health system transformation in support of the Medicaid waiver, legislation, and implementation.

*Timelines*: The calendar was essential to our success. We tied communications timelines to the state policy-making calendar, taking into account legislative, budget, and federal deadlines. We leveraged every opportunity to communicate key messages linked to activities relevant to stakeholders.

*Create a newsroom*: OHA staff created content that was widely distributed through the state. Staff reported on public meetings for agency and partner newsletters, visited health care settings, and interviewed people who would be affected by the changed system, shot photographs and videos, and engaged in social media.

**EXHIBIT 4.1 For Oregon Doctor, the Future of Coordinated Health Care Is Already Here**

*Gideonse says shift to prevention and better management of chronic diseases improves patient health and satisfaction—and keeps costs down.*

*May 16, 2011*—**As Oregon policymakers consider legislation to improve how the state delivers health care** to hundreds of thousands of Oregon Health Plan clients, some local doctors are already pointing the way to better care at lower costs.

Family doctor Nick Gideonse says he was concerned that rising costs and mounting paperwork were making family medicine unsustainable. But a 2006 visit to the *Southcentral Foundation's Native Primary Care Center* in Anchorage offered a model of coordinated care with an emphasis on prevention and better management of chronic diseases.

"Here I really saw a model that could again feel successful, that put us in alliance with our patients," Gideonse says. "Things like reaching out to patients who weren't necessarily on our schedule, being in a mutually responsible relationship, using basic quality tools to work towards goals that you'd like to achieve, to balance supply and demand—that all these things were really possible. And that was tremendously refreshing to see."

Back in Portland, Gideonse oversaw big changes at his own clinic, including moving to a team-based approach, setting up regular group sessions to help patients manage chronic diseases, and saving room in doctors' schedules to allow patients to make same-day appointments so they can get help right away and avoid expensive emergency room visits. Hospitalization rates and charges have declined significantly for Gideonse's clinic and others like it since 2007, according to a study by CareOregon.

Gideonse who also teaches at Oregon Health & Science University says now he sees hope for the future of family medicine—and so do his students.

"We have had greater than a 50 percent increase in the number of medical students choosing family medicine—I'd say one of the main primary care specialties—and explicitly saying that they see a survivable future in working in primary care."

The changes did not happen overnight, but 5 years later Gideonse says he is happy with the result. "I know I feel much more effective in my work and much more supported. And I would never go back."

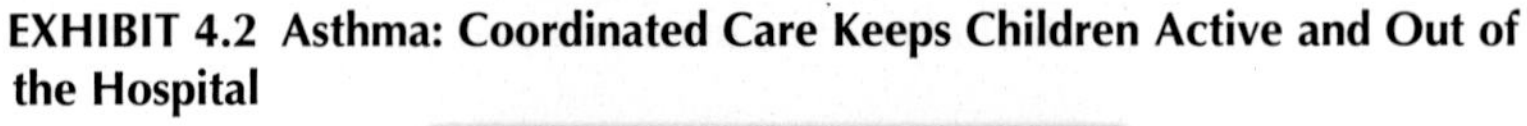

**EXHIBIT 4.2 Asthma: Coordinated Care Keeps Children Active and Out of the Hospital**

*Coordinated care means that Malik, 8, no longer makes twice-monthly visits to the emergency room.*

**Malik Wilkerson is an 8-year-old third-grader who has asthma.** A dust mite, cigarette fumes, or a chemical additive in scented soap can trigger a severe reaction in his nose, throat, windpipe, or lungs. He starts to wheeze and gasp for breath.

Asthma attacks used to send Malik to the emergency room once or twice a month, says his mother, Maydean Wilkerson. "He'd get to the point where he couldn't breathe."

Today, that almost never happens. The primary care team at Portland's Northeast Health Center has helped Malik bring his asthma under control and drastically reduce his trips to the hospital. The new, coordinated approach involves reduced exposure to household asthma "triggers"; a home nebulizer to convert his medication into a quicker-acting mist; and a portable miniinhaler to help him overcome wheezing attacks at school or on the playground.

Asthma is the most common chronic disease among children—and especially prevalent in low-income households. Although asthma can be well controlled by medication and preventive measures, Oregon children went to hospital emergency rooms 550 times for asthma attacks in 2006, according to the state Asthma Program. The average cost of a hospital visit for asthma was $12,000. The Oregon Health Plan paid more than $5.5 million in 2007 for hospital care of asthma.

Malik is an example of how coordinated care can reduce that expense while improving health.

His mom enrolled in Multnomah County's Healthy Homes program, which helps children with asthma get control over their disease through careful choice of medications and reduction of environmental hazards in the home.

Healthy Homes visited the Wilkerson home to search for asthma triggers and recommend preventive steps: Natural cleaning agents. Mold prevention. Nonallergenic bedding and pillow cases. A special vacuum cleaner that captures dust without releasing any into the air.

Since Malik started using the nebulizer and special bedding and since his mom substituted vinegar-based cleaners for scented detergents, he has made only two visits to the hospital for asthma. These lower-cost interventions are a good deal for everyone, his mother says, including taxpayers.

"In the long run it saves money by keeping Malik out of the hospital."

---

*Audiences*: We identified audiences, both by category and geography. While core messages stayed the same and message discipline was essential, we highlighted information that met the needs and concerns of a particular audience. For example, the costs and benefits of Medicaid were provided for each county or local jurisdiction, making it more immediate and real at the community level. Also, different audiences often have unique concerns. For example, in Oregon, the behavioral health system was concerned about the ways in which a coordinated care model would change their funding streams. The physical health care system, in addition to having concerns about reimbursement rates, needed to know how they would be held accountable and what incentives for improvement would be. Community advocates wanted to know how the new model would affect Medicaid members' access to care. The public health community was eager to participate early and often to affect the upstream conversation. And agency staff, too, was anxious to understand how their jobs may be changing to adapt to a new vision for the health care system.

## Special Note About Communications With Oregon Health Plan Clients

It was important that during this time of transformation, Medicaid clients had information available about what health reform would mean to them. Thanks to a series of informal focus groups with people waiting for assistance at a local social service agency, we learned quickly that clients first and foremost wanted to know how or whether their benefits would be changed. Would their current treatments be covered? Were their health services at risk? Coming out of that, our number one message for clients was that nothing would change in their benefits. That was central to all of our communications with them. We repeated that assurance with every publication, all over the website, and in public settings. Also, we timed our direct client communications until after the policy had passed so that we did not cause any unnecessary uncertainty. Our messages focused on the fact that all benefits would be protected and that not much would change for most Oregon Health Plan clients while highlighting how coordinated care would be value added to people with complex health needs. We also worked with client stakeholder groups to ensure that they had this information as well.

**EXHIBIT 4.3 Health System Transformation Outreach and Communications Plan Summary**

**Key messages**

- Opportunity, not crisis
- Vision of better health—aspiration—give people something to believe in
- Not reinventing the wheel
- Prevention
- Focus on chronic care—20% of population drives 80% of cost. Most OHP clients will not see a major change.
- This is a way to reduce waste and inefficiency

**Stage 1: July 2011 to February 2012. Building toward the legislative session**

- *Stakeholder outreach and public engagement*:
  - On-the-ground contact with providers/plans/hospitals/counties through individual and group meetings
  - Statewide public meetings, September/October
  - Monthly Oregon Health Policy Board and House Bill 3650 work group meetings as opportunities for input, outreach, and education
- *Lawmakers*:
  - Meetings, conversations, outreach, materials and engagement, and earned media in key districts
- *Media*:
  - Constant drumbeat of media outreach and story pitches using researched messages about what Oregonians want from health system transformation to build interest and demand in coordinated care
  - Editorial board meetings statewide
  - Concurrent campaign to create demand
  - New website: www.health.oregon.gov and ongoing social media
  - Toolkit of materials
  - Videos and produced presentations about patient-centered coordinated care geared toward providers and clients
- *Sourcing funding for additional research*:
  - Client/member information needs
  - Diverse communities' information needs
  - Provider information needs

**Stage 2: February 2012 to July 2012. Building toward launch of first CCO**

- Continued meetings and outreach with providers/plans/hospitals/counties
- More direct communications for clients to build interest, create demand, and ease concerns about change
- Continued drumbeat of earned media and editorial boards
- Ongoing social media
- Updated materials in preparation to launch and more videos and produced presentations

**Goals—Four keys to success**

- Provide timely, clear and accurate information and input into OHP health system transformation
- Build coalitions to support and sustain CCOs and drive innovation to improve health and lower cost
- Educate OHA staff about CCOs
- Build public support and understanding of how health system transformation will lower costs and improve care

**Assumptions—Communications will include messages about**

- Prevention
- Improving health equity and reducing health disparities
- Primary care health homes
- Community-level accountability for improving health
- Services that are person-centered, provide choice, independence

**Strategies**

- Disciplined, clear messages
- Why, when, how, affect on particular audiences
- Over-communicate through every available channel—from the e-bulletin to face-to-face meetings
- Public input—OHPB meetings
- Stories—OHA/NWHF story bank project
- Engage directly with key constituencies at the right time in the right way
- Track contacts—oha.comms.outreach

**Audiences**

- Clients/families
- Client advocates
- Providers
- Provider associations
- Stakeholders
- Tribes
- Counties
- Lawmakers
- Statewide and local mass media
- Local community opinion leaders

**Spokespeople**

- Local community and health care leaders
- Governor Kitzhaber
- Lawmakers
- Bruce Goldberg, OHA leadership and staff
- Oregon Health Policy Board members
- Stakeholder supporters—providers, members, advocates
- OHP client/CCO members and advocates.

## Communications Channels

### *Mass Media—Earned Communications*

We saw the media as an ally in helping the public understand health system transformation. We developed press kits with localized information, did TV and radio talk shows, had face-to-face visits with every major newspaper and broadcaster, and made several rounds of editorial board visits. We worked hard to make sure that the media understood what we were trying to do and how it would affect people in their local communities. This effort paid off in several front-page stories and editorial endorsements of health system transformation.

### *Owned Communications*

Nearly all of the OHA communications staff was repurposed to focus on Health System Transformation and on creating frequent and clear direct communications to stakeholders, clients, staff, and the general public. It was, in essence, a newsroom.

*Health.Oregon.Gov*. We created a new website that was the hub for all communications activities. It was short and easy to remember and we created a graphic that our stakeholders used on their own sites to help promote health system transformation. There were fact sheets, presentations, and real life stories of how a transformed health system could work.

*Pushed communications*. We also had numerous and frequent outlets for communications including a weekly director's message that repeated the same messages on a regular basis, a monthly newsletter, and special updates to staff.

*Public engagement*. We also leveraged all existing opportunities for public engagement and created new ones tied to milestones. We took advantage of standing meetings, such as advisory committees, legislative hearings, board meetings, or tribal consultations, to stay on course with message discipline throughout the waiver process, telling the story of transformation, and updating stakeholders. We established public work groups made up of key stakeholders to provide forums for stakeholders and media to learn first-hand about the opportunities and challenges of the state's health reform work. We held regular webinars tied to key milestones. Media visits and stakeholder visits with local health systems were scheduled to support those events and they received strong local media coverage. In all, there were more than 90 public facing opportunities to learn and/or comment to the state about health reform over the 13-month period between January 2001 and February 2012. By the time, the public comment period opened at CMS for the 1115 waiver, interested Oregonians were well-versed on the state's health reform vision.

*Statewide Town Halls*. One of the most successful public outreach efforts was a series of town halls throughout the state. For a state with the population of Oregon, eight is a large number, especially given that most of the citizenry live in a few metropolitan areas. But while our population is small, our geography is large and we really wanted to have a presence in as many parts of the state as we could manage to do in the time we had, particularly in areas with significant health systems serving rural Oregon. The governor's policy staff, Oregon Health Policy Board members, and agency leadership split up the state. We made sure to always have two of them at each meeting.

We selected venues that were large enough so that everyone who was interested could come and we promoted them heavily through the local media and our coalition partners, who turned out their members.

The agenda reflected both our aspirational messaging and our focus on local innovations. Our speakers from the governor's office and the agency described health system transformation and then a local health care leader would give an example of how they were already transforming care in their local community. For example, in one community, a local neurosurgeon talked about how care coordination after surgery reduced readmissions for his patients.

After the presentations, there was a group discussion that allowed people to talk about what they wanted from a transformed health system. Again, we focused on aspirational questions so people could begin to envision with their local community what might be possible.

In every town we also did an editorial board visit with the local newspaper and made sure that reporters came to the town halls and reported on them. We also did our own write ups about the meetings and shared them with coalition partners and agency staff. The town halls created important energy and momentum for health reform and made people feel as if they were part of a movement.

*Spokespeople*: In Oregon, senior state officials and policymakers made telling the story of the state's health system transformation a top priority. Top messengers included the Governor and health policy staff, Oregon Health Authority Director, the Oregon Health Authority Chief Policy Officer, the state Medicaid Director, the Addictions and Mental Health Director, the Chair of the State's Health Policy Board, and legislative champions. Whether to rotary clubs, trade associations, health systems, media interviews, keynote speeches to groups of hundreds of people, the executive and agency leadership gave their time to lay out the state's vision and answer all questions. It was an extraordinary effort and time investment, but one that paid off.

In addition, we identified third-party validators such as health care providers across the state. They informed and participated in local town hall meetings and with local media.

*Materials*: OHA created toolkits with simple fact sheets, presentations, videos from the Governor's office and agency leadership, talking points, and newsletter articles, and made them widely available on the website created for health reform. We also created content specifically drafted for stakeholders and partners to use within their own organizations, including newsletter articles, presentations, and web content (see Exhibits 4.4 and 4.5).

*Accountabilities*: Without ongoing attitudinal research, the effectiveness of communications efforts cannot be reported by measuring changes in attitudes or understanding. But there are valid and helpful activities measurements that we tracked and reported to agency leadership and the governor's office. These included the type and number of media stories, social media metrics, public meetings, and our own story bank. We analyzed our activities weekly to see where we were lacking in quantity or geography.

---

**EXHIBIT 4.4 Oregon Health Plan Transformation—Why, What, When, and How**

**Why—Taking the opportunity to remake our health care system to one that brings better health, better care, and lower costs. The system we have is not working and our state cannot afford to wait any longer.** House Bill 3650 proposes a redesign to the Oregon Health Plan with the goal of better coordinated care to improve access to primary and preventive care and to break down barriers between physical, mental, and other types of care. Today, OHP serves some 600,000 Oregonians and by improving health and lowering costs we can create a more sustainable system.

**What—Common sense approach that uses best practices from Oregon and elsewhere. Remove barriers that keep providers from working together for their patients' optimum health. Reward prevention and local innovation. Reduce waste and inefficiency.** Oregon's providers are on the front lines caring for our state's most vulnerable citizens. The vision of House Bill 3650 includes a plan for the creation of community-based CCOs to remove barriers that prevent them from working together.

Key elements include: Local control, local coordination; health equity; global budgets, flexibility, and shared savings; metrics/performance measures; primary care health homes.

CCOs would focus on patient outcomes, would integrate physical health, mental health, and oral health, would make health equity a key focus, would be required to manage costs within a global budget that is tied to the consumer price index, and would have local flexibility to allocate resources. CCOs allow for a system where hospitals, clinics, mental health providers, community health workers, nurses—where everyone is working together for better health. The vision also includes more primary care health homes, which use a team-based approach for patients, and better management of chronic illnesses, which account for up to 80% of health care costs.

**When—While nothing is changing today, local communities are talking about how they can form CCOs. Change will begin in 2012 if the legislature approves plan.** Under House Bill 3650, with approval by the Legislature, the first CCO would launch in July 2012. Prior to that, there are key elements of House Bill 3650 that need to be fully developed and brought back for legislative review at the February 2012 session.

**How the change will happen—Public input and working together on the reforms already started. While CCOs are a new type of organization, the vision**

**that created them comes from reforms already underway in Oregon.** In order to be successful, we need to bring together the best thinking in the state and build on the work already done by Health Fund Board, Health Policy Board. Four work groups appointed by the Governor will inform the CCO plan that will be presented to the legislature. In addition, the Health Policy Board holds monthly meetings and will be holding additional statewide public meetings around the state in the fall. Stay informed at www.health.oregon.gov.

---

**EXHIBIT 4.5 Sample Newsletter Article**

**The budget brought a challenge and an opportunity**. Providers will be taking difficult reductions and we all must do what we can to keep the system stabilized through these reductions. We must also remake our system to be more sustainable to bring better health and lower costs. The high costs are hurting individuals, families, and businesses, while eating up dollars that could be used for education, public health, or other important services. As evidenced by these budget cuts, we cannot afford our current health care system any longer.

House Bill 3650, proposed by Governor Kitzhaber and passed with strong bi-partisan support, will change the way health care is delivered and provide greater access to health care coverage in our state.

House Bill 3650 proposes a redesign to the Oregon Health Plan with the goal of better coordinating care to improve access to primary and preventive care and to break down barriers between physical, mental, and other types of care. Today, OHP serves some 600,000 Oregonians. Those of you on the front lines of service know first-hand the limitations of our current system and the subsequent effect on health and costs.

The vision of House Bill 3650 includes a plan for the creation of community-based CCOs that would focus on patient outcomes and health equity, would integrate physical health, mental health, and oral health services, would be required to manage costs within a global budget, and would have local flexibility to allocate resources.

We know we can get better health, better care, and lower costs because across the state providers are taking innovative approaches to delivering care. We will be most successful if we begin by focusing on the patients who will most benefit from better care—those individuals with chronic conditions who account for up to 80% of health care costs.

*Here are two examples: (Use whichever works best)*

During a year-long pilot program, Mosaic Medical in Bend identified the 100 Medicaid patients with the highest medical bills in early 2010—by and large because of frequent ER visits. Mosaic started coordinating their care with a team to stay in regular contact with patients. Since then, the total health care costs for Mosaic's 6400 Medicaid patients decreased by more than $621,000 in 2010, thanks to just 6 months of reduced reliance on the emergency room for nonemergent care.

CareOregon conducted a pilot project of team-based care involving 18 clinics and 41% of their 140,000 members. Median inpatient hospitalization rates for Medicaid adults decreased by statistically significant levels beginning in late 2008 by between 16% and 18%. And markers of good clinic care, such as diabetes or hypertension control, have steadily increased, while patients increasingly perceive that the clinics are successfully meeting their needs.

Under House Bill 3650, with approval by the Legislature, the first CCO would launch in July 2012. Before that, there are key elements of House Bill 3650 that are to be fully developed and brought back for legislative review at the February 2012 session.

That is a fast timeline and we have already started. The Governor has appointed four work groups on CCO criteria, global budgeting, metrics methodology, and Medicaid and Medicare dual eligible populations, to provide input on some of the key elements. The groups will meet through November and their work will inform the final plan. In addition, the Oregon Health Policy Board has monthly public meetings and will be holding special evening meetings in locations around Oregon this fall.

You can keep up on all the work happening and sign up for regular updates at health.oregon.gov.

---

### *Get Feedback and Communicate About Communicating*

Stakeholders have great expertise in the populations they serve. We asked for their help in identifying audiences, setting priorities, and targeting messages. We also had formal and informal communications tables that helped us solicit feedback and advice on outreach plans and communications materials. We worked hard to develop a communications process that is, in essence, a model of the state's transformation efforts: team-based, audience-centered, and broadly accountable. We also shared our communications plans broadly so people could see the work underway.

### *Agency Partnerships*

Depending on the state structure, not all the health-related activities are likely to be under one agency. In addition, human services and other related agencies will be affected by improvements in the Medicaid program. Staff and stakeholders of those agencies also need information about health reform. OHA stood up a multiagency joint communications cabinet to build allies, inform the communications plan, and create more opportunities for dissemination.

### *Internal Communications to Agency Staff*

Direct, frequent, and interactive communications with agency staff were critical to the success of moving forward health reform. After all, they are telling the story of health reform every day, with or without direction. We engaged them as ambassadors and also made sure there were opportunities for them to learn how a changing health system affected their work and the agency. Our tactics for the internal communications plan included:

- Prioritizing health reform in existing internal communications channels wherever possible.
- Ensuring division and program leadership understood and used the core messages in their communications to staff.
- Where possible, engaging public employee unions in health reform and staff communications.
- Creating a dedicated Intranet site.
- Producing internal presentations, leadership brown bags, e-mail updates.
- Branding health reform internally with materials such as posters or flyers.

## LESSONS LEARNED

The communications and outreach work did not stop after CCOs were launched, and we took the lessons learned forward into the new era. Client communications intensified as OHP enrollees received notification about their new CCO. The next year, OHA began an early expansion of Medicaid, enrolling eligible Oregonians into coverage through the federally approved "fast-track" program, bringing coverage to hundreds of thousands of people.

Essential for OHA's communications and outreach program from 2009 to 2014 were putting first and foremost the stories of real people when talking about health reform, which allowed us to translate complex health policy changes into simple and clear examples. Using those examples and the research-tested messages, we were extremely disciplined in our approach. We made sure our messages were widely available to the coalitions we built or were part of, and we tracked how the messages were used and amplified. And to make all of this possible, we prioritized health reform communications and found ways to invest in it because even the best policies will fail if people do not understand or believe in them.

Section II

# Implementation of the Coordinated Care Model: Key Components

Chapter 5

# Delivering Health Care Through the Coordinated Care Organization Model

**Bob Dannenhoffer**
*Douglas Public Health Network, Roseburg, OR, United States*

As health care transformation in Oregon was taking shape, a key question was how to shape and form the delivery system, that is, how the actual care would be delivered to the patients within their local community. The choice of delivery system was key, as the transformation plan called for the delivery system to engage patients and providers to meet the triple aim of improved population health, improved patient care, and a moderation of costs with the imperative that Oregon's per capita Medicaid costs would grow at 2% less than the national average. While there were many models of Medicaid delivery systems in Oregon and in other states, there was no obvious model for success that would meet the triple aim and would also meet the specific objectives planned for health care transformation in Oregon.

The development of the delivery system was important, as the high ideals and specific objectives of transformation could not be achieved unless the system performed. This chapter includes my recollections of the history of how Coordinated Care Organizations (CCOs) evolved and reviews their structure and organizational models. For the purpose of this chapter, the term delivery systems will be used to describe the systems of risk aggregation, care coordination, and the actual delivery of care.

## THE CCO BACKGROUND STORY

In the spring of 2011, the Oregon Health Authority (OHA) leadership presented a white paper that outlined a plan for a Medicaid delivery system that could carry out transformation and meet the triple aim. This white paper outlined many of the key features of transformation and called for a delivery system of regional health authorities in the state that would serve as the aggregators of risk and the payer for services in their jurisdictions.

**Health Reform Policy to Practice. DOI: http://dx.doi.org/10.1016/B978-0-12-809827-1.00005-7**

Transformation of the delivery system was seen as a key to transformation, as the fee-for-service delivery model was seen to encourage the production of lots of "widgets," but not necessarily improvements to health outcomes.

During the discussions on transformation, Governor Kitzhaber had given speeches that decried a system that would pay for repeated hospital admissions for congestive heart failure, but not for the air conditioner that might prevent these exacerbations. Thus, "Kitzhaber's air conditioner" became an oft repeated phrase. This proposal was quite bold, transformational, and inspirational. But the roadmap from this inspiration to a workable delivery system was uncharted. While there was early work in the New England states, in the commercial environment in Colorado, and managed care models in other states, there was no other well developed Medicaid model from which to choose.

It would be charitable to say that initial reaction to this white paper was mixed. In such a broad and sweeping proposal, almost everyone could find something to dislike. Smaller communities were worried that they would be swallowed up by the largest systems in the region. In several of the proposed regions, there were competing hospitals or medical systems and the high degree of cooperation required seemed unlikely. There was great concern of who would be in charge of these regional health authorities. A shared governance model was originally proposed and in many areas of the state, the dominant health care systems, hospitals, and providers did not have the sufficient level of trust to make that happen.

The state budgetary shortfalls following the great recession of 2008 made some sort of health care transformation mandatory. This budgetary reality made it necessary to find consensus on a model and it was not clear that consensus could be achieved. At the time, the Oregon House of Representatives was evenly split with 30 Republicans and 30 Democrats, with cospeakers and cochairs of the relevant committees, and thus a political as well as a programmatic consensus was necessary. Unlike the national discussion of the Accountable Care Act (ACA), in Oregon, there was truly a bipartisan consensus for reform.

In a brilliant move, the OHA called together 60 health care stakeholders including providers, patient advocates, managed care executives, community members, legislators from both parties, and OHA senior staff to meet in a seemingly endless series of meetings on Tuesday nights in the state capital of Salem. A vigorous and entertaining discussion of the form, structure, and function of the delivery system ensued. Perhaps more importantly, most of the health care leaders in the state were present and their concerns were able to be voiced and heard. Many friendships were made and strengthened. With excellent facilitation and many breakout groups, key concepts were debated and considered (Table 5.1). Slowly, and somewhat grudgingly, a consensus was reached on almost every decision point. With so many of the leaders participating, it would later be difficult to foil the consensus that was needed.

**TABLE 5.1 Key Discussion Themes of Medicaid CCO Delivery Model Development**

- Adopt the goals of improving the health of all Oregonians, increasing quality, reliability and availability of care, and reducing costs.
- Care is integrated and coordinated, including physical health, mental health, addictions treatment, oral health, home and community-based services, and long-term care services and support.
- Consumers get the care and services they need, coordinated locally with access to statewide resources when needed.
- Care and services delivered through ACOs using alternative payment methodologies that shift the focus to prevention, improve health equity, and utilize person-centered primary care homes, evidence-based practices, and health information technology.
- The ACO is a single integrated organization that accepts responsibility for the cost within its global budget and for delivery, management, and quality.
- Essential elements of an ACO model include:
  - Work cooperatively with community partners to address public health issues.
  - Health equity is prioritized and disparities are reduced.
  - Actively engages consumers in making its decisions that impact the populations served, the communities where it is located, and decisions about how integrated care is delivered.
  - Person-centered, providing integrated care and services designed to provide choice, independence, and dignity.
  - Individuals have a consistent and stable relationship with a care team that is responsible for comprehensive care management and service delivery, including comprehensive transitional care.
  - Local access to care, including use of community health workers and nontraditional settings.
  - Referral to community and social support services, with access to statewide resources when needed.

Source: Legislative Concept paper: preliminary synopsis for discussion purposes. Health Systems Transformation Team, Oregon Health Authority, February 2011.

At the outset, the range of possibilities for the delivery systems was very broad and included the use of regional health authorities, the possibility of bidding out to insurers, Accountable Care Organizations (ACOs), a single payer system, or the use of locally developed delivery systems. Almost all stakeholders had their advocates and all had significant detractors. In the beginning, none of these models were the clear front runner.

## THE REGIONAL HEALTH AUTHORITY CONCEPT

The most transformational concept proposed was for regional health authorities. Although the initial reception was mostly negative, the regional health authority concept received a full hearing at the Tuesday night meetings. This concept called for three to seven larger regional health entities which would contract with the state and then with providers in their geographic areas to provide care. These organizations would aggregate risk and become the single payer for Medicaid services in their defined regions. The board of these regional health authorities would include county commissioners, consumers, local citizens, and representatives of the delivery system.

The regional health authorities would be the purchasers of care for up to 25% of the population and would be heavily focused on transforming the health delivery system, to include population health and the social determinants of health. The size and buying power of these authorities suggested that they would have power to transform the delivery system. It was anticipated that they might later expand to provide care for state employees and retirees, thus increasing their leverage on the provider community and becoming the largest purchaser of care in each of their regions.

There were obvious potential advantages of such a system including

- regionalization and local accountability of services,
- transparency of cost and quality data,
- rationalization of costs and services over a larger geographic area, and
- public input as to the delivery of care.

The benefits for transformation were obvious. However, there were no existing models in Oregon or in other states. These regional health authorities would require considerable cooperation among the several counties in the regions, a big undertaking in those regions where the individual counties are geographically large, politically independent, and in many cases, politically divergent. For example, neighboring Lane and Douglas counties in West Central Oregon would likely be in the same region, but are politically quite different, with President Obama receiving 60% of the vote in Lane County in 2012, but only 35% in Douglas County. As Oregon has 36 counties, each of the regional health authorities would be spread over multiple counties, further complicating the issue.

This concept would deal a blow to many of the existing Managed Care Organizations (MCOs), many of whom were locally prized, financially stable, and politically well connected. Commercial insurers were concerned that these regional health authorities might assume care for state employees and retirees, a business line long held by the commercial insurers.

The regional health authority would also require considerable cooperation among hospitals and provider groups, most of whom were competitive and some of whom were on very unfavorable terms. For example, in Lane County, the two hospital systems had been involved in a protracted legal battle that made it unlikely that they would be cooperative partners. This organizational model was also feared by some small-town interests to tremendously favor the largest hospital or health system in the region, eclipsing smaller and perhaps competitive systems in the region.

Perhaps most importantly, the timeline for transformation was very tight—these new regional health authorities would need to be up and running in less than 18 months. Given that time frame, it was seemingly impossible to pull together the degree of cooperation, the intergovernmental agreements and the development of the boards, staff, and systems needed to enact regional health authorities in the few short months allotted. Thus, despite the great advantages of regional health authorities, this model was dropped, but many of the goals were retained, at least in part, by the CCO model.

## CONSIDERING OTHER DELIVERY MODELS

A statewide fee-for-service system was only briefly considered. At the time of transformation, Oregon had a "fee-for-service" Medicaid system that covered a little less than 20% of the Medicaid population, mostly in areas without fully capitated care or for those who were tribal members. The bulk of the patients were enrolled in 13 Fully Capitated Health Plans. Most of these plans developed in the mid-1990s and were outgrowths of local Individual Practice Associations (IPAs) or hospital systems. These Fully Capitated Health Plans had been reasonably financially successful and had good provider participation and moving back to a fee-for-service system was seen as a step backward. Also, during the years of managed care, many of the competencies needed for a fee-for-service single payer system no longer existed within the state bureaucracy and building those competencies in such a short time seemed both impractical and politically impossible. Lastly, a fee-for-service system had no obvious method to control costs. Thus, a statewide fee-for-service system dropped from consideration.

The possibility of bidding out the delivery system to commercial insurers was also considered. In 2010, Chris Dudley, the former Portland Trailblazer star and Republican candidate for governor suggested that if elected, he would consider this strategy for Medicaid delivery. He lost by less than 23,000 votes to Governor John Kitzhaber and there was some sentiment to consider this proposal. However, unlike other states in which large commercial insurers have dominated the Medicaid market, the large insurers had only a limited presence in the Medicaid market in Oregon. When the fully capitated plans were begun in the 1990s, several insurers were initially part of the system, but dropped out after a few years, leaving no big insurers and no national plans in the mix. It was felt that commercial insurers could not provide the amount of community involvement required and more importantly, none of the big insurers expressed much interest. While the commercial insurers would later be important parts of several delivery systems during these discussions, bidding this out to the insurers did not receive much consideration.

The popular federal model of the ACO was briefly considered. In the excitement over the ACA, ACOs were seen as a very early experiment with some promise to improve quality and accountability, but the complexity of member assignment, the lack of local control, the uncertainty about cost saving, and real concerns about the durability of this concept led to abandonment of the ACO model. Nonetheless, accountability for care, strong primary care, and a quality focus were concepts that were strongly desired by Oregon.

## THE EMERGENCE OF THE CCO MODEL

Conceptualizing the new model was made easier by naming it. Representative Tim Freeman proposed a name that would meet the goals of transformation and proposed the term "Community Coordinated Care Organization," later shortened to Coordinated Care Organization (CCO). The name and concepts of strong primary and coordination of care were quickly embraced, but the form was still unsettled.

Oregon had a reasonably robust and successful managed care plan in place since the mid-1990s. This system was delivered by 13 MCOs or Fully Capitated Health Plans reflecting popular terminology and structure of the mid-1990s. The system geographically covered most, but not all of the state, had very little community involvement and focused almost entirely on physical health. The MCOs received a capitated payment for their services and then contracted with local providers to provide the services. This system had been, in general, financially successful. Despite very low reserve requirements, none of the plans had gone insolvent. The plans were operationally stable, with little change in service areas and great continuity in plans for many years. There was experimentation within these plans, including some innovative payment models and some forays into population health, but the degree of innovation required of the MCOs was low and the amount provided for experimentation by the MCOs was correspondingly low. There was little community input required and not much was sought. There were few required quality measures and no real way to judge quality of care.

Many of these MCOs started as projects of the local IPAs. In their local areas, they were generally popular with their providers, begrudgingly accepted by the hospitals and generally embraced by their communities for maintaining local control. Their local roots gave them outsized political influence, further aided by a powerful advocacy group, the Coalition for a Healthy Oregon (COHO; www.cohoplans.org) that represented many of the plans. In other parts of the state, the Fully Capitated Health Plans were associated with hospital systems or local insurers. In the Portland area, there were several competitive large health care systems.

Given the short time frame, the reasonable performance of the previous MCOs, and the political difficulty of eliminating the MCOs, the MCOs were seen by some as critical to the new system of CCOs. However, continuing the MCO status quo would not serve the interests of greater community involvement, payment innovation, the addition of mental and physical health, and focus on population health or quality. Thus, a consensus developed that MCOs might serve as the basis for the new delivery system, but that changes in governance, transparency, scope, and quality would be required. As there was no other obvious and politically acceptable alternative, this consensus strengthened.

With a relatively blank slate, the Tuesday evening group undertook some basic questions of structure and function, including coordination of care, the inclusion of for-profit models, the degree of community involvement desired, the board structure for the CCOs, the risk arrangements with the CCOs, and the required levels of reserves.

The previous MCOs dealt only with the delivery of medical care. Dental care organizations were parallel systems for dental care and mental health care was provided by mental health care organizations, frequently in conjunction with county sponsored mental health programs. There was an overwhelming sentiment that splitting the patient into three separate and frequently uncoordinated care systems violated many of the principles of transformation and the CCOs were required to incorporate mental and dental health in addition to physical health. This coordination of care seemed intuitive yet was rarely achieved in other models, including other Medicaid delivery models or commercial insurance models. Coordinating this care meant that dental care organizations and mental health organizations would need to fold into or contract with the new CCOs. Despite the many changes and compromises that would be needed to accomplish this coordination, the benefits of coordination won the day and coordination of physical, mental, and dental health was achieved.

There was sentiment that the CCOs should be not-for-profit entities and that any revenue in excess of expenses should be invested back into the community. As many of the predecessor MCOs were for-profit entities, they resisted any move to require not-for-profit status. Oregon already had reasonably good experience with for-profit MCOs that covered a large portion of the state and as there was no obvious alternative not-for-profit entities in those areas, the decision was made that any Oregon corporation, regardless of type could act as a CCO. This allowed Limited Liability Corporations (LLCs) to become contractors. LLCs by themselves do not have profit or not-for-profit status, but pass through distributions to their owners, who may or may not be not-for-profit. In many instances, there were several owners of the contracting LLCs, sometimes with differing status and thus CCOs could be wholly not-for-profit, partially not-for-profit or wholly for-profit.

The plan called for considerable community and provider input with a requirement that each of the CCO boards has a doctor, a mental health or addictions professional, and members of the community on the governing board. The existing MCOs objected that others would have control of the money they had invested, and subsequent legislation called for a majority of the board to be controlled by the group that bore a majority of the financial risk. In practice, community involvement has been inconsistent, with excellent community representation in some areas, but in others, the community and provider members see board membership as a token appointment with the board activities controlled by the risk bearing owners.

In addition to the composition of the board, the plan called for the formation of a Community Advisory Council (CAC) within every CCO. The CAC was designed to be broadly representative of the community and the chair of the CAC was to have a board seat on the CCO Board. The CAC was charged with developing a community health assessment (CHA) and a community health improvement plan (CHIP). As might be expected, the degree of community involvement varies by CCO, but in general the first round of CHAs and CHIPs were rather robust. Again, the performance varies greatly by CCO, with some CACs empowered and others seen as a useless appendage. As we now enter the second round of CHIPs, it will be instructive to see whether the original enthusiasm can be sustained.

## Risk and Funding Arrangements

The risk arrangements for the CCOs were only briefly discussed as the plan for the CCOs was to have a budget that was predictable and sustainably increased annually, but at a rate slower than that of medical inflation. A fee-for-service or "cost plus" model between the state and the CCOs would have made cost control nearly impossible. As most of the CCOs (or their predecessor organizations) had long experience and considerable success with capitated payments, a global capitation method of funding the CCOs was retained. The capitated model has served the state well and most of the CCOs did well, as evidenced by positive margins for almost all and considerable profits by some.

However, in subsequent years and with the addition of the ACA Medicaid expansion population, rates changed significantly, but not always predictably, not always positively, and in some cases, perhaps not sustainably. This has led to discord, with one of the CCOs suing the state over its rates.

The only specified payment relationship between the CCOs and their providers was for hospitals, in which the governor interceded and a rate was set as the floor for hospital reimbursement. For other providers, payment amounts or methods were not specified and all manners of payment models were used including fee-for-service, subcapitation, quality adjusted payments, and many variants thereof. The large number of CCOs and the many payment schemes and rates led to a rapid evolution of many payment models, although at this early stage, there is no clear leading model.

The previous MCO arrangement called for risk reserves that were low for the amount of premium received by the MCOs and in stark contrast to the reserves required for commercial insurers. While this was a potentially risky arrangement for the state, in the 18 years of MCO operation, not a single MCO failed or needed to be taken into receivership. However, in a prudent step, the plan and subsequent legislation called for CCO reserves to increase over the ensuing 5 years. This requirement to add to reserves was a struggle for some of the CCOs. A lingering question is ownership of those reserves in the event that a CCO no longer provides services.

## Service Areas

The service areas for CCOs were another area for discussion. While there was some desire to have competition in all areas and some desire to follow county borders, this proved not to be feasible, as the natural flow of patients and commerce did not always follow county borders and in some rural areas, more than one CCO was not feasible. Thus, CCOs made themselves available to designated zip codes. Amazingly, every zip code had at least one CCO, most zip codes have a single CCO, and a few have more than one. In zip codes with more than one CCO, it is the patient's choice as to which CCO they want to join. Fig. 5.1 shows the distribution of CCOs in the state as of 2016.

There was broad agreement that improving primary care, choice of providers, achieving equity in care, smooth transitions in care, and holistic care were key features of transformation and needed to be included in any legislation and eventually all were included in legislation.

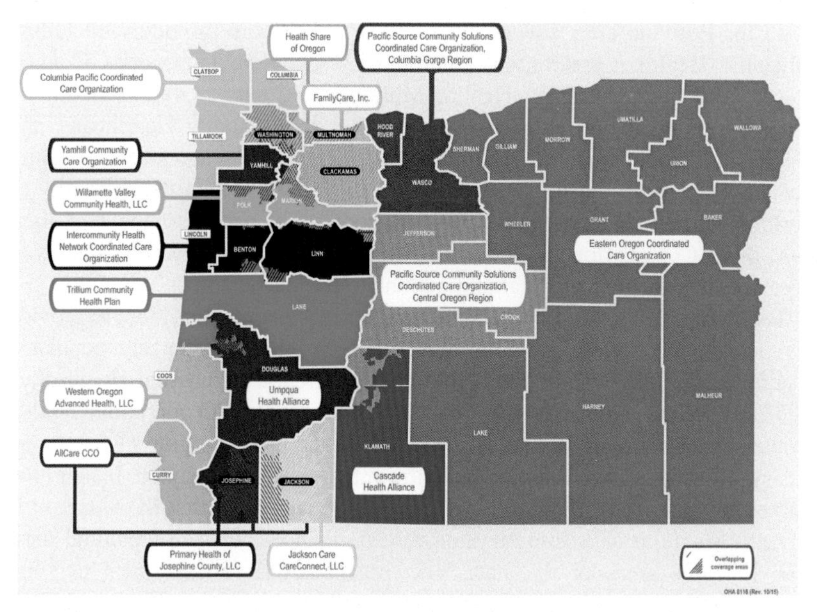


**FIGURE 5.1** Coordinated Care Organization service areas. *From Oregon Health Authority. Available from: https://comm.ncsl.org/productfiles/83403380/CCO_Service_Area_Map.pdf*

## Legislation Drives CCO Development

At the conclusion of the Tuesday night meetings, there was a remarkable degree of consensus among the participants. The Committee on Health Care Transformation took the task of writing the legislation that would become House Bill 3650 (http://gov.oregonlive.com/bill/2011/HB3650/). Remarkably, the final product differed considerably in detail from the original white paper, although the underlying principles of community and provider engagement were retained and perhaps strengthened. When debated on the house floor in the spring of 2011, the bill was passed in the evenly split House 59-1 and in the Senate 22-7 and signed by Governor Kitzhaber on July 1, 2011. Given the desire to begin the CCO model by the end of 2012, an expedited application process was constructed. There was furious activity over the summer of 2012 to nail down final decisions (such as the hospital default payment rate) and finalize contracts. The first contracts offered were seen as unworkable and in a remarkable feat of cooperation, the state Medicaid director and interested parties worked late nights and weekends to fashion a workable contract. The lengthy CCO applications were approved in August 2012 and the first CCOs began seeing patients on October 1, 2012.

In southwest Oregon, the MCOs evolved into the new CCOs, generally with new names, with similar leadership, similar business structures, and similar staff, but now with requirements for increased community participation, greater level of reserves, and the addition of dental and mental health care responsibility.

In the Portland area, several of the large health care providers, including Adventist Health, CareOregon, Central City Concern, Clackamas County, Kaiser Permanente, Legacy Health, Multnomah County, Oregon Health & Science University, Providence Health & Services, Tuality Healthcare, and Washington County joined forces to form HealthShare, a not-for-profit entity that would grow to be the largest CCO in the state. Despite the difficulty of working with such large and powerful groups with strong leaders of their own, this CCO has done well financially and has been innovative, thanks in large part to excellent medical and administrative leadership.

PacificSource Health Plan worked with various county providers to provide coverage in the rapidly growing central Oregon area. In sparsely populated eastern Oregon, the insurer Moda and Greater Oregon Behavioral Health, Inc. (GOBHI) joined to provide care as Eastern Oregon CCO for about 50,000 people spread out over 49,732 square miles. In the capital region of Salem, an uneasy alliance of a local IPA and the major hospital resulted in a group marked by external turmoil and lawsuits. While it appears that the situation is currently stable, it calls into question forced alliances between unnatural partners. In some areas of the state, new organizations formed, such as rural Yamhill county, where new and innovative partnerships were developed leading to coverage in areas not previously served by the MCOs (Table 5.2).

**TABLE 5.2 Oregon CCO Overview**

| CCO | Business Model | Membership (as of November 2016) | Website |
|---|---|---|---|
| All Care CCO | For-profit corporation, a project of the local IPA[a] | 46,600 | www.allcarehealthplan.com |
| Cascade Health Alliance | LLC[b], a project of the local IPA and hospital | 15,438 | www.cascadehealthalliance.com |
| Columbia Pacific CCO | LLC, formed by CareOregon[c] with diverse governmental and health care leadership | 23,063 | www.colpachealth.org |
| Eastern Oregon CCO | LLC, owned by a behavioral health organization, an insurer (Moda Health), local hospitals, and physician groups | 45,403 | www.eocco.com |
| FamilyCare, Inc. | Not-for-profit corporation | 113,147 | www.familycareinc.org |
| Health Share of Oregon | Not-for-profit corporation made up of 11 health care organizations and counties | 206,874 | healthshareoregon.org |
| Intercommunity Health Network CCO | Not-for-profit corporation associated with Samaritan Health System | 50,712 | www.samhealth.org |
| Jackson Care Connect | LLC formed by CareOregon,[c] an assumed business name of Jackson County CCO | 27,446 | www.jacksoncareconnect.org |
| Pacific Source Community Solutions CCO, Central Oregon Region | Domestic business corporation, associated with PacificSource, an Oregon independent, not-for-profit insurer | 46,193 | www.communitysolutions.pacificsource.com/ |

*(Continued)*

**TABLE 5.2 (Continued)**

| CCO | Business Model | Membership (as of November 2016) | Website |
|---|---|---|---|
| Pacific Source Community Solutions CCO, Columbia Gorge Region | Domestic business corporation, associated with PacificSource, an Oregon independent, not-for-profit insurer | 11,865 | www.communitysolutions.pacificsource.com |
| PrimaryHealth of Josephine County, LLC | LLC formed by Grants Pass Management Services, associated with a multi-specialty medical group | 10,222 | www.primaryhealthjosephine.org |
| Trillium Community Health Plan | Domestic business corporation associated with Centene, a for-profit national insurer | 84,497 | www.trilliumchp.com |
| Umpqua Health Alliance | LLC owned by Architrave, a venture of the local IPA and hospital | 25,155 | www.umpquahealthalliance.org |
| Western Oregon Advanced Health, LLC | LLC formed by a local IPA | 18,627 | www.woahcco.com |
| Willamette Valley Community Health, LLC | LLC formed by local IPA, hospital, and governmental groups | 91,704 | www.WVCHealth.org |
| Yamhill Community Care | Assumed business name for a domestic not-for-profit corporation with diverse representation by government, providers, and hospital | 21,934 | www.yamhillcco.org |

[a] *Independent Physician Association.*
[b] *Limited Liability Corporation.*
[c] *CareOregon—a 501(c)(3) public benefit nonprofit organization.*

## CCO SUCCESSES AND FAILURES

By the fall of 2016, the CCO process is old enough to begin to judge its successes and failures. As this is a system with 16 very different CCOs, the results are mixed and somewhat in the eye of the beholder. Objectively and to the good, the CCOs did a very good job of providing an insurer option in every part of the state. The CCOs have been a stable delivery system with no CCO financial failures and minimal change in coverage areas. The leadership of the CCOs has changed some in the interval, but at least half of the original CCO CEOs remain, giving remarkable continuity to the system. This is in stark contrast to the leadership at the state, which has seen a nearly total turnover of key staff since the formation of the CCOs.

The relationship between the state and the CCOs has been one of the most interesting stories. Initially, the CCOs and the OHA were on a "shared journey" toward transformation. The degree of collaboration and cooperation was remarkable, including a monthly informal meeting between CCO CEOs and top leaders at OHA, including the Governor on occasion, to discuss the progress of transformation. Many saw this as quite helpful in smoothing out the very many rough edges of such a rapid and complicated process. Others saw this as inappropriate closeness between the state and its contractors. In any case, the state was a fairly lenient overseer and gave the plans a fair amount of room to grow and innovate. This relationship has changed with new players on both sides and will bear further watching.

The CCOs did a good job of providing coverage for the many new members enrolled as a result of Medicaid expansion under the ACA in early 2014, when the Medicaid population increased significantly. While there were issues with primary care access, the problems were less pronounced for CCO members than what was occurring in other segments of the health care market such as Medicare.

Financially, the plans did well. None of the plans failed financially or needed to be "rescued." The financial success of the plans was perhaps too robust, with several of the plans reaping profits of greater than 10%, calling into question undue profits derived from the use of Medicaid funding. The decision to allow for-profit companies is one of the decisions that continues to be debated. Large profits by several for-profit plans in 2014 and 2015 further fueled this debate. In 2015 Eugene-based Trillium/Agate was bought out by the Missouri-based for-profit Centene corporation for over $90,000,000. There were multimillion dollar payouts to investors and to member physicians, leading to local concerns that a plan for Medicaid members should have generated such profits. At this time, the profits derived by the for-profit plans since formation of the CCOs is in excess of $200,000,000. It is anticipated that the inclusion of for-profit entities will continue to be revisited in the upcoming legislative sessions.

The CCOs were successful in engaging providers to address the quality incentive metrics, with almost all insurance plans meeting most or all of the required metrics resulting in remarkable decreases in emergency department (ED) utilization and hospital readmission. This improvement is a testament to the high degree of provider engagement that was required to improve quality measures such as developmental screening and colorectal cancer screening. The improvement on the quality metrics that are not incentivized is less impressive. The incentivized metrics provide only a snapshot of a small portion of care and whether overall quality improved is subjective. There has been no suggestion from providers, patients, or metrics that quality has decreased. The increase in primary care utilization and decrease in ED use is impressive and the anecdotal stories are encouraging, but the Medicaid system is in many ways tied to other societal and health system trends, so that a definite cause and effect relationship to the CCO model is hard to prove. Nonetheless, the CCO model has likely met the goal of improved care while living on a fixed budget.

The CCOs were somewhat less successful in improving patient satisfaction or access as measured by the CAHPS surveys, although this is probably due to larger community trends, such as a shortage of primary care providers. That neither access nor satisfaction was highly incentivized might also contribute.

Community involvement was variable throughout the state and the durability of the community involvement is still a question mark. The lack of a reliable measure of community involvement makes any objective measure difficult and the subjective measures are just that. The integration of dental and mental health services is similarly a mixed story with some notable successes, but other areas where little improvement has been made. Public health has been more involved with the CCOs, but again the picture is mixed and hard to measure.

One experimental feature of the model was the addition of "innovator agents." As suggested by Atul Gawande's analogy of farm agricultural agents ("Testing, Testing," *The New Yorker*, December 14, 2009), these were envisioned as mid- to high-level state employees who would be the sole point of contact between the state and the CCOs and would serve as agents who could help the CCOs to transform and to spread best practices. As with many features of the transformation plan, innovator agents have been variably successful in different CCOs. The use of nontraditional health care workers, navigators, and the allowance of spending for nonmedical items such as air conditioners was agreed to and is discussed elsewhere. A transformation center was established and each of the CCOs engaged in numerous transformational ventures supported by the center.

## CONCLUSION

Overall, the CCO experiment has been a qualified success. The triple aim of improved care, improved population health, and a stabilization of costs has been met. As we now come to the end of the 5-year waiver, it appears that the CCO model is popular and likely to be retained. No better models have surfaced and other possibilities at the time of the CCO creation, such as the ACO model, have continued to show variable success. However, there will certainly be calls for changes to the for-profit status of CCOs, limits on CCO profitability, and calls for more community involvement. In addition, we may also see a call for consolidation of the many CCOs.

The learnings from the CCO experiment are many and may be instructive to those looking to transform their systems. Some of these include:

- The involvement of many stakeholders at an early stage was messy, but a well-thought-out proposal followed by a robust discussion phase was very useful to arrive at a working group consensus, so that the number of decisions necessary at the legislative level was manageable.
- The consensus model was crucial to allow CCOs to grow and develop with a minimum of second guessing, in contrast to the federal ACA legislation that faced legal and political challenges at every step.
- Sticking firm to the principles of community involvement, coordination of care, and the triple aim was crucial.
- History is important—the political differences between counties, the competitive feuds between providers, and the previous history of successful MCOsshaped decisions.
- Heterogeneous delivery systems can be challenging but may be necessary to have local delivery systems.
- Success is multifactorial and will be in the "eyes of the beholder." A clear definition of success in the formative stages will be useful, especially for more difficult issues such as transformation or community involvement.

Chapter 6

# Primary Care as a Cornerstone of Reform

**Jeanene Smith[1] and Nicole Merrithew[2]**
*[1]Health Management Associates, Portland, OR, United States, [2]CareOregon, Portland, OR, United States*

Primary care has been the foundation for Oregon's health reform efforts for several decades. Policies and programs have been designed to prioritize care toward prevention and early intervention. Oregon's primary care infrastructure began a dramatic transformation in 2009 with the creation of the Patient-Centered Primary Care Home (PCPCH) Program, in conjunction with several key federal initiatives, and the adoption of the Coordinated Care Model (CCM) in the Medicaid and state employees' health programs. Early evaluation results demonstrate a positive impact of the PCPCH model on health processes and outcomes. This chapter identifies and describes the role of some of the policies and programs that worked synergistically to move Oregon's primary care system forward, beginning with a brief history of Oregon primary care policy, and providing perspective on barriers and lessons learned.

## POLICY DRIVERS AND LANDSCAPE

### A History of Primary Care Reform

Primary care has been an integral feature of Oregon's health policy initiatives and landscape for the past few decades. Developed in the mid-1980s as a cornerstone of the Oregon Health Plan, Oregon's Prioritized List of Health Services used a strong primary care perspective to ensure Medicaid coverage of common conditions treated in the community, in addition to prioritizing coverage based on academic evidence. In the early 2000s, initiatives of state policy decision-making bodies such as the Oregon Health Council and the Oregon Health Policy Commission called for strengthening primary care to enhance prevention and reduce utilization of expensive care in hospitals and emergency rooms. This policy focus was coupled with strong leadership in

**Health Reform Policy to Practice. DOI: http://dx.doi.org/10.1016/B978-0-12-809827-1.00006-9**

the Legislature and in the state health agency by primary care providers, which enhanced the education of other policymakers as to the importance of primary care in delivery system transformation. Support for primary care transformation took a leap forward in 2008 when the Oregon Health Policy Board charged the state with providing access to patient-centered primary care for all its covered lives including Medicaid and the Children's Health Insurance Program, state employees, and Oregon educators. In addition, the Board urged the state to successfully spread adoption of the model outside of publicly sponsored health coverage such that 75% of all Oregonians had access to this kind of care. This was codified in legislation in House Bill (HB) 2009 in July 2009, giving birth to Oregon's PCPCH model.

## The Coordinated Care Model and PCPCH

A second piece of legislation in 2011, Senate Bill (SB) 3650, was also critical to primary care reform in Oregon as it provided the vehicle for integrating the PCPCH into a statewide health system transformation using the CCM. The PCPCH lies at the core of the CCM and shares its Triple Aims' goals: a healthy population, extraordinary patient care, and reasonable costs. PCPCHs achieve these goals through a focus on wellness and prevention, coordination of care, active management and support of individuals with special health care needs, and a patient- and family-centered approach to all aspects of care. Fig. 6.1 illustrates the relationship of the PCPCH to the

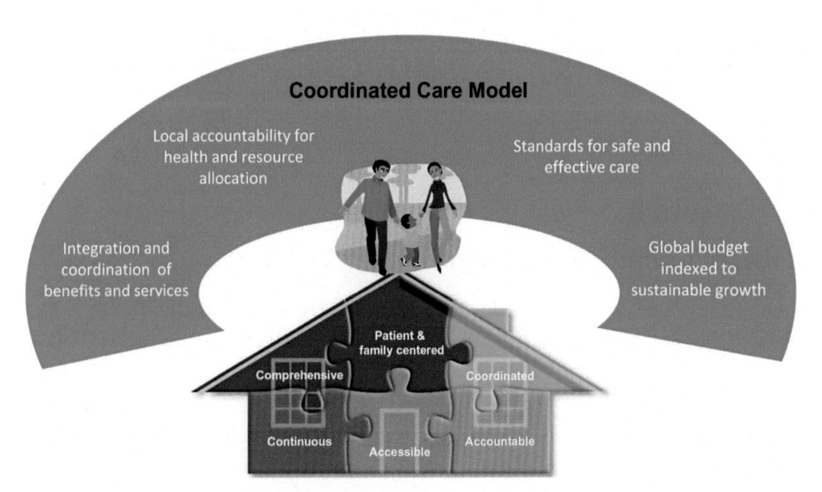


**FIGURE 6.1 Primary care home at the core of Oregon's Coordinated Care Model.** *Modified from Oregon Health Authority. Patient-Centered Primary Care Home Program. 2014–2015 Annual Report. Available from: http://www.oregon.gov/oha/HPA/CSI-PCPCH/Documents/2014-2015%20PCPCH%20Program%20Annual%20Report.pdf.*

CCM, where the PCPCH serves as a hub of primary care service delivery and coordination.

The CCM was first introduced in Oregon's Medicaid program in 2012 to better integrate and coordinate all of a member's benefits and services, with an expectation that care would be centered in PCPCHs. Spread of the CCM beyond Medicaid to the state employee population resulted in placing both CCOs and commercial health plans on new CCM-based contracts that included the requirement to use PCPCHs recognized by the state to the greatest extent possible in their networks and maximize PCPCH enrollment of members. The Public Employees Benefit Board (PEBB), the governing body for the public employees' program, even reduced the benefit copays for primary care visits if a member was seen in a PCPCH. These requirements were critical to the success of PCPCH model adoption and spread across primary care clinics, which will be discussed later in this chapter.

## FROM LEGISLATION TO PRACTICE

### Patient-Centered Primary Care Home Model Development

Oregon drew heavily upon expertise and local experience in developing and implementing the PCPCH model through convening a stakeholder advisory committee in 2010 to create a common set of PCPCH Standards among payers. The committee included practicing clinicians, policymakers, patient advocates, insurers, and the major medical professional societies. The objective in creating the Standards was to reduce administrative burden for the primary care providers rather than having to conform to each insurer or CCO requirements as the delivery system transformation moved toward value-based payments.

This work arose as the nation was looking at the Patient-Centered Medical Home (PCMH), a model initially developed by pediatricians focused on coordination of care for children with special needs. The National Committee for Quality Assurance (NCQA) had developed certification standards to define the PCMH, and many provider groups had embraced this team-based model of care. Due to its national recognition, many state and federal initiatives pointed to the NCQA definition. However, the 2010 public stakeholder committee responsible for Oregon's PCPCH development reviewed various PCMH approaches, including the 2008 NCQA definition, and felt that those standards did not go far enough to encourage improved health outcomes and change in delivery of care.

Ultimately, the committee chose to design a model from the perspective of the patient and family, framing the Standards around six key attributes outlined in Table 6.1. The person-centered language was an instrumental step for the multistakeholder committee to create a common vision for this new model of primary care: patient- and family-centered care. By framing the model

**TABLE 6.1 Patient-Centered Primary Care Home Attributes**

| Attribute | Patient-Centered Description |
|---|---|
| Access to Care | "Health care team, be there when we need you." |
| Accountability | "Take responsibility for making sure we receive the best possible health care." |
| Comprehensive, Whole-Person Care | "Provide or help us get the health care, information, and services we need." |
| Continuity | "Be our partner over time in caring for us." |
| Coordination and Integration | "Help us navigate the health care system to get the care we need in a safe and timely way." |
| Person and Family-Centered Care | "Recognize that we are the most important part of the care team—and that we are ultimately responsible for our overall health and wellness." |

Source: Oregon Health Authority. Patient-Centered Primary Care Home Program. 2014–2015 Annual Report. Available from: http://www.oregon.gov/oha/HPA/CSI-PCPCH/Documents/2014-2015%20PCPCH%20Program%20Annual%20Report.pdf.

around language such as "be there when we need you" and not health insurance quality measure language such as "third next available appointment," the Committee's perspective focused on the needs of the person and their family when encountering the health care system. The key attributes were fully endorsed by the committee (Oregon Health Authority, 2010).

The Standards committee also considered the importance of integration across behavioral health and physical health. The committee was cochaired by a physician and a director of a mental health clinic to bring a broad perspective to the work. There were several discussions about patients with behavioral health needs, particularly those with a diagnosis of severe and persistent mental illness, who have a stronger relationship with a behavioral health provider than a physical health provider. The group agreed that the Standards needed to be flexible enough to allow continuation of those relationships, either through behavioral health providers building on-site primary care capacity or through close relationships with medical providers. The committee decided to be agnostic about the type of clinic eligible to be a PCPCH, which ultimately resulted in several behavioral health providers and even several women's health clinics (for the course of a woman's pregnancy) becoming certified.

Another area of focused discussion for the group included the concept of the medical neighborhood as well as population health. The first iteration of the Standards, and subsequent updates, incorporates aspects of these concepts by encouraging the primary care providers to look outside the walls of their clinics to improve the health of their patients. As one committee member stated, she would rather not diagnose and manage diabetes in her patients but work with community partners to reduce the risk of developing diabetes in the first place.

Each attribute was assigned several corresponding measures, which divided PCPCH recognition into three tiers according to the complexity of the measure. To fully engage primary care providers across Oregon, the stakeholder group felt that it was critical to acknowledge the spectrum of transformation capacity and meet providers where they were. As one rural physician committee member stated, "We need to make a model where most the clinics in the state should be able to be a Tier 1 (the least advanced level)." It was also acknowledged that to encourage providers to progress along the transformation spectrum, the expectations and requirements should become more stringent over time. Since its inception, the advisory committee has continued to update and refine the measures. Beginning in 2017, the PCPCH model will have five tiers, spreading the criteria over a wider spectrum. The intent in doing so is to encourage further evolution of the clinics in their efforts to improve care coordination and health outcomes, as well as to more accurately categorize a clinic's PCPCH maturity. The refined model will also place a greater emphasis on integration of behavioral health services in primary care, a critical component of a robust primary care home, and Oregon's CCM.

## Statewide Implementation

After the stakeholder advisory group had completed its work on PCPCH development in 2010, the state had a framework for primary care transformation but no identified mechanism for adoption and spread of the model. The Medicaid and state employee benefits transformation efforts toward the CCM were just completing development and not fully implemented, so state incentives were not yet in place. However, several federally sponsored opportunities unveiled themselves somewhat simultaneously, all providing levers for PCPCH adoption support.

First, Oregon partnered with Alaska and West Virginia on a Children's Health Insurance Program Reauthorization Act (CHIPRA) Demonstration, which had a focus on PCMH implementation (the summary of the CHIPRA quality demonstration grants is available at: https://www.medicaid.gov/chip/chipra-quality-demonstration-grants-summary.html). Given that Oregon had recently completed its PCPCH model development, there was a strong desire to have clinic participants gain recognition as a PCPCH and use grant funding to support technical assistance for the clinics to adopt the model. This provided the impetus for the state to move the model from theory to implementation and build a mechanism for identifying clinics that met the standards outlined in the model. Next, a Health Resources and Services Administration (HRSA) State Health Access Planning (SHAP) grant provided additional funding to further develop the state's system for recognizing clinics that met the model and provided some additional funding for technical support for model implementation (http://www.hrsa.gov/statehealthaccess/).

Following the SHAP grant, Section 2703 of the Affordable Care Act (ACA) allowed Oregon to align its PCPCH model with the federal government's vision for "Health Homes." This provision enabled the state to provide enhanced funding to clinics that were recognized by the state as a PCPCH. Although only available for 2 years, this was the first payment support available to clinics adopting the model. Beginning in 2012, Oregon's State Innovation Model (SIM) grant from the federal Center for Medicare and Medicaid Innovation (CMMI) supported both the PCPCH recognition programs and expanded technical assistance to gain even broader adoption of the model and movement along the primary care transformation spectrum (https://innovation.cms.gov/initiatives/state-innovations/). Finally, Oregon's participation in CMS' Comprehensive Primary Care Initiative (CPCI) set the stage for multipayer primary care reform (https://innovation.cms.gov/initiatives/comprehensive-primary-care-initiative/).

These federal initiatives were invaluable to assist the state in establishing the PCPCH model in the primary care delivery system as the new Medicaid CCOs were emerging. They enabled the state to build the recognition and technical assistance supports necessary when there was not any funding specifically allocated in the state budget for primary care transformation or PCPCH implementation as Oregon was slow to emerge from the recession. The state applied and shaped the activities under these funding opportunities to fuel its own policy direction while fulfilling each grant's requirements.

## The Outreach and Engagement Strategy

As mentioned previously, there was a strong sentiment within the Oregon provider community that the state needed to meet providers where they were and actively support their movement along the transformation spectrum. Recognizing that the provider community is the fundamental element to this kind of reform, the state embarked on what would become known as the "outreach and engagement strategy." After the initial Standards Advisory Committee finalized its recommendations for the PCPCH model in early 2010, the state surveyed providers and the key Medicaid delivery system stakeholders to better understand which of the PCPCH standards were currently in practice and which were most meaningful or had the most potential to positively impact clinical practice. In doing this, the state hoped to glean which pieces of the model providers may be the most invested in working toward, regardless of the difficulty.

Based on this input and recognizing that implementation of the entire model as articulated by the advisory committee would likely be overwhelming for most providers, the state pared back the number of measures for initial implementation, and focused on those holding the greatest value and level of feasibility to the provider community. The intent remained to increase the rigor of the model over time as practice workflows shifted and provider capacity increased. This approach cultivated support from the major professional societies such as the Oregon Academy of Family Practice, the Oregon Medical Association, the Oregon Nurses Association, and the Oregon Primary Care Association; all of which were critical to successful implementation and spread of the model.

To recognize clinics that had met the PCPCH standards, the state developed a centralized recognition process, similar to other certification or accreditation programs that exist nationwide. The fundamental difference, however, between the state's PCPCH program and other accrediting bodies is the active support built into the verification, or audit, process. Teams of primary care practice coaches partner with regionally based clinicians to conduct site visits to recognized PCPCHs, creating an experience of both regulatory verification and best-practice sharing.

The PCPCH application is largely attestation-based; therefore, these teams are deployed to verify that the structures and processes that the clinic attested to in the application are in place. However, as opposed to taking a strict regulatory approach, the teams provide technical assistance, education, and support to clinics that have discrepancies between what was stated in the application and actual operations. The teams are also available to provide general practice coaching and assistance on workflows, data reporting and interpretation, and a myriad of other issues that clinics regularly encounter, as well as point to additional technical assistance resources the clinic could seek out.

Key to the success of this approach is the inclusion of a locally based clinical champion that has experienced practice transformation in their own primary care clinic. This peer-driven approach has led to extensive provider collaboration and peer mentoring relationships extending beyond just the site visit itself. By coupling a simple application process, free of cost or extensive administrative reporting, with a supportive technical assistance infrastructure, the state successfully spread implementation of the PCPCH standards to most of the primary care clinics across the state. As of September 2016, more than 660, or over 70%, of Oregon's primary care clinics had achieved recognition (Oregon Health Authority, 2016a).

## Technical Assistance Partnerships

Oregon is fortunate to have a wealth of expertise around primary care transformation spread throughout various groups, including its academic medical center and community and provider organizations. These entities have relationships with providers that are stronger than those that the State has and

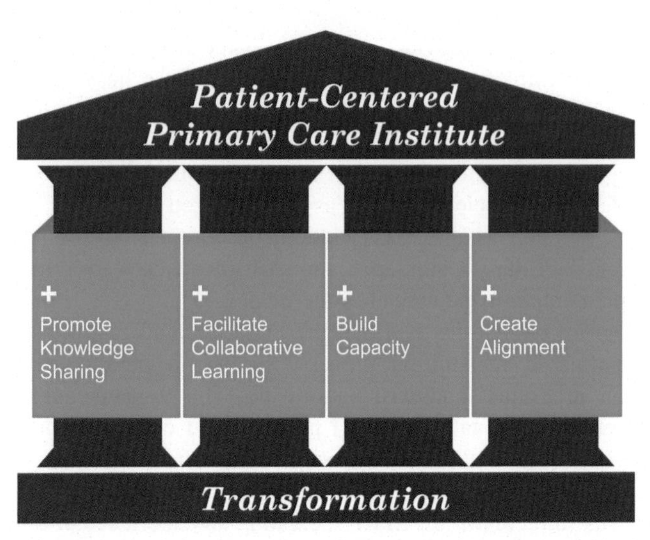


**FIGURE 6.2** Patient-Centered Primary Care Institute pillars. *Image available from: http://pcpci.org/about-institute.*

play an active role in supporting the primary care delivery system in Oregon. Many have practice coaches or other human resources in communities across the state to deploy to primary care clinics needing transformation support. Recognizing the need to partner, the state convened several of these stakeholders along with primary care providers and asked how to best use the limited funding for technical resources it had through the federal initiatives noted earlier.

From these discussions, the concept of the Patient-Centered Primary Care Institute (PCPCI) arose. The Institute furthers primary care transformation by convening organizations that provide clinical technical assistance to share best practices and it also connects primary care clinics to these organizational resources. In a state with numerous, often disconnected, primary care improvement efforts, it serves as a resource hub for these various activities. This collaborative effort has produced an extensive database of coaching materials, webinars, and other resources that are publicly available free of charge, due in part to the state's investment in the PCPCI. The Institute has also supported hands-on technical assistance to multiple cohorts of clinics (Patient-Centered Primary Care Institute, 2016) (Fig. 6.2).

## Sustainability of Primary Care Transformation

As with any systemic overhaul, financial support to the primary care clinics is critical for not only initial change, but also sustainability of reformed activities. Primary care transformation in Oregon so far has been fueled by three key payment incentive initiatives: Health Home payments through

Section 2703 of the ACA; inclusion of PCPCH enrollment as a metric of the state's quality incentive pool for the Medicaid CCOs; and Oregon's participation in CMS' multipayer CPCI. The ACA allowed the state to provide per member per month payments to recognized PCPCHs for their Medicaid population; however, this incentive was limited to 2 years of federal funding. Still recovering from the recession, Oregon could not afford to support continuation of those payments without federal funding, so they ended. CCOs are provided an incentive through the state quality pool to enroll members in recognized PCPCHs, and many CCOs pass this incentive directly through to the PCPCHs and/or pay through alternative payment models. However, the level of funding associated with this is not enough to fully sustain transformation work. Lastly, through the CPCI, PCPCHs are provided financial support from five payers across the state, in addition to Medicare to meet practice transformation milestones. This is also a time-limited opportunity, although a 5-year extension of this initiative is underway as Oregon was selected as a region in the next round of CPC development, renamed as CPC Plus (https://innovation.cms.gov/initiatives/comprehensive-primary-care-plus).

Due to the fragmentation and limited nature of the funding opportunities to support this work, the state and the Oregon Health Leadership Council convened a multistakeholder meeting[1] in July 2013 with the objective of creating a shared vision for a multipayer strategy. The group gained consensus on signing a voluntary agreement stating that all major payers in Oregon would:

1. use a common definition of primary care home based on Oregon's PCPCH Program and
2. provide variable payments, or other payment models, to those primary care practices in their network participating in state's PCPCH program, based on each practice's PCPCH points total and their progress toward achieving outcomes, which lead to the Triple Aim (Center for Evidence-Based Policy, 2013).

While the agreement held promise and was considered by some participants as a success, others held reservations. Among the concerns were a potential risk for non-PCPCH recognized clinics to be financially penalized, lack of

1. Participants included representatives from Lifewise, Tuality Health Alliance, Oregon Health Authority, First Choice Health, Oregon Pediatric Society, CareOregon, Oregon Association of Hospitals and Health Systems, Umpqua Health Alliance, Oregon Medical Association, Childhood Health Associates of Salem, Providence Health Plans, PacificSource Health Plans, Oregon Academy of Family Physicians, Moda Health, Oregon Health Leadership Council, Oregon Nurses Association, Regence Blue Cross Blue Shield, Health Net of Oregon, CIGNA, Aetna, and Kaiser Permanente.

clarity on quality measure alignment, and a potential for additional misaligned administrative requirements given the open-ended nature of agreement.

Unfortunately, the agreement did not result in meaningful change in primary care payment immediately. This, coupled with a charge from the Governor to strengthen the primary care infrastructure, resulted in proposed 2015 legislation, which would have required all major payers to implement some form of primary care alternative payment. The objective of the legislation was to sustain support for primary care transformation activities. Amidst concerns from payers, language in the legislation changed such that the final bill, SB 231, only requires payers to report on the percentage of their total expenditures that are allocated to primary care (Senate Bill 231, 78th Oregon Legislative Assembly, 2015). This did not result in immediate financial support to primary care providers, however, it supported the sentiment that the multipayer agreement had not stimulated meaningful primary care payment reform. Spending allocated to primary care ranged from 3% to 31% across payers (Oregon Health Authority, 2016b). On average, CCOs allocated the largest proportion of spend to primary care, 13.1%, while PEBB and Oregon Educators Benefit Board (OEBB) allocated the smallest proportion, 7.8%. Commercial health insurers and Medicare Advantage plans allocated 9.9% and 8.5%, respectively. The wide variation in spend could be due to multiple factors including organizational priorities, a relatively short timeframe for data collection, a lack of extensive data validation, or demographic variables within each covered population include age, health status, or distance to medical care.

The report also created transparency in the level of support that each payer organization directs to primary care. This transparency will allow for better informed primary care payment policy development in the future. That said, to date, there is not a broad-based sustainability solution for primary care reform. In addition to the report, SB 231 created a state-convened multistakeholder learning collaborative charged with developing and sharing best practices in technical assistance and methods of reimbursement. Recommendations from this group were released in November 2016. Heading into the January 2017 legislative session, discussions continue as to whether further directives will be created.

## Evaluation Results

The primary care reform road has not been short and has, at many times, been rocky. But perseverance and commitment to better health by Oregon's providers, the state, and stakeholder partners appear to be paying off. Evaluation results indicate that, even with fragmented financial support, the PCPCH approach to care produces improvement in health care processes, outcomes, and costs.

A recent evaluation found that PCPCH-designated clinics have accomplished significant transformation, resulting in greater effectiveness and efficiency, both within primary care and the larger health care system. Looking across the clinics using both public and private claims data, total service expenditures per person were reduced by 4.2% or approximately $41 per person per quarter. Effects were more significant the longer clinics were designated as a PCPCH, doubling by the third year of recognition. There was a $13 savings in other services, such as specialty care, emergency department, and inpatient care for every $1 increase in primary care spending across the recognized PCPCH clinics. The report estimates that over $240 million was saved in the first 3 years of the recognition program (Gelmon et al., 2016a).

An earlier study found that relative to non-PCPCH practices, preventive procedures increased by 5.0%, specialty office visits decreased by 6.9%, pharmacy claims decreased by 11.4%, primary care visit expenditures decreased by 3.2%, and specialty office visit expenditures decreased by 6.6% (Wallace, 2014). The study also found decreases in emergency department utilization and inpatient costs for people seeking care in PCPCHs relative to those cared for in nonrecognized practices; however, these results were not statistically significant. Furthermore, a study conducted by the Oregon Healthcare Quality Corporation found that PCPCHs had significantly higher rates of screening, brief intervention, and referral to treatment for alcohol and substance use; diabetes HbA1c testing; diabetes LDL testing, diabetes kidney disease testing; adolescent well care visits; and chlamydia screening (Oregon Health Care Quality Corporation, 2015).

While primary care practice and culture can vary greatly from region to region, the PCPCH standards appear to be achievable regardless of clinic demographics. A 2016 study found little variation in ability to achieve the PCPCH standards among practices grouped by population density or organizational structure, as defined by practice ownership and independence (Gelmon, Sandberg, Merrithew, & Bally, 2016b). There was, however, variation by practice size with clinics having 0–2 practitioners demonstrating the most difficulty. These results, coupled together, further support the outreach and engagement approach that the state took in spreading and supporting clinic implementation of the PCPCH model. If realistic standards for clinics are created, combined with good communication that the standards will increase in rigor over time and hands-on support is provided, clinics can reform their workflows regardless of demographic and, in turn, produce systemic improvements.

## LOOKING BACK AND MOVING FORWARD

### Challenges

As with any major endeavor, hindsight is always 20/20. Many challenges, both at the clinic and state policy level, were encountered that could potentially have been avoided or lessened had they been anticipated or better planned for. The key challenges were:

- *Development of policy without adequate inclusion of operational knowledge*: The best, most well-intentioned, theoretical plans may not be fully realized if logistical details are not considered early in the policy development process. Translation of policy to practice will be most successful if there is clearly a bridge between policy development and operational implementation. An often-overlooked nuance, the professional skill-set required for each of these pieces is distinct. Knowledge of both policy and operations should be included from development through implementation. This requires an intentional connection between policy making and operational implementation at the legislative, state agency, health plan, and clinic level.
- *Existence of data reporting and payment barriers due to unsophisticated data systems at the state, health plan, and clinic levels*: The state and many health plans did not have mechanisms in place to easily and consistently provide payment at the clinic level nor provide data back to the clinics. Many claims databases and payment processes in place are aggregated to the medical group or system level, especially for the Medicaid program. Variance and lack of aggregation of data at the payer level can also lead to a dearth of actionable data at the clinic level. Even for those payers who have found successful mechanisms for providing clinics with data, technical assistance is often required for clinics to translate the data. These facets can all be barriers to clinic-level improvement.
- *Variance in support needed at the clinic level depending on clinic demographics*: While evaluations to-date have not demonstrated a significant difference in a clinic's ability to achieve PCPCH recognition due to demographics, the support needed to implement processes to meet particular standards does vary based on clinic culture, location, size, and population served. Technical assistance must be tailored to the clinic's needs and using local peer-to-peer support is invaluable. This support can often revitalize clinicians experiencing burnout or change fatigue, a sentiment common in an era of accelerating reform.
- *Need for an established, adequate payment structure to support and sustain reformed primary care practice*: Without a long-term funding strategy, processes implemented and structures changed using short-term payment bonuses may disappear or be stunted. Sustainable payment reform and shared, meaningful incentives are critical to sustain and advance the cultural changes in the clinical setting. While there is promising work underway through Oregon's participation in CPC Plus and the SB 231 Primary Care Payment Reform Learning Collaborative, it is still unknown whether these opportunities will yield a concrete long-term strategy. As previously mentioned, time-limited opportunities and collaborative agreements without action will not sustain changes necessary for true primary care transformation.

## Synergies

While these barriers are important to identify and share, of perhaps greater significance are the partnerships and activities that facilitated success. Overall, the key to advancing primary care reform in Oregon has been a combination of several factors.

- *Leveraging opportunities for financial support*: Oregon intentionally captured sequential federal opportunities for financial support, which allowed the state to use each one as a springboard to accelerate its primary care reform activities. The successes experienced to-date may not have been possible without the alignment and sequencing of activities supported by the CHIPRA Demonstration, SHAP grant, ACA Section 2703, CPCI, and SIM grant, which allowed time for state support and local incentives to emerge. Examining any new opportunities either at the federal level, or at the state or county level can assist other states and regions to support primary care reform activities, as well as looking to new accountable care efforts on a local level. Upcoming incentives and value-based payment targets expected by Medicare under MACRA could also enhance payment structures as the public and private payers align, with alternative payment for medical homes qualifying for points under the new Merit Incentive Program (more information about MACRA available at: https://www.cms.gov/Medicare/Quality-Initiatives-Patient-Assessment-Instruments/Value-Based-Programs/MACRA-MIPS-and-APMs/MACRA-MIPS-and-APMs.html).
- *Creating partnerships between regulatory bodies, providers, and payers to move in the same direction*: As with Oregon's broader health reform, garnering support from the legislature, state agencies, and health care providers was critical to both development and implementation of the PCPCH program and other primary care reform activities. The PCPCH program was initiated as a foundation to support the broader transformation work toward the CCM across the state. Its importance was emphasized as accountability for adopting and spreading the model was integrated into Medicaid CCO and state employee benefit contracts. Somewhat unique to PCPCH, however, was the deliberate partnership between the state and practicing physicians to carry-out what could be traditionally viewed as a regulatory function, the PCPCH site visit. The site visits continue to function as a source of practice coaching/technical assistance, peer-to-peer mentoring, best-practice spread, and regulatory audit all at the same time.
- *Developing a unique program that places health care improvement over pure regulation*: Using the site visit, structure as a vehicle for primary care improvement has demonstrated success both at the program and individual clinic level. This would not have been possible if the state had chosen to focus solely on ensuring that clinics were meeting particular standards without providing support for making changes in their practice.

The entire structure of the PCPCH standards also allowed for engagement of clinics that would have otherwise been uninterested or motivated. This alters the usual role of the state as a regulator, and expands the role to be a supportive partner that encourages successful adoption of a model, benefiting both the state and the providers by focusing on improving the outcomes of the patient. By creating standards that acknowledged a wide spectrum of primary care practice capacity and providing support through the practice transformation process, true partnerships were created. Maintaining these partnerships will be essential for continuous improvement to Oregon's primary care.

## CONCLUSION

Primary care has been the foundation for Oregon's health reform efforts for several decades. Oregon's primary care infrastructure began a dramatic transformation in 2009 with the creation of the PCPCH Program, in conjunction with several key federal initiatives, and the adoption of the CCM in the Medicaid and state employees' health programs. Early evaluation results demonstrate a positive impact of the PCPCH model on health processes and outcomes, and new studies are documenting cost savings that increase over time with the PCPCH model. Policies and programs have worked synergistically to move Oregon's primary care system forward, although there have been bumps along the way. Perhaps more important than state regulation, leveraging and alignment of supportive payment structures, statewide partnerships, provider and community focus on health care improvement, and a unique approach to state regulatory function have been key to Oregon's success to have primary care as a cornerstone to reform.

## REFERENCES

Center for Evidence-Based Policy. (2013). Multi-payer strategy to support primary care homes. November 5, 2013.

Gelmon, S. et al. (2016a). Implementation of Oregon's PCPCH Program: Exemplary practice and program findings. Available from: http://www.oregon.gov/OHA/HPA/CSI-PCPCH/Documents/PCPCH-Program-Implementation-Report-Final-Sept-2016.pdf. Accessed 06.11.16.

Gelmon, S., Sandberg, B., Merrithew, N., & Bally, R. (2016b). Reporting mechanisms in Oregon's patient-centered primary care home program to improve performance. *The Permanente Journal*, *20*(3), 15–115.

Oregon Health Authority. (2016a). Patient Centered Primary Care Home Program. PCPCH recognition information for payers. Available from: http://www.oregon.gov/OHA/HPA/CSI-PCPCH/Pages/Recognition-Oregon-Payers.aspx. Accessed 02.09.16.

Oregon Health Authority. (2016b). Primary care spending in Oregon. A report to the Oregon state legislature. February 2016.

Oregon Health Authority. Standards and measures for Patient Centered Primary Care Homes: Final report of the Patient Centered Primary Care Home Standards Advisory Committee

February 2010. Available from: http://www.oregon.gov/oha/HPA/CSI-PCPCH/SACDocuments/2010%20Standards%20Advisory%20Committee%20Report.pdf. Accessed 06.11.16.
Oregon Health Care Quality Corporation. (2015). Information for a healthy Oregon: Statewide report on Health Care Quality. Available from: http://q-corp.org/sites/qcorp/files/qCorp-statewide-report-2015-postpress-corrected-singlepages_WEB-FINAL%20BBF%202.pdf.
Patient-Centered Primary Care Institute. (2016). Available from: http://www.pcpci.org/resources.
Senate Bill 231, 78th Oregon Legislative Assembly. (2015). Available from: https://olis.leg.state.or.us/liz/2015R1/Downloads/MeasureDocument/SB231.
Wallace, N. (2014). Patient Centered Primary Care Home (PCPCH) evaluation: Cost and efficiency. Mark O. Hatfield School of Government, Portland State University. Available from: http://www.oregon.gov/oha/HPA/CSI-PCPCH/Documents/2014%20PCPCH%20Cost%20and%20Efficiency%20Evaluation.pdf.

Chapter 7

# The Path to Integrating Medical, Behavioral, and Oral Health Care: Oregon's Experience With Change

**Deborah J. Cohen[1], Jennifer D. Hall[1], Daniel A. Reece[2] and Eli Schwarz[1]**

[1]*Oregon Health & Science University, Portland, OR, United States,* [2]*Oregon Health Authority Transformation Center, Portland, OR, United States*

*It is very expensive to give bad medical care to poor people in a rich country.*

– Paul Farmer

*Better is possible. It does not take genius. It takes diligence. It takes moral clarity. It takes ingenuity. And above all, it takes a willingness to try.*

–Atul Gawande

The human body perceives symptoms from its different parts holistically. I may have a toothache due to dental disease, but this may have been due to early negative experiences with a dental visit, which created anxiety about visiting the dentist. The dental disease symptoms are now exacerbated because I also suffer from type II diabetes. While the human body experiences these symptoms as a whole, our health care system, in its current, traditional form, responds to my symptoms separately, and by different professionals (e.g., dentist, primary care or emergency room doctor, behavioral health expert). Each professional may attempt to isolate the symptoms and treatment plans within their own expertise without much connection to the others.

Oregon's medical, behavioral, and oral health care delivery systems—like many other states in the United States—have evolved in silos with separate funding streams, state agencies, provider licensing and credentialing methods, payment and billing systems, and data gathering and reporting requirements designed to meet different regulatory and billing needs.

**Health Reform Policy to Practice. DOI: http://dx.doi.org/10.1016/B978-0-12-809827-1.00007-0**

However, an important strategy for accomplishing the triple aim of improving the patient experience of care, improving the health of populations, and reducing per capita costs of health care is to reduce the silos and fragmentation that are rife in the US health system. What is the evidence that suggests that removing the silos that separate medical, behavioral, and oral health care can be an effective strategy for achieving the triple aim?

## EVIDENCE SUPPORTS INTEGRATION

There is mounting evidence that integration of behavioral health (care that addresses mental illness, substance use conditions, health behavior change, life stressors and crises, stress-related physical symptoms, and health care utilization) (Peek & The National Integration Academy Council, 2013) and medical care is essential for improving quality of care, access to treatment, and patient outcomes and reducing costs and fragmentation (Butler et al., 2008; Collins, Hewson, Munger, & Wade, 2010; Croghan & Brown, 2010; Institute of Medicine, 2001; Laderman, 2015; Miller et al., 2014; Stange, 2002).

There is also robust research documenting the benefits of integration. Behavioral health needs are prevalent and often identified in primary care (Kessler et al., 2005; Prince et al., 2007); however, many individuals with a behavioral health issue do not receive behavioral health treatment (Kessler et al., 2005), either because patients never make an appointment after receiving a referral (Fisher & Ransom, 1997; Hoge, Auchterlonie, & Milliken, 2006), or because patients cannot access outpatient behavioral health services due to provider shortages or health insurance barriers (Cunningham, 2009). Left untreated, behavioral health problems increase morbidity and health care utilization (Dickey, Normand, Weiss, Drake, & Azeni, 2002; Scott & Happell, 2011). Likewise, individuals with severe and persistent mental health needs often do not have their general medical needs met as they receive most of their care in specialty mental health settings. This can place them at higher risk for medical problems, such as hypertension, coronary heart disease, and diabetes (Druss, 2007; Nasrallah et al., 2006; Salsberry, Chipps, & Kennedy, 2005).

Similarly, evidence shows the benefits of integrating oral health as part of general health care (Institute of Medicine, 2011; US Department of Health and Human Services, 2000; US Department of Health and Human Services Health Resources and Services Administration, 2014). Integrating oral health into primary care can help increase access to preventive care, treatment, and promote overall health. The two dominating oral diseases, dental caries and periodontal diseases, are highly preventable and very common; left untreated, dental problems accelerate diabetes and cardiovascular disease complications and may also be associated with respiratory diseases and adverse pregnancy outcomes (Bobetsis, Barros, & Offenbacher, 2006). In spite of overall improvements in the oral health, disparities in oral health are significant with a high burden of oral diseases prevalent in the most vulnerable populations. Prevention and early detection are foundational to oral health care and to the role of primary care. Integrating oral health into primary care addresses a currently unmet need for identification, prevention, and control of oral diseases (US Department of Health and Human Services Health Resources and Services Administration, 2014). Expected additional benefits are improved whole person care, better overall health outcomes, and reduced health care costs by avoiding systemic complications of dental diseases on other chronic diseases, and when dental problems are presented in hospital emergency rooms (Sun et al., 2015).

While evidence shows the benefits of integrating behavioral health with medical care and oral health with medical care, Oregon is unique in its attempt to integrate all three. In this chapter, we explore the silos that exist between medical, behavioral, and oral health care and describe the steps that Oregon has taken to reduce those silos and work toward a health system that coordinates medical, behavioral, and oral health.

## OREGON'S EVOLUTION OF CHANGE

The evolution of change in the state of Oregon has been an iterative process, with behavioral health and medical integration, perhaps, further along on that journey than oral health integration. This story starts with each sector as separate silos of care, and it continues with the passing of legislation to reduce fragmentation through the implementation of Coordinated Care Organizations (CCOs) in 2012.

### Delivering Care in Separate Silos

Oral health, behavioral health, and medical care developed and evolved separately, even though Medicaid managed care was applied to these health care delivery sectors since the institution of the Oregon Health Plan (OHP) in 1989. These separate delivery systems were reinforced by different (1) technical foci (mind, body, mouth), (2) types of trained professionals delivering the care, (3) funding streams, and (4) state organizational structures. House Bill 2009 created the Oregon Health Authority (OHA) to oversee all the state's health-related programs, including addictions and mental health, public health, and the OHP (Medicaid). Separate departments were created to oversee operations for each sector, and each maintained different policies and standards for accountability and billing, many of which were influenced by federal regulations and restrictions. For example, mental health services were often paid through federal block grants; these funds were carefully monitored and separate from medical care funding, and the local counties administered funds in their region. This was not the model for medical care.

General dentistry was separate from other health care sectors because of the lack of provisions for dental care in Medicare. In addition, professional participation in Medicaid was limited as a result of poor reimbursement rates for general dental practitioners and inconsistent eligibility criteria for dental services for adults in the OHP (Oregon Medicaid Advisory Committee: Oregon Health Work Group, 2016). Thus, the primary care practices, community mental health centers, dental practices, and substance use treatment programs that deliver medical, oral, and behavioral health services to patients developed quite separately.

State and national quality improvement also affected dental, behavioral, and medical professionals in different ways, creating an imbalance in resources and foci on improvement, which has further separated these service areas. For example, primary care practices in Oregon participated in state and federal programs directed at changing care delivery structures to become Patient-Centered Primary Care Homes (PCPCHs), which focused attention on better coordination and patient-centered care. Other federal initiatives for primary care (e.g., Comprehensive Primary Care Initiative and Meaningful Use) were encouraging practices to implement care coordination, care management and risk stratification, and practices were choosing to take federal incentives to implement electronic health records, and taking steps to "meaningfully use" these tools to promote better care quality. Some of this work included an emphasis on creating patient registries and tracking quality measures.

Dental and behavioral health organizations were not directly affected by these programs, and electronic information infrastructure in these organizations had less sophistication and capacity. Dental organizations had record systems, but they were largely designed for medico-legal and billing purposes, and rarely included diagnosis codes like those included in medical records systems, as dentists primarily use procedural, Current Dental Terminology (CDT) codes. Behavioral health organizations and dental clinics have been slower to adopt Electronic Health Record (EHR) systems. There are multiple reasons for this. First, there has been little incentive for smaller, single-owner behavioral and dental clinicians to adopt EHRs, as there simply was a less urgent need for these tools among independent practices. Second, EHRs are expensive. Even among larger behavioral health clinics who are connected to medical practices, and who see the value of these systems struggle to afford EHRs (Cifuentes et al., 2015). With the absence of federal funding to motivate the purchase and implementation of EHRs, few behavioral and oral health organizations use these tools. Where EHRs are used by these organizations, firewalls are in place to prevent or stringently restrict the flow of clinical information between medical, behavioral, and dental clinicians. Even those practicing within the same organization and using the same information system may have information sharing restrictions. Although there was a regulatory basis for many of these restrictions [e.g., 42 CFR Part 2, Health Insurance Portability and Accountability Act (HIPAA)] particularly related to substance use treatment, the flow of information between clinicians caring for the same patients has been, in our opinion, excessively restrictive.

## Pockets of Innovation

Despite these challenges, there were pockets of innovation emerging among primary care practices, hospitals, community mental health centers, and dental practices across the state. Some examples include a community foundation supporting a pilot psychiatric consultation service for primary care providers; a county mental health center embedding a primary care clinician from a local Federally Qualified Health Center (FQHC) into their center; and using motivational interviewing to encourage oral health prevention among pregnant women and young mothers. While these efforts showed promising results, these small pilots were not being sustained and spread due to lack of system support and financing.

There is good reason for this. For example, a primary care practice must make an investment to integrate a behavioral health clinician in the practice (e.g., salary, benefits, office space) (Wallace et al., 2015). For the practice, there is no return on this investment because the savings accrued from better managing diabetes and depression, such as reduced emergency department utilization does not return to the practice, but to the payer. Health care transformation needed to occur at a macrosystem level—in Oregon's case the state and CCO level—to start the reorganization of and compensation for integrated care.

## Steps Toward Integration

It was in this environment that, in 2012, the state of Oregon passed Senate Bill (SB) 1580 that supported the creation and implementation of 16 regional CCOs (Oregon Legislative Assembly, 2012). CCOs are Oregon's approach to Medicaid Accountable Care Organizations. Through Medicaid Section 1115 demonstration waiver (Centers for Medicare & Medicaid, 2012), the Centers for Medicare & Medicaid Services (CMS) agreed to provide $1.9 billion to Oregon to support transformation, and in exchange, Oregon agreed to reduce Medicaid spending growth by 2% (Centers for Medicare & Medicaid, 2012; McConnell, 2016; Oregon Legislative Assembly, 2012), while not diminishing care quality as defined by 33 quality measures. Quality measures include metrics for medical, behavioral, and oral health. Exceeding the cost cap carried penalties for both the state and the CCOs. Unlike most Accountable Care Organizations in other states, each CCO accepted full financial risk for the members they served in the form of a global budget, and CCOs were intended to oversee the integration of medical, oral, and behavioral health services for more than 600,000 Medicaid members.

Through this legislation, Oregon created a regulatory framework linking historically distinct and separate health care providers and establishing the foundation for an integrated delivery system. The integration of medical and behavioral health care began in 2012 and efforts to start integrating dental care started in 2014, when CCOs began contracting with nine statewide Dental Care Organizations (DCOs), which previously contracted directly with the state, but were now contracting with CCOs. With the new Medicaid delivery structure, both the state and the CCOs were "at risk" and incentivized to reach the triple aim. This led to state-led and CCO-led efforts to support the defragmentation of medical, behavioral, and oral health care, which we describe later.

## OREGON INTEGRATION EFFORTS

Both the state and the CCOs took important steps to lead and support efforts to integrate behavioral, medical, and oral health care. At the state level, this included developing quality metrics, which were tied to financial incentives for CCOs, providing technical assistance, passing legislation to formalize integration, and reorganizing the OHA Medicaid office to reduce administrative silos. CCOs included representatives from the three disciplines on their governing boards and clinical leadership committees and identified administrative people to instigate integration efforts. They were able to identify funding for integration experiments or clinical pilot studies, and financed training to educate CCO affiliates in integration so they would have the knowledge and capacity to implement these programs on the ground.

### State-Led Integration Efforts

The state engaged in efforts to support behavioral, medical, and oral health integration, which include reorganizing the OHA to reduce silos, identifying a set of metrics to measure and incentivize integration, providing technical assistance and resources to the CCOs, and implementing legislation that further supports integration.

### OHA Reorganization

In 2015 OHA started the process of reorganizing its internal structure to better support integration and connection between departments at the state level. The refined organizational structure brought Mental Health and Addiction Services into the same division as medical services, which had been previously separate from the Medicaid program. The reorganization maintained the Oral Health Unit as part of the Maternal and Child Health Section of the

Public Health Division, and subsequent to the reorganization, OHA recruited a State Dental Director (see Section "Legislation"), who reported to the Chief Medical Director in the Health Policy and Analytics Division. The PCPCH program, which certifies clinics and provides them with guidance on meeting standards, such as integrated care standards, was merged with the Transformation Center, a department within the Oregon Health Authority, which aims to test, accelerate, and spread effective delivery system innovations among the CCOs. This merge facilitated closer coordination of technical assistance efforts.

## Metrics

State-led efforts to support accountability for improved health outcomes through quality measurement have been an important feature for supporting integration. Examples of state-developed process measures for medical, behavioral, and oral health care are shown in Table 7.1. The metrics for behavioral health were developed in 2012, and included assessment of screening adults for depression and substance use and postpsychiatric hospitalization follow-up. Oral health metrics were added in 2014, after the Dental Quality Metrics Work group made its recommendations to the Metrics and Scoring Committee. At the time, oral health lagged significantly behind the rest of the health care system in developing quality measures both in Oregon and nationally, where there was a lack of accepted oral health quality metrics (National Quality Forum, 2012). The first metric (i.e., children receiving dental sealants) prioritized prevention of oral diseases in children, with more comprehensive measures to be added later. Quality metrics were aimed at establishing defined benchmarks and performance and improvement targets. Metrics specifically targeted quality in a range of health service domains and were linked to financial incentives, which were intended to provide a focus for the CCOs and their partner health care organizations to foster quality improvement.

## Technical Assistance and Resources

The state has further supported integration of medical, behavioral, and oral health care by offering learning opportunities and technical assistance through its Transformation Center (http://www.oregon.gov/oha/HPA/CSI-TC/Pages/index.aspx). The Transformation Center supported CCOs with integrating medical, behavioral, and dental care through a range of offerings, which include: hands-on consultation and/or financial support; access to information and resources; learning collaboratives which involve peer-to-peer learning, information sharing, subject matter expert education, quality improvement strategies, and networking opportunities which include

**TABLE 7.1 State-Led Resources for Supporting Integration**

| Resource | Behavioral Health Integration | Oral Health Integration |
|---|---|---|
| **Metrics** | | |
| | Alcohol and drug misuse (SBIRT)<br>Depression screening and follow-up<br>Psychiatric hospitalization postdischarge follow-up<br>Mental, physical, and dental health assessment within 60 days of placement into foster care for children >4 years old | Dental sealants on permanent molars for children<br>Mental, physical, and dental health assessment within 60 days of placement into foster care for children >4 years old |
| **Consultation and Financial Support** | | |
| Technical Assistance | Through the technical assistance bank, CCOs can request up to 35 hours of technical assistance from in-state content experts on topic areas, which include assistance with behavioral health integration processes and work flows; integrated psychiatry; telehealth; workplace/clinic design; dental integration into primary care; oral public health and prevention; shared use of electronic health records; alternative payment methodologies; and expanding the range of settings in which dental care can be reimbursed.<br>http://www.oregon.gov/OHA/HPA/CSI-TC/Pages/Technical-Assistance-Bank.aspx | |
| Grants and Financial Support | The Transformation Center provided grant awards to support innovation and further CCOs' efforts to transform health care delivery in Oregon. Many CCOs used these funds to jump-start and/or expanded behavioral health integration activities in their local communities.<br>The Oregon Health Authority, in partnership with the Oregon Office of Rural Health, provided grant opportunities to CCOs implementing projects that improved care coordination; increased individuals' access to their own health data and engagement in their care; expanded system capacity; and achieved efficiencies in health care delivery. | |
| Council of Clinical Innovators Fellowship program | This fellowship program created a statewide network of leaders from a wide range of health disciplines: physicians, behavioral health consultants, pharmacists, expanded practice dental hygienists, etc. | |

(*Continued*)

**TABLE 7.1 (Continued)**

| Resource | Behavioral Health Integration | Oral Health Integration |
|---|---|---|
| **Information Resources** | | |
| Integration Resource Library | The state is providing an online repository of information that is organized around both clinical and operational topics and includes links to both in-state and national online resources, research articles, video site visits, expert interviews, and recorded webinars. http://www.pcpci.org/search/resources | |
| Information Sharing Strategies | The state convened the Behavioral Health Information Sharing Advisory Group to examine information sharing barriers and recommend solutions. http://www.oregon.gov/oha/HPA/CSI-BHP/Pages/Behavioral-Health-Info.aspx | |
| **Workforce Development** | | |
| | The state conducted a workforce development assessment to determine its readiness to meet the demand for primary care behavioral health providers by examining data from state professional licensing boards. http://www.oregon.gov/OHA/HPA/HP-HCW/Documents/Workforce%20Compentencies%20for%20New%20Systems%20of%20Care.pdf | Local university initiated a campus-wide interprofessional education program for students from Schools of Medicine, Dentistry; and Nursing, Physician Assistant, and Pharmacy programs to gain experience working in interprofessional groups. |

(*Continued*)

**TABLE 7.1 (Continued)**

| Resource | Behavioral Health Integration | Oral Health Integration |
|---|---|---|
| **Learning Collaboratives** | | |
| CCO Summits and Innovation Cafe | Learning collaborative which aims to support and spread innovative health system models (e.g., complex care, behavioral health integration, telehealth) through peer-to-peer learning, information sharing, and networking. | |
| Quality and Health Outcomes Committee | Monthly statewide learning collaborative of medical directors and quality improvement staff to discuss health outcomes, metrics, and measures, and promote integration, efficient working relationships, and data-driven decision making. Topics relevant to behavioral health and/or oral health integration are discussed. | |
| Learning Collaborative/ Workgroups | The state supported a Behavioral Health Home Learning collaborative which included periodic face-to-face gatherings, online learning sessions, and on-site practice coaching for behavioral health organizations that wished to develop primary care services. | The state contracted with external consultants to conduct situation analysis and establish parameters for further integration of oral health care across the state. Specific workgroups have addressed dental care access issues and an oral health strategic plan. |
| **Trainings** | | |
| | CCOs facilitated behavioral health trainings for clinicians including:<br>Mental Health First Aid<br>SBIRT<br>ACEs | The First Tooth program training provided to medical clinicians to implement preventive oral health services for infants and toddlers. |

large-scale transformation summits and an innovation café event. See Table 7.1 for descriptions of these initiatives.

## Legislation

Oregon's CCOs strived to balance statewide standards and measure with local control over the means to achieve those goals. Transformation Center technical assistance, for instance, provided information and resources, while leaving implementation up to CCOs. Until 2015, the state did not have a highly refined model for integrating behavioral care. It became clear that people at all levels of the CCO organization were unclear about the core components of integration and how to implement and pay for them. With leadership from people "on the ground" integrating care, the state clarified these questions when the Legislature passed SB 832. The bill put into statute the following definition of integrated care:

> *Care provided to individuals and their families in a patient centered primary care home or behavioral health home by licensed primary care clinicians, behavioral health clinicians, and other care team members, working together to address one or more of the following: mental illness, substance use disorders, health behaviors, developmental risks or conditions, stress related symptoms, preventive care, and ineffective patterns of health care utilization. (Oregon Legislative Assembly, 2015b)*

SB 832 also described the range of licensed behavioral health clinicians and unlicensed team members who can work in integrated settings, and directed OHA to use the bill's definition to develop additional integration standards for certified primary care homes and behavioral health homes.

Following the passage of SB 832, OHA convened a PCPCH Standards Advisory Committee that developed requirements for behavioral health screening and referral coordination. Under these standards, practices would receive points for demonstrated comanagement of care between medical and behavioral health clinicians or collocated care, with the highest awarded points for practices with on-site "integrated behavioral health services, including population-based, same day consultations by providers/behaviorists specially trained in assessing and addressing psychosocial aspects of health conditions." The definition was important because it made clear that integration was a team-based approach that required licensed behavioral health clinicians. Other related team members were vital but needed the support of licensed clinicians. The other critically important component of SB 832 was its enabling language to support billing for behavioral health services in primary care settings. By passing legislation that both named and defined integrated health care, the state laid a foundation of standards, which CCOs and practices could adopt and use to develop alternative payment methodologies.

In the same legislative session, SB 672 (Oregon Legislative Assembly, 2015a) established the Dental Director position within OHA. The legislature's decision to put the position into statute reflected an understanding that oral health is inseparable from overall health. The Dental Director, as dictated by SB 672, shall:

> *Provide recommendations and guidance to the authority and other state agencies, individuals, and community providers on how to prevent oral diseases and measures to take to improve, promote and protect the oral health of the residents of this state, with a focus on reducing oral health disparities among underserved populations.*

One of the first major projects of the Dental Director was to develop statewide oral health integration policies that achieve better outcomes and help Oregon achieve its policy and operational delivery system transformation goals for oral health care integration into the CCOs. The final report and recommendations, which are underway, will define how the state will proceed with its integration efforts.

## CCO-LED INTEGRATION EFFORTS

Oregon CCOs have encouraged efforts to support behavioral, medical, and oral health integration through its governance structure, by supporting experimentation and piloting integrated care in its practices, by providing educational and training opportunities to local stakeholders, and by collaborating with other CCOs to discuss and coordinate common approaches to integration.

### Governance Structure

Broad representation of medical disciplines and community representatives on CCO advisory and decision-making bodies was important to integration efforts. CCOs had active leaders with behavioral health integration experience on its boards, and with the addition of new contracts with DCOs in 2014, oral health leaders were brought to the table and included in CCO decision making. Some CCOs also have individuals on staff who oversee integration efforts. In addition to board members and staff members to lead implementation, CCOs are required to have Community Advisory Councils (CACs), which are comprised of OHP members and representatives from community organizations. CACs encouraged CCOs to support a holistic approach to health, including behavioral health needs and social determinants of health.

### Experiments and Pilots in Practices

CCOs were supporting integrated behavioral health and medical care by experimenting with different integration models in its practices. This experimentation was important to innovation and largely funded—initially—with grant funds, as there were start-up costs for integration and challenges using the existing billing structure to generate revenue for this new model of care. These models included integrating behavioral health services into primary care, integrating primary care into community mental health centers, and providing psychiatry consultation. CCOs and its affiliated practices had difficulty determining how to bill for integrated services; therefore, integration efforts were primarily funded through short-term state general funds allocated by the legislature in 2013, often referred to as "transformation" grant funds (see Table 7.1). Twelve of the 16 CCOs chose to use these funds to support hiring additional or new types of staff in primary care or behavioral health homes, and to remodel clinic space to better support the collaboration that is critical to integration (Cohen et al., 2015; Gunn et al., 2015). Grant funds allowed CCOs to jump-start integration efforts and circumnavigate barriers to establishing new contracting relationships and financing integrated care; however, heavily relying on grants put program sustainability at risk.

CCOs were also experimenting with alternative payment methods for integrated care using the global budget. Examples of alternative payment approaches include: a tiered primary care payment system, where higher tier primary care practices—those with more integrated services—received additional payment; paying for integrated services by distributing incentive payments back to practices for meeting its quality benchmarks; and asking the state to provide reimbursement and "open up" additional billing codes to finance integration through more traditional fee-for-service payment models.

To date, CCOs and DCOs have only started to pilot programs to integrate oral health services and reduce overall health care costs associated with untreated dental disease. State efforts to integrate oral health started after the state transformation grant awards were issued. However, since that time, 8 of the 16 CCOs have added specific oral health strategies in their transformation plans, which include efforts to eliminate or minimize barriers to dental care for all members; dental screenings for comorbid severe and persistent mental illness (SPMI) and diabetes populations; and value-based payments for dental care (Oregon Health Authority, 2016).

Several CCOs contributed community benefit funds to plan and implement integration projects with DCOs. CCOs and DCOs have also initiated some early pilots to integrate dental hygienists into primary care settings and are exploring the use of integrated EHRs, tele-dentistry, and care coordinators to prevent emergency department (ED) utilization for nonemergent dental issues. There are also plans to connect dental practices to the Emergency Department Information Exchange (EDIE), so clinicians can follow up after one of their patients visits the ED.

## Training and Education

CCOs supported behavioral health integration in its practices by providing opportunities to engage, train, and disseminate information about integration to clinicians and stakeholders through conferences, learning collaboratives, training sessions, and educational programs. Large group learning topics included implementing Screening Brief Intervention Referral to treatment (SBIRT) for substance use disorders, educational sessions on tobacco and addiction behaviors, Mental Health First Aid, and screening for and treating adverse childhood experiences (ACEs). CCOs also supported integration through training sessions and educational programs that either were internally led or were consultant-led or sent clinicians to external training programs to prepare their workforce for integration. Training efforts educated clinicians and clinical staff on integration models, brief interventions and therapeutic practices, medical vocabulary, and information on common diagnoses behavioral health clinicians were likely to encounter in primary care settings. For example, one CCO stakeholder from a local university initiated a campus-wide interprofessional education program that was mandated for freshmen enrolled in the Schools of Medicine, Dentistry, and Nursing programs. This program heightened students' knowledge and awareness of multidisciplinary collaboration.

To further support dental integration, several CCOs engaged in programs like the First Tooth program, which educates primary care clinicians in the prevalence and impact of oral disease, trains them in risk assessment, fluoride varnish application, and how to integrate oral health preventive services into their practice. These programs provide on-site support and build collaboration between dental and medical clinicians as well as increase access to oral health preventative services for the pediatric population.

### CCO Collaborations

To further support integrated care in its communities, some CCOs have also started to collaborate with each other. These CCOs have organized clinicians across networks to coordinate common approaches to integration, with common data reporting processes and reimbursement strategies. Thirteen CCOs have joined CCO Oregon (www.ccooregon.org), which is a collaborative that is comprised of clinicians, health care leaders, and CCO stakeholders from across the state. This member organization sponsored an interdisciplinary work group called the Integrated Behavioral Health Alliance of Oregon, which led efforts to better define integration. Many of their recommendations were adopted into legislation. This group developed recommended standards for primary care practices providing integrated behavioral health and recently developed a set of recommended measures to assess integrated care outcomes. CCO Oregon also developed the Oral Health Work Group. That group developed a framework for assessing oral health access and advocated for an expanded set of oral health quality metrics to be adopted for children and adults.

## FUTURE OPPORTUNITIES

Although Oregon has made great progress in reducing the silos between medical, behavioral, and oral health care, there is still more work to be done. Opportunities remain to further refine and foster the spread of integration across the state.

Oregon's emphasis on health system transformation started with its Medicaid population, and integrated care experimentation began in FQHCs and Community Health Centers, as these practices served a greater proportion of Medicaid beneficiaries. Access to integrated care is now more available to Oregon's Medicaid population than it is for patients with commercial insurance or Medicare. State leaders in Oregon recognized that for the health system to truly transform, this model of care delivery must be extended to other payers and populations. To accomplish this, Oregon's leaders have started to spread the coordinated care model to its state employee population, which includes individuals working for state agencies and educators. Reducing costs and improving care quality for these employees could improve the state's budget (e.g., health care costs are a large portion of the school budget) and free more funds for other important initiatives. State leadership is also engaged in efforts to develop alternative payment models for integrating primary care across all payers in the state.

Additional support for integrated behavioral, medical, and dental care at the federal level is also needed. It is important that the federal government avoids creating unintended barriers to local, state innovation. For instance, Oregon's Medicaid waiver allowed CCOs to pay for nonbillable services without requiring encounter and billing data. Creative strategies for improving the cost-quality curve emerged from this change. However, federal actuarial reviews continued to require billing data to prove that the capitated rates were actuarially sound, and CCO leaders fear that the federal government will question the actuarial soundness of their actions. CCO leaders have expressed that CMS requirements limit their flexibility in a number of ways, including providing integrated services to patients. Thus, advocacy with federal leaders is needed to further support local innovation related to integration (Kroening-Roche, Hall, Cameron, Rowland, & Cohen, 2016; Mendelson, Goldberg, & McConnell, 2016).

Oregon's initial Medicaid waiver is scheduled to expire in 2017, but the state has proposed a renewed waiver, which would extend through 2022, allowing for more opportunities to expand upon initial efforts. The newest waiver continues to emphasize actions that support integration of medical, behavioral, and oral health care and proposes extending integrated, coordinated care beyond the walls of health facilities. It also involves moving away from process measures to outcome-based metrics for measuring performance and quality incentives, and it prioritizes population health by focusing on social determinants, particularly housing, and improving health equity for vulnerable populations. In this extension period, we expect that more work will be done to mature and improve dental care integration. Access to dental care remains a challenge for considerable numbers of low-income and rural populations who carry the heaviest dental disease burden, and we anticipate further spread of dental and medical integration beyond the Medicaid population, which to-date has been limited.

In addition to expanding its waiver, Oregon has also applied to participate in the Certified Community Behavioral Health Clinic (CCBHC) demonstration project, which is administered by the Substance Abuse and Mental Health Services Administration through the Excellence in Mental Health Act (National Council for Behavioral Health, 2016). These clinics will provide a comprehensive range of mental health and substance use services, particularly to vulnerable individuals with complex needs. Although primary care integration was not a central feature of the national CCBHC program, Oregon proposed to adopt the behavioral health home standards from SB 832 as a component of its CCBHC certification standards. If approved, 14 certified clinics, some with multiple locations, would begin the CCBHC program in 2017.

## CONCLUSION

Health care leaders and policy makers in Oregon have taken steps to reduce health care fragmentation by integrating medical, behavioral, and dental care for its Medicaid population, and work is being done to further develop and evolve these changes as well as to spread them to patients with commercial insurance. Integration efforts are at an early stage, but the innovative work to-date shows promise. Others embarking on such a change should note that Oregon's global approach, which includes financial levers that incentivize all parts of the Medicaid health care system to quality through better integrated and coordinated care, is a critical element of this change strategy. By addressing financial, organizational, and human resource components, and underpinning these efforts with legislation and quality measurements, the Oregon health care transformation is a unique experiment. For researchers and policy makers, it is a natural experiment worthy of further study, as these efforts can inform others across the nation. Although early indications of success are promising, other countries have spent decades to restructure and tweak systems of care, and it is imperative that Oregon takes time to monitor and evaluate outcomes appropriately in the years to come, and to make the changes that will be necessary to continue to evolve and improve this new system.

## REFERENCES

Bobetsis, Y. A., Barros, S. P., & Offenbacher, S. (2006). Exploring the relationship between periodontal disease and pregnancy complications. *The Journal of the American Dental Association*, *137*(Suppl.), 7s−13s.

Butler, M., Kane, R. L., McAlpin, D., Kathol, R. G., Fu, S. S., Hagedorn H., & Wilt, T. J. (2008). Integration of mental health/substance abuse and primary care. Available from: http://www.ahrq.gov/research/findings/evidence-based-reports/mhsapctp.html.

Centers for Medicare & Medicaid. (2012). Services waiver list and expenditure authority.

Cifuentes, M., Davis, M., Fernald, D., Gunn, R., Dickinson, P., & Cohen, D. J. (2015). Electronic health record challenges, workarounds, and solutions observed in practices integrating behavioral health and primary care. *The Journal of the American Board of Family Medicine*, *28*(Suppl. 1), S63–S72.

Cohen, D. J., Davis, M., Balasubramanian, B. A., Gunn, R., Hall, J., deGruy, F. V., III, & Miller, B. F. (2015). Integrating behavioral health and primary care: Consulting, coordinating and collaborating among professionals. *The Journal of the American Board of Family Medicine*, *28*(Suppl. 1), S21–S31.

Collins, C., Hewson, D.L., Munger, R., & Wade, T. (2010). Evolving models of behavioral health integration in primary care. Available from: http://www.milbank.org/wp-content/uploads/2016/04/EvolvingCare.pdf.

Croghan, T. W., & Brown, J. D. (2010). *Integrating mental health treatment into the patient centered medical home*. Rockville, MD: Agency for Healthcare Research and Quality.

Cunningham, P. J. (2009). Beyond parity: Primary care physicians' perspectives on access to mental health care. *Health Affairs (Millwood)*, *28*(3), w490–w501.

Dickey, B., Normand, S. L., Weiss, R. D., Drake, R. E., & Azeni, H. (2002). Medical morbidity, mental illness, and substance use disorders. *Psychiatric Services*, *53*(7), 861–867.

Druss, B. G. (2007). Improving medical care for persons with serious mental illness: Challenges and solutions. *The Journal of Clinical Psychiatry*, *68*(Suppl. 4), 40–44.

Fisher, L., & Ransom, D. C. (1997). Developing a strategy for managing behavioral health care within the context of primary care. *Archives of Family Medicine*, *6*(4), 324–333.

Gunn, R., Davis, M. M., Hall, J., Heintzman, J., Muench, J., Smeds, B., ... Cohen, D. J. (2015). Designing clinical space for the delivery of integrated behavioral health and primary care. *The Journal of the American Board of Family Medicine*, *28*(Suppl. 1), S52–S62.

Hoge, C. W., Auchterlonie, J. L., & Milliken, C. S. (2006). Mental health problems, use of mental health services, and attrition from military service after returning from deployment to Iraq or Afghanistan. *JAMA*, *295*(9), 1023–1032.

Institute of Medicine. (2001). *Crossing the quality chasm: A new health system for the 21st century*. Washington, DC: National Academy Press.

Institute of Medicine. (2011). *Advancing oral health in America*. Washington, DC: The National Academies Press.

Kessler, R. C., Demler, O., Frank, R. G., Olfson, M., Pincus, H. A., Walters, E. E., ... Zaslavsky, A. M. (2005). Prevalence and treatment of mental disorders, 1990 to 2003. *The New England Journal of Medicine*, *352*(24), 2515–2523.

Kroening-Roche, J., Hall, J., Cameron, D., Rowland, R., & Cohen, D. J. (2016). Integrating behavioral health under an ACO global budget: Barriers and progress in Oregon. *American Journal of Managed Care, in press.*

Laderman, M. (2015). Behavioral health integration: A key component of the triple aim. *Population Health Management*, *18*(5), 320–322.

McConnell, K. J. (2016). Oregon's Medicaid Coordinated Care Organizations. *JAMA*, *315*(9), 869–870.

Mendelson, A., Goldberg, B., & McConnell, K. J. (2016). New rules for Medicaid managed care—Do they undermine payment reform? *Healthcare: The Journal of Delivery Science and Innovation, in press.*

Miller, B. F., Petterson, S., Brown Levey, S. M., Payne-Murphy, J. C., Moore, M., & Bazemore, A. (2014). Primary care, behavioral health, provider colocation, and rurality. *The Journal of the American Board of Family Medicine*, *27*(3), 367–374.

Nasrallah, H. A., Meyer, J. M., Goff, D. C., McEvoy, J. P., Davis, S. M., Stroup, T. S., & Lieberman, J. A. (2006). Low rates of treatment for hypertension, dyslipidemia and diabetes in schizophrenia: Data from the CATIE schizophrenia trial sample at baseline. *Schizophrenia Research*, *86*(1–3), 15–22.

National Council for Behavioral Health. (2016). Excellence in Mental Health Act. Available from: https://www.thenationalcouncil.org/topics/excellence-in-mental-health-act/.

National Quality Forum. (2012). Oral health performance measurement: Environmental scan, gap analysis & measure topics prioritization. Available from: http://www.qualityforum.org/Publications/2012/07/Oral_Health_Performance_Measurement_Technical_Report.aspx.

Oregon Health Authority. (2016). Oregon Coordinated Care Organizations transformation plans. Available from: http://www.oregon.gov/oha/HPA/CSI-TC/Pages/Oregon-CCO-Transformation-Plans.aspx.

Oregon Medicaid Advisory Committee: Oregon Health Work Group. (2016). A framework for oral health access in the Oregon Health Plan. Available from: http://www.oregon.gov/oha/HPA/HP-MAC/Pages/Oral-Health-Work-Group.aspx.

Peek C. J., & The National Integration Academy Council. (2013). Lexicon for behavioral health and primary care integration: Concepts and definitions developed by expert consensus. Agency for Healthcare Research and Quality, AHRQ Publication No.13-IP001-EF, Rockville, MD.

Prince, M., Patel, V., Saxena, S., Maj, M., Maselko, J., Phillips, M. R., & Rahman, A. (2007). No health without mental health. *Lancet*, *370*(9590), 859–877.

Salsberry, P. J., Chipps, E., & Kennedy, C. (2005). Use of general medical services among Medicaid patients with severe and persistent mental illness. *Psychiatric Services*, *56*(4), 458–462.

Scott, D., & Happell, B. (2011). The high prevalence of poor physical health and unhealthy life-style behaviours in individuals with severe mental illness. *Issues in Mental Health Nursing*, *32*(9), 589–597.

Oregon (State). Legislature. Assembly. An act to approve Oregon Health Authority proposals for Coordinated Care Organizations (S 1580). Reg. Sess. (2012). 78th Oregon Legislative Assembly. Available from: https://olis.leg.state.or.us/liz/2016R1/Downloads/MeasureDocument/SB1580/Introduced.

Oregon (State). Legislature. Assembly. An act to establish the Dental Director position within the Oregon Health Authority. (S 672). Reg. Sess. (2015a). 78th Oregon Legislative Assembly. Available from: https://olis.leg.state.or.us/liz/2015R1/Downloads/MeasureDocument/SB672/Enrolled.

Oregon (State). Legislature. Assembly. An act to require Oregon Health Authority to prescribe rule standards for integrating behavioral health services and physical health services in patient centered primary care homes and behavioral health homes. (S 832). Reg. Sess. (2015b). 78th Oregon Legislative Assembly. Available from: https://olis.leg.state.or.us/liz/2015R1/Downloads/MeasureDocument/SB832/Enrolled.

Stange, K. C. (2002). The paradox of the parts and the whole in understanding and improving general practice. *International Journal of Quality in Health Care*, *14*(4), 267–268.

Sun, B. C., Chi, D. L., Schwarz, E., Milgrom, P., Yagapen, A., Malveau, S., ... Lowe, R. A. (2015). Emergency department visits for nontraumatic dental problems: A mixed-methods study. *American Journal of Public Health*, *105*(5), 947–955.

US Department of Health and Human Services. (2000). Oral health in America: A report of the Surgeon General. Available from: http://www.nidcr.nih.gov/DataStatistics/SurgeonGeneral/sgr/ack.htm.

US Department of Health and Human Services Health Resources and Services Administration. (2014). Integration of oral health and primary care practice. Available from: http://www.hrsa.gov/publichealth/clinical/oralhealth/primarycare/.

Wallace, N. T., Cohen, D. J., Gunn, R., Beck, A., Melek, S., Bechtold, D., & Green, L. A. (2015). Start-up and ongoing practice expenses of behavioral health and primary care integration interventions in the Advancing Care Together (ACT) Program. *The Journal of the American Board of Family Medicine*, *28*(Suppl. 1), S86–S97.

Chapter 8

# Aligning Financial Models With Health Care Delivery

**Kristen Dillon**[1] **and Kelly Ballas**[2]
*[1]Family Physician, Hood River, OR, United States, [2]Financial Consultant, Portland, OR, United States*

The architects of health system transformation in Oregon understood the crucial importance of implementing financial models that supported their goals. In the United States, the financial risk for health care costs has typically remained with insurance companies or government entities, and payment for healthcare services has relied primarily on fee-for-service models. The incentives in these models have led to high levels of spending on health care services as well as fragmentation and inefficiencies in care. In response, Oregon state leaders designed the Coordinated Care Organization (CCO) model to include a global budget as well as a requirement to implement Alternative Payment Methods (APMs) that moved away from fee-for-service payments. A global budget with a single point of accountability and a sustainable rate of growth gave CCOs the freedom to be innovative in developing payment models that would support transformation. In addition, the model included accountability of the CCOs for health outcomes among their members.

This chapter focuses on Oregon's efforts to achieve financial alignment across the healthcare delivery system as well as the history and programs that created the context in which the reforms took place. This work started with a vision for health care in the state and then moved forward through collaboration between state government, the state legislature, and the federal government. The CCO model as implemented has allowed significant changes in the financing of the state's Medicaid programs and accomplished improvement in the payment models for care and the spectrum of services covered. During the early years of this work, the national implementation of the Affordable Care Act had numerous impacts on the work in Oregon, many of them unexpected or still evolving.

**Health Reform Policy to Practice. DOI: http://dx.doi.org/10.1016/B978-0-12-809827-1.00008-2**

## STATE GOVERNMENT AND MEDICAID LANDSCAPE PRIOR TO COORDINATED CARE ORGANIZATIONS

Prior to 2012, Oregon had a long history of Medicaid innovation, beginning with implementation in 1994 of the first state plan waiver (Medicaid 1115 waivers: https://www.medicaid.gov/medicaid/section-1115-demo/about-1115/index.html), which established the Oregon Health Plan. This state plan waiver, known as an 1115 Waiver after the law that permits states to alter the terms of the Medicaid program, was the first in a long series of agreements extending to the present time (Table 8.1).

Under this first waiver in 1994, Oregon's Medicaid program transitioned from one administered by the state to a managed care program with much of the risk and administration of benefits contracted to other organizations. Managed Care Organizations (MCOs) were formed in Oregon to provide physical health services to Medicaid beneficiaries. These were typically founded by Independent Practice Associations (IPA) or insurance companies, and most parts of the state were served by a single MCO. Mental health funds went to Mental Health Organizations (MHO), which were separate from MCOs and accountable for paying for behavioral health services to the Medicaid population. MHOs were most commonly run by Community Mental Health Programs (CMHPs) or a nonprofit entity with leadership from county government, which in Oregon is responsible for public mental health services. Management of oral care budgets was delegated to Dental Care Organizations (DCO). The six largest DCOs were for-profit corporations owned by dental providers or shareholders. They each served large areas of the state with multiple DCOs serving a single region.

While 20 years of managed care Medicaid in Oregon had improved care and allowed some control of cost, the limitations of the arrangement had also become apparent. The separation of funding streams was inefficient and failed to support integration and coordination of care, and services were often duplicated in different settings. In response, health reform combined virtually all Medicaid funding into global budgets with a fixed rate of increase, making CCOs the single point of accountability for the cost of most Medicaid benefits.

Prior to CCO implementation, the state made important changes to its own structure, the most significant being the formation of the Oregon Health Authority (OHA) in 2009, with the eight-member Oregon Health Policy Board as its policy-making and oversight body. The OHA assumed responsibility for all of the health care paid for by the state, including the Oregon Health Plan, which had previously been administered by the state's Department of Human Services, and the purchasing of health insurance for state employees and the state's educators. Medicaid reform in Oregon was given an extra boost in visibility and importance because the agency

**TABLE 8.1 Timeline of Payment Reform Strategy**

| | | |
|---|---|---|
| 2010 | Oregon Action Plan for Health | Identified goals for payment reform: equity, accountability, simplicity, transparency, and affordability |
| 2011 | House Bill 3650, Senate Bill 204 | Mandated cost controls for Medicaid and use of APMs for state sponsored health plans |
| 2012 | Oregon Health Policy Board Implementation Proposal | Projected cost savings and proposed global budget for CCOs |
| 2012 | Medicaid Waiver Request to Centers for Medicare and Medicaid Services | Requested federal matching funds for state expenditures, securing $1.9 billion for health system transformation |
| 2012 | CCO Request for Applications | Required financial projections to ensure solvency |
| 2012 | CCO Certification and Transformation Plans | Included contractual requirement for CCOs to use APMs in provider contracts |
| 2012 | State Innovation Model Grant | Required state to demonstrate the use of payment for value models in all state sponsored health plans |
| 2013 | Coordinated Care Model Alignment Committee | Developed strategies to spread APMs throughout the broader health care marketplace in Oregon |
| 2014 | Public Employees Benefits Board Request for Proposals | Required that plans have experience with APMs to be eligible to offer health insurance to state employees |
| 2015 | Hospital Transformation Performance Program | Established bonus payments to hospitals for meeting or improving performance metrics |

implementing the change had become the purchaser of health care for 40% of Oregonians. The work of the state's Coordinated Care model alignment committee continued for years to bring the state's health care purchasing under the same structure, or at least the same models and philosophy. While CCOs have not become the sole delivery model for state-purchased benefits, some CCOs have earned the bid to be a health insurance option for teachers in parts of the state, and the principles inherent in CCOs are being integrated into the public employee programs.

In 2011 the state legislature passed legislation (Oregon House Bill 3650: https://olis.leg.state.or.us/liz/2011R1/Downloads/MeasureDocument/HB3650/Enrolled), which described key elements of the CCO structure. Also during 2011, the Oregon Health Authority published the state's Action Plan for Health (http://www.oregon.gov/oha/Pages/Action-Plan-Health.aspx) and implemented a 15% rate cut on MCOs to compensate for budget shortfalls. In this time of significant change, the legislation, the Action Plan report, and later significant budget cuts signaled the state's intention to take aggressive action on cost control and health system change.

One year later, the OHA delivered a business plan for CCOs to the Legislature and gained approval in February 2012 to move forward with health system transformation. The target start-up date for the initial round of CCOs was July 1, 2012, only 4 months after the legislation that defined the model. Immediately following legislative approval, the OHA submitted a waiver request to CMS to enable the reforms and received CMS approval of the waiver within months. The first three CCOs were approved in July and certified for an August 1 start-up. By the end of 2012, 15 of the eventual 16 CCOs had been certified and begun operations (Chapters 5 and 1: Delivering Health Care Through the Coordinated Care Organization Model and The Oregon Narrative: History of Health Reform in Oregon).

## FUNDING FOR TRANSFORMATION

In addition to federal funding for continuing to operate the Medicaid program, the state of Oregon received a commitment of $1.9 billion from the Federal government's Centers for Medicare and Medicaid Services (CMS). The money came as a CMS match to state spending on health care programs and was for spending on Oregon's Medicaid program over 5 years beginning in 2012. The state programs that were used as a match for the federal money were the Oregon's insurance pool for high-risk individuals, health care training programs at public universities and community colleges, state-funded addictions and mental health services, and other healthcare-related activities. This funding was made available on the condition that Oregon control its Medicaid spending over the 5-year period to an average annual increase of 3.4%, or two percentage points lower than the national Medicaid spending trend at the time. The money was used to augment the funding for CCOs to support the additional cost of implementing health system transformation.

In addition to the matching funds, Oregon received a grant for $45 million from CMS under their State Innovation Models (SIM) initiative (https://innovation.cms.gov/initiatives/state-innovations/). This grant provided 5 years of funding to establish a center for system transformation within the OHA, the Transformation Center, which monitored implementation of system transformation and supported the work through activities such as a technical assistance bank of consultants to work with CCOs and educational programs for consumer members and professionals involved with CCOs.

## GLOBAL BUDGET

The creation of a global budget covering all Medicaid services sat at the center of the CCO model. Global budgets established a single point of accountability for cost containment, health care delivery, and quality of services across sectors of care that had historically been segregated among multiple budgets for physical health, mental health, oral health, transportation, and other services. Historically, each of these budgets had its own administrative structure, authorization criteria, and provider relationships. Rarely did the multiple institutions paying for the care for an individual patient interact with each other, and failures in one area easily led to costs being incurred in another. CMS and the OHA agreed that all possible services should be paid for by CCOs within the global budget. CCOs incorporated medical and behavioral health services at the outset and added oral health and nonemergent transportation services in early years. After this phase of implementation, the state has continued to retain direct payment responsibility for nursing home care, targeted case management, and in-home custodial care services as well as administering all benefits for the 10% of OHP members who remained on fee-for-service coverage through the state instead of being enrolled in CCOs.

In addition to creating an integrated point of accountability, the process of global budgeting allowed the state to build in a structure to contain the rise in costs to 3.4% and to shift the insurance risk of paying for care to the CCOs. Once annual budgets were negotiated, CCOs bore full risk for paying for care for their members. There were a few exceptions, including a process to protect CCOs from losses due to spending on members with a defined list of very high-cost conditions.

To complement the transition to global budgets for CCOs, the OHA collaborated with CMS to create a quality measurement strategy to protect member access to services and improve health care quality. Over the years after implementation, several other factors intervened to complicate the plan for CCO global budgeting, including the implementation of the Affordable Care Act and a transition by CMS to examining the actuarial soundness of CCO rates based on historic use patterns instead of allowing a fixed rate of rise in the global budget.

## QUALITY MEASUREMENT STRATEGY

In addition to setting up a global budget for each CCO, Oregon created a quality measurement strategy that included a long list of measures of system performance, a subgroup of measures with financial incentives, and a parallel program for hospitals administered directly by the OHA. As part of the 2012 Medicaid waiver, the state and CMS agreed on over 50 quality measures that were tracked by the state and reported publicly (Oregon's Measurement Strategy: http://www.oregon.gov/OHA/HPA/ANALYTICS/MetricsScoringMeetingDocuments/MeasurementStrategy.pdf), many of which had been in place to monitor quality under Oregon's prior Medicaid waivers. CMS required these quality measurements as a way to ensure that quality of care did not decline under the CCO model of care. In order to avoid financial penalties under DSHP funding to Oregon, the state needed to maintain and then show improvement in the quality measures over the years from 2012 to 2017, which has been successful to date.

In addition, the OHA created a Quality Pool of money to be paid annually to CCOs. Starting at 2.5% of CCO revenue and rising annually to a target of 5%, the Quality Pool money was distributed back to CCOs annually based on their ability to meet customized targets for a subset of 17–18 quality measures, called Quality Incentive Measures (CCO Incentive Measures: http://www.oregon.gov/oha/HPA/ANALYTICS/Pages/CCO-Baseline-Data.aspx).

Finally, in 2015 the OHA established a hospital quality incentive program, the Hospital Transformation Performance Program, with separate funding. The program sought to align the hospital incentives with the CCOs, as well as improve services in the hospital that were largely independent of CCO span of control. This program was built in parallel with the one for CCOs, with overlapping goals and some alignment of the incentives.

## IMPACTS OF THE AFFORDABLE CARE ACT

The implementation of the Affordable Care Act came during the first full calendar year of operation of the CCOs. While it supported the state's work overall by significantly increasing the number of people covered by OHP, starting in 2014 it also significantly disrupted financial and care models. Three major areas of ACA impact were the significant increase in membership under expanded Medicaid eligibility rules, the uncertainty around rate--setting for new adult members, and the disruptive effect of implementing retroactive rules around minimum levels of medical spending.

During 2014, overall enrollment in the Oregon Health Plan and CCOs nearly doubled, with adults going from 40% to 60% of state Medicaid members. The increase came with the move of Oregon Kids Care, previously a separate program funded with Children's Health Insurance Program money, into the OHP pool and the expansion of income-based eligibility to nondisabled adults without children.

For the OHA, adding the expansion population of adults to CCOs created significant uncertainty in rate-setting for this group. The state and CMS had little actuarial data on the expansion population given that this group had historically been largely uninsured. For CCOs, this meant bringing on many new members who had little experience with using health care benefits, many with potential costly medical or behavioral health concerns and a significant reservoir of unmet dental needs.

The ACA established minimum amounts that insurance companies must spend on health care, as distinct from administrative expense and profits. The fraction of spending that goes to health care services and other activities to improve health is known as the medical loss ratio (MLR). The ACA sets a minimum MLR of 80%–85% for insurance plans, and insurers who spend less than this amount must rebate unspent money back to those who paid the premiums for the insurance. Although not required in the ACA legislation, a minimum MLR is now being implemented through regulation into government-funded health insurance programs such as managed care Medicaid programs. In Oregon, a minimum MLR expenditure of 80% was implemented for the expansion population during early 2015, to apply retroactively for the 18 months ending December 31, 2015. CCOs and the OHA were challenged to define and agree on a process for this calculation. The calculation of MLR was particularly difficult in situations where the CCO had created payment models that included capitated or global budget arrangements to pay providers for patient care.

In the absence of recognized valuation models for transformational care, such as Primary Care Medical Homes, community health worker services, and team case management services for highly complex members, CCOs face strong pressure to return to fee-for-service payment methods. Leaders in CCOs struggled to understand how to fund care improvement and transformation that by definition is not fairly captured in encounterable services. Having a rigorous and consistent method for monitoring MLR will increase accountability for CCOs going forward, but the process of implementation has drained the time and energy of all parties involved and has strained the relationships between the OHA and CCOs.

## ESTABLISHING PAYMENT RATES TO CCOs

CCO premium payments in the form of global budgets were initially set using an actuarial method that relied on past utilization and costs, with adjustment for regional factors where appropriate. As part of the application process to OHA, CCOs submitted encounter data to the state, which formed the historic cost basis, risk adjustment, and cost trends used by actuaries to develop rates. Other geographic factors that impacted the initial rates included the presence of cost-based hospitals in a CCO region, which were more expensive on a unit basis than larger hospitals.

In the first year of CCO operations, some CCOs suffered significant operating losses and were able to negotiate rate increases based on population-health status or local health care costs. By early 2015, CMS was insisting on the state's rates being actuarial sound, to the point that CMS refused to approve the rates in contracts that the OHA had already executed with CCOs for the year. CMS performed a retroactive review of rates, leading to significant changes in some cases. A lawsuit between one large CCO and OHA resulted after significant payment reductions were recommended for the CCO.

The future of how CCOs' revenue will be set continues to be unsettled. The switch to an actuarial rate-setting model that relies on historic utilization and spending impedes innovation. CCOs that succeed in care transformation and cost savings run the risk of seeing their revenue for patient care decline year over year. Certainly, many CCOs aspire to reinvest the savings into measures that support care transformation or future long-term health outcomes, but budgeting and developing these programs takes time. In the meantime the budget surpluses that would be used to fund these investments can instead become a flag for cuts in the global budget.

## FINANCIAL RESERVES, SOLVENCY, AND RISK

In most communities, new corporations were formed to be certified as CCOs, and many of the new corporations formed included MCOs or insurance companies as major partners. The CCO implementation brought a great deal of new revenue into these organizations, along with increased responsibilities. There was concern in the legislature that CCOs would not have adequate financial reserves to ensure their ongoing viability. Financial failure of a CCO would interrupt care to beneficiaries, damage the providers of care that had outstanding claims in the region, and impose a burden on the state. The enabling legislation contemplated this scenario and charged the Department of Consumer and Business Services (DCBS), the state department that includes the state insurance commissioner, with supporting the OHA in determining the viability of a CCO applicant. The legislation also asked DCBS to consider the possibility of a special insurance license for CCOs and a regulatory role similar to their role in overseeing viability of commercial health insurers in the state.

During the latter part of the certification period, the OHA decided to use DCBS expertise only for initial certification of CCOs. Because the OHA had previously monitored MCOs for financial solvency and performance, it allowed CCOs to choose between two options. The first option was reporting to OHA using the existing MCO financial solvency reporting and reserve requirements. The second was reporting to DCBS using the standard format and rules developed by the National Association of Insurance Commissioners (NAIC) as used by DCBS to monitor and regulate commercial insurers in Oregon. The OHA reporting and reserve requirements were less burdensome to CCOs than the NAIC requirements, and all but one CCO chose to report to OHA. This approach has been successful to date.

A second concern related to the long-term viability of a CCO had to do with the increased insurance risk for CCOs with small populations. This risk is higher for smaller CCOs because they cannot dilute the unpredictable expenses of high-cost members across a large population. To mitigate this risk, CCOs are required to maintain a reinsurance policy that will absorb the costs from a single member that rise above a certain threshold, but the deductibles and premiums for these policies are high. The insurance risk faced by small CCOs also increased when CMS implemented minimum MLR requirements, which to date have been calculated approximately annually. Small CCOs are much more likely to have uncontrollable fluctuation in their spending from year to year. They face the risk that they will be required to rebate excess premium in years with low spending and absorb losses in years where spending is higher, even when this fluctuation is due to the known unpredictability in health care spending across a small population rather than actual system performance.

## ALTERNATIVE PAYMENT MODELS

### Implementation

Alternative Payment Models (APMs) for health care services were a central feature of the financial reforms in the CCO model and one that the state itself had to delegate to the CCOs to implement (Table 8.2). CCOs had a single global budget and also accountability for quality outcomes for their population, which created a strong incentive to put payment models in place that support high-value, high-quality care.

In addition, the required governance structure for CCOs mandated participation of all major health care providers, which in many cases created relationships among organizational leadership that had not existed previously. Finally, each CCO's Transformation Plan required it to create a plan for implementing APMs.

The fee-for-service payment model, historically dominant in health care in the United States, pays health care providers for each health care service or supply provided to a patient. Providers of services benefit financially by providing a greater volume and higher intensity of services. This system fails to account for services that take place over time or do not have recognized billing codes, such as primary care medical home services, other types of care coordination, and encounters provided via alternative means such as telephone or email. The model lacks incentives for control of overall spending and coordination among care providers. It also fails to value patient experience and outcomes and creates an incentive to provide more services than may actually be of benefit to a given individual.

The initial formation of CCOs took place over a matter of months, during which CCOs had to incorporate themselves and establish contracts with an adequate provider network. In most cases, preexisting payment models were

**TABLE 8.2 Definitions of Alternative Payment Models**

| Payment Models |
|---|
| 1. Per-Member-Per-Month (PMPM) payments for nonvisit functions<br>The provider receives additional payment for providing services such as care coordination that are not reimbursed under the fee-for-service payment model. |
| 2. Performance incentive payments<br>The provider may earn additional payments if the practice meets specified quality or financial performance targets. Payments are paid retrospectively after the CCO has been able to assess the practice's performance level. These "pay-for-performance" programs may be funded by dollars withheld from CCO payment to providers or from supplemental dollars from the CCO. |
| 3. Shared savings<br>Services are paid on a fee-for-service basis. At the end of the year, the cost of services is compared against a predetermined annual budget. If the cost of services is below the budget amount, the provider will share savings with the CCO. Shared savings arrangements may have an integrated quality component, such that the level of shared savings distributed to the provider is tied to the provider's performance on quality measures. |
| 4. Shared risk<br>Services are paid on a fee-for-service basis. At the end of the year, the cost of services is compared against a predetermined annual budget. If the cost of services is below the budget amount, the provider will share savings with the CCO. If the cost of services is above the budget amount, the provider will share losses with the CCO. |
| 5. Case rates<br>A case rate is a flat amount, such as a per diem rate, that covers a defined group of procedures and services, usually provided by a single provider at one point in time, such as a case rate for colonoscopy. |
| 6. Prospective and retrospective bundled payments<br>A bundled payment is a flat amount paid for an individual patient receiving a defined set of services (e.g., joint replacement; maternity care) or for a specific condition (e.g., adult asthma) that often covers services provided across the continuum of care by multiple providers for a specified period of time. |
| 7. Capitation: full or partial<br>Fixed dollar payments for a defined set of services are paid to a provider for each person cared for by the capitated provider. Payments are based on estimated costs of all or most health care services that the person may need (full capitation) or for a smaller set of services such as behavioral health or primary care services (partial capitation). Payments are prospectively paid, usually via a PMPM payment, and usually involve the provider accepting full risk for the capitated services. |

Source: Adapted from OHA Coordinated Care Model Alignment Work Group documents. Available from: http://www.oregon.gov/oha/HPA/HP/Pages/Coordinated-Care-Model-Alignment-Work-Group.aspx.

left in place and became carve-outs within the global budget of each CCO. DCOs typically continued to receive a global budget for all dental services provided to their members, with a variety of payment methods from the

DCO to its dental providers. The OHP behavioral health network had been primarily county-based CMHPs. These programs in many cases continued to receive a global budget intended to cover the cost of all mental health services and with full upside and downside risk for expenditures such as inpatient care. In addition, many MCOs had contracted with physical health providers in ways that shared financial risk or provided a capitated budget for some services, and again these contracted arrangements were often absorbed into the CCO.

Because the CCO model required a governance board that included all major providers of health services, it brought institutions and people into relationships that had not existed before 2012. These relationships, including some provision of trust, shared vision, and good will, appear to have been crucial for the communities that have been able to implement innovative payment models (Table 8.3).

In addition, CCOs were required to submit biannual Transformation Plans (CCO Transformation Plan Reports: http://www.oregon.gov/oha/HPA/CSI-TC/Pages/Oregon-CCO-Transformation-Plans.aspx). The Transformation Plans were created by CCOs in response to an outline of requirements and formed an amendment to each CCO's contract with the OHA. Each CCO was required to commit to a 2-year implementation timeline with interim written reports on areas that included APMs, Integration of Physical and Behavioral Health, and support for Patient-Centered Primary Care Homes (PCPCH).

In implementing new APMs, one of the most common strategies employed by CCOs was per-member, per-month payments to primary care in support of PCPCH, the state's model for primary care medical home design. One large urban CCO also implemented augmented payments to primary care clinics in exchange for a commitment to take on new members as the CCO struggled with primary care access for the expansion population. One rural CCO integrated a preexisting global budget contract for hospital services that also included cross-sharing of quality withhold returns with outpatient clinics not affiliated with the hospital system. And in another example, a CCO developed a model with budgets for each sector and a process to share any savings from the overall global budget. Current ongoing efforts include developing payment models for highly specialized integrated settings such as maternity care clinics for pregnant women with drug addiction and complex care clinics for patients with high-cost medical and behavioral health conditions.

## Results

Each CCO was allowed to define and implement its own APMs, without constraint on acceptable models. This created wide leeway for innovation but also set up for challenge and uncertainty. Because of the difficulty in developing aligned payment models and continued incentives for providers

**TABLE 8.3 Summary of APMs Used by Oregon CCOs**

| | PMPM[a] | Performance Incentives | Shared Savings | Shared Risk | Case Rates | Bundled Payments | Capitation—Full[b] | Capitation—Partial[c] |
|---|---|---|---|---|---|---|---|---|
| **Number of CCOs Reporting APM by Type[d]** | | | | | | | | |
| Count of CCOs | 2 | 15 | 2 | 2 | 3 | 1 | 0 | 11 |
| Percent of CCOs Using APM | 13% | 100% | 13% | 13% | 20% | 7% | | 73% |
| **Number of CCOs Reporting APM by Type and Sector** | | | | | | | | |
| Primary Care | 3 | 11 | | | | 1 | | 7 |
| Other Physical Health[e] | | 5 | 1 | | 1 | | | 3 |
| Dental Care | | 3 | 1 | | | | | 5 |
| Behavioral Health | 1 | 5 | | | 1 | | | 7 |
| Hospital Services | | 4 | 1 | | | | | 1 |
| Provider Networks | | 2 | 1 | 2 | | | | |
| Prescription Medications | | | 1 | | | | | |
| Other[f] | | 2 | | | 1 | | | 3 |

[a] *Per-member per-month payment for activities beyond in-person patient encounters.*
[b] *In full capitation payment models, the recipient assumes responsibility for the full cost of care within and outside its institution.*
[c] *In partial capitation models, the recipient assumes responsibility for only those services it provides.*
[d] *Data from 15 of Oregon's 16 CCOs are incorporated in this table. One CCO did not give the OHA permission to publicly share their APM information in the original committee document.*
[e] *Includes office care by subspecialty physicians and inpatient physician services.*
[f] *Examples include Physical Therapy and Nonemergency Medical Transportation.*
Source: Adapted from OHA Coordinated Care Model Alignment Work Group documents; based on data as of March 2015.

to resist change, overall progress in moving to payment for value has been limited.

As part of CCO creation, the OHA required implementation of APMs and produced materials describing them but did not define acceptable models. This led to a wide diversity of strategies with a wide spectrum of incentives and qualitative boundaries. In addition, the OHA did not define allowed rates for providers nor the actuarial process that would be used to calculate MLRs for APMs, which became a point of conflict after CMS retroactively implemented minimum MLR requirements on CCOs. CCOs found themselves working to create aligned payment models and adequately compensate providers without knowing at what level provider payments would be deemed excessive.

Difficult aspects of implementing APMs included defining and measuring the services that were to be covered through population-based payments, administering APMs using computer systems built for fee-for-service payments, and writing contracts that freed providers to innovate while maintaining accountability. Many APMs seek to provide more flexibility for care providers to meet the patient's needs. Within this flexibility, CCOs struggled with mitigating the unintended consequences of payment change, for example, the need to closely monitor access, quality, and utilization in a model that shifts to a capitated payment for a large group of services.

Under the legislation that established CCOs, each CCO was required to have a letter of support from the Local Mental Health Authority, a responsibility held by county government and typically accomplished through a Community Mental Health Program (CMHP) run or overseen by county government. Later legislation required that each CCO contract with all Dental Care Organizations (DCOs) serving its geographic region. In order to ensure DCO and county support, CCOs in nearly all cases simply continued the payment arrangements then in place. This took the form of subdelegated budgets that transferred all financial risk and left all savings with the DCO or CMHP. Many of these contracts lacked provisions to ensure that members were receiving an adequate quantity of essential services and to monitor qualitative aspects of care such as member experience, access, quality of care, and outcomes. In addition, the carve-out of mental health budgets to mental health agencies retarded the integration of behavioral health services into other settings such as primary care clinics.

The default payment model for any major health care provider that refused to contract with a CCO was fee-for-service at Medicaid rates. House Bill 3650 addressed the need for a process to resolve disputes between CCOs and providers regarding reimbursement for services and directed the OHA to develop a process, including the use of an independent third party arbitrator. Specifically, the legislation states that "a healthcare entity may refuse to contract with a CCO if the reimbursement established for a service provided by the entity under contract is below the reasonable cost to the entity for providing the service." It also prohibits a provider from unreasonably refusing to contract with a CCO.

In general, CCOs struggled to negotiate APM contracts with hospitals, related to generally low Medicaid rates, the prospect for hospitals of being forced to change in ways that would reduce their revenue long term, and an imbalance in bargaining power between CCOs and Oregon's relatively consolidated hospital systems. In addition, many CCOs were negotiating with financially strapped rural hospitals that were the sole hospital provider in large geographic areas of the state. As a third issue, in Oregon most commercial insurers and Medicare pay hospitals using fee-for-service or similar volume-based models. This makes hospitals reticent to enter APM agreements with CCOs, a single, low-revenue payor, when their clinical care processes and financial systems are set up for the state's dominant fee-for-service model.

Medicaid level funding did not give CCOs adequate leverage in negotiations with healthcare providers. While a very large piece of the budgets go to pay for hospital care, hospitals typically lose money on Medicaid and were averse to taking risk without the possibility of significant reward. This varied, of course, depending on health system leadership and local conditions, but CCO-provider contracts were voluntary agreements with little incentive for many health systems to accept risk, change clinical care models, or accept overall payment reduction. As a result, limited change has been accomplished in how most Oregon Health Plan dollars are paid out to providers.

When negotiations were unsuccessful and led to stand-offs between CCOs and providers or legal action, conflict resolution between CCOs and provider groups became problematic. The strategies laid out by the legislature were neither complete nor effective enough, and the OHA typically left CCOs and providers to work out an agreement or leave things unchanged.

## LEGACY PAYMENT MODELS

With the implementation of financial reforms related to CCOs, several preexisting payment structures also remained in place in Oregon. These legacy payment models included cost-based reimbursement for rural hospitals and per-visit payments to safety net clinics.

Oregon's rural hospitals have benefitted from a revenue support program implemented by the state legislature, similar to the federal Critical Access Hospital program. For 32 hospitals located in qualifying communities, the program pays a cost-based fee-for-service rate for care provided to Medicaid beneficiaries. These payment rates, agreed to by the state and each hospital, remained in place, and CCOs were required to pay the rates with their global budget dollars. This caused problems in the original rate-setting for CCOs, with many rural CCOs suffering significant losses in the first year due to high hospital costs. In addition, augmented fee-for-service payments created a powerful incentive for hospitals to maintain a high volume of hospital-based care, including emergency department visits and hospital inpatient days, which is contrary to efforts to prioritize health care provided in less costly clinic settings.

Through Oregon's implementation of CCOs, the Prospective Payment System (PPS) for Federally-Qualified Health Centers (FQHCs) and Rural Health Centers also remained in place. These payment models incentivize high volume, face-to-face care in outpatient clinics. While increasing the access and use of outpatient services was certainly a goal of Oregon's health care reform, the PPS rate only compensates clinics for certain in-person encounters conducted by designated clinicians. This means that clinics operating under PPS receive no additional compensation for implementing newer models of care such as team-based management of a patient population, visits with nurses and community health workers, and care provided by phone, email, or telemedicine.

Beginning in 2010, FQHCs were offered the opportunity to join the Alternative Payment and Care Methods program (APCM) (http://depts.washington.edu/payeval/docs/oregon-report.pdf). APCM was established through an 1115 Medicaid waiver that predated the CCO model and had already enrolled the first centers before CCOs were formed. FQHCs have continued to join the program while also participating in CCOs, with most Centers in the state now enrolled in APCM. Through this program, FQHCs can voluntarily transition to a capitated, per-member per-month payment that is based on a clinic's historic utilization and PPS rates. As long as they maintain quality outcomes and patient experience, FQHCs can transition their care to the model that best meets the needs of their patients.

## FLEXIBLE SERVICES

As Governor John Kitzhaber promoted the idea for CCOs, he commonly told a story about an unnamed middle-aged woman. She lived with multiple chronic medical conditions and experienced frequent hospitalizations. An observant care provider detected that the woman's tiny apartment became overheated in summer, a significant contributor to her fragile health. Providing the patient with an air conditioner allowed her to maintain appropriate temperature and hydration, resulting in many thousands of dollars of savings on emergency department and hospital costs.

In establishing CCOs, Oregon desired to have flexibility to pay for services or supplies that were likely to be less costly alternatives to those covered by Medicaid, often referred to by CMS as "in-lieu of" services. These were referred to in Oregon's health care reform as "flexible services" and were intended to be paid for with Medicaid dollars in order to avoid use of higher cost health care services. Handrails can help frail people avoid falls. Money for childcare might allow an individual to make an important medical appointment. The concept of flexible services as an element of the

Oregon Health Plan received strong leadership support within the state and captured the imagination of Oregonians, even those working outside health care.

While CCO contracts permitted CCOs to adopt policies and budgets to use money in this way, the dream of an "air conditioner fund" has yet to be fully realized. This is the result of several factors related to implementation of the flexible funds model: limited support from CMS, disincentives in how flexible spending was handled in the rate-setting process, and difficulty defining the administrative process and scope of options for flexible spending.

While CMS did grant permission in the initial 1115 waiver for flexible services, they created a strong disincentive to the spending by adding the condition that the costs would not be counted as a medical expense. This posed a problem for CCOs because reducing medical expense by providing less expensive substitutes would have the impact of reducing the revenue available to CCOs in future years. Oregon has since developed a waiver extension request to CMS that would allow CCOs to account for flexible services as a medical expense in exchange for greater accountability for evaluating their effectiveness.

In the early years of CCOs, there was little use of flexible spending, and OHA remained silent on flexible services other than adding them as a permitted activity in contracts. This led to investigation by advocacy groups for people with disabilities. In the 2016 CCO contracts, CCOs were no longer simply permitted, but instead required, to implement flexible services spending. The rules on flexible services that were initially released in January 2016 provided some parameters for CCOs to write their policies and also went beyond spending directed at individual needs to include a provision for population health interventions with these dollars to be overseen by the Community Advisory Council of each CCO.

Under the 2016 state requirements, each CCO was required to develop its own policy and process for implementing flexible services spending, including a way to track spending, tie spending to specific members, and accept grievances over decisions. These policies are being implemented slowly, due to the nature of administering the process and the financial disincentives inherent in the current rules around accounting for flexible spending.

Communication around flexible services within the state has become quite confusing as CCOs increasingly identify that in order to improve health and reduce costs, they must address social needs in their members and their members' communities. Several CCOs have provided significant funding for purchase of housing units, nutritious food, and other nonmedical items intended to improve health. While these activities reflect what many felt flexible services were intended to do, they have so far been funded out of administrative dollars or community benefit funds that derived from surplus revenue in prior years.

## CONCLUSION

Financing and payment models get in the way of health system transformation when they are fragmented or fail to adequately pay the costs of new models for care delivery. In Oregon, the OHA and the newly formed CCOs enacted substantial financing changes, but successful adoption of APMs has been more challenging. Oregon's financial reforms were largely successful in removing funding silos at the state level and permitting CCOs to apportion funding as needed to meet their populations' needs. There does continue to be tension between a global budget paradigm that incentivizes CCOs to innovate and save money and a rate-setting paradigm that bases future payments on the historic value of services provided to members. When it comes to the implementation of APM, the results have been mixed. The fast pace of CCO implementation and voluntary nature of provider–CCO contracts have caused many CCOs to emulate the payment methods that were in use by the state prior to 2012, with all their inherent strengths and weaknesses. In addition, as CMS implemented Minimum Loss Ratio requirements for Medicaid managed care programs, CCOs faced the risk of implementing APMs without a way to receive assurance in advance from CMS regarding how MLR would be calculated. A more proscriptive list of allowed APMs, or at a minimum a safe-harbor list of models with MLR calculation rules, would ease the growing concern among CCOs that they will be required to pay rebates to OHA on money already paid to providers under APMs. Finally, the state has put extensive quality measurement in place, but the system of measures appears to be allowing care in areas that is still plagued by poor access, low utilization, or low quality. APMs have significant promise as a tool in health care reform, but Oregon's experience shows the importance of implementing them with great clarity around the technology needed to administer the payments, the actuarial process that will be used to value the care provided, and how each model's incentives align or compete with the overall goals of health care transformation.

Chapter 9

# Engaging the Community to Promote Health

**Chris DeMars**
*Oregon Health Authority Transformation Center, Portland, OR, United States*

From the beginning of Oregon's health system transformation experiment—launched by former Oregon Governor John Kitzhaber along with a cast of Oregon legislators and leading health policymakers—the goal was achieving health for all Oregonians. In 2007, Institute for Healthcare Improvement founder Dr. Don Berwick coined the Triple Aim of lower costs, better care, and improved health as the vision for the health care system. The concept rapidly caught on in Oregon health circles, and became the overarching goal for Oregon's health care reform efforts. During the years between his second and third term as Governor—which started in January 2010—Dr. Kitzhaber spoke frequently about the need to reform Oregon's health care system in pursuit of the Triple Aim. He contended that there were sufficient funds in the health care system to achieve health; the system simply needed to be designed more efficiently to deliver improved quality for individuals and improved health for populations. His "air conditioner" story characterizing a redesigned system has become Oregon health policy lore. He would describe an elderly woman with congestive heart failure whose apartment building was sweltering in the summer, which exacerbated her condition and often resulted in visits to the emergency room. He reflected how the dysfunctional health care system would cover hundreds if not thousands of dollars in emergency room visits, but would not pay for a $120 air conditioner that could prevent the need for her to go to the emergency room altogether. He argued that Oregon's health care system needed to be redesigned so that an air conditioner could be purchased with health care dollars, thereby decreasing costs and improving care and health.

Around this time, the inextricable link between health and the social factors required to promote health was becoming more widely accepted both nationally and locally, to the point where it is now commonly recognized that "upstream" social factors and community supports such as housing, edu-

**Health Reform Policy to Practice. DOI: http://dx.doi.org/10.1016/B978-0-12-809827-1.00009-4**

cation, and jobs have a greater impact on long-term health outcomes than medical care (Centers for Disease Control and Prevention, 2015). The last decade has seen growing acceptance that, to achieve health, the health care system must engage with the community to address the social determinants of health (DeCubellis & Evans, 2014; Magnan, Fisher, & Kindig, 2012; Shortell, 2013). Improving population health requires partners across many sectors—including public health, health care, community, and business—to integrate investments and policies across the social determinants (Kindig, 2015).

Furthermore, beginning with the Medicaid population to foster ties between the health care system and the community is a logical place to begin this work. Vulnerable, low-income populations—which disproportionately include people of color—are most significantly affected by the lack of access to social services that address the social determinants of health, often resulting in chronic complex medical, behavioral health, and/or supportive service needs. Not only does this situation contribute to health disparities (Centers for Disease Control and Prevention, 2015), but it has been noted that effectively serving these so-called "super-utilizers" will be essential to controlling US health care costs. It has also been noted that doing so may require the health care system to reach beyond its traditional boundaries into social services (DeCubellis & Evans, 2014).

Oregon's reformed health care delivery system—which began with a focus on the Oregon Health Plan or Medicaid population—was intentionally designed to allow and even promote this type of cross-sectoral community engagement and partnership. In 2010, the Oregon Health Policy Board, which serves as the policy-making and oversight body for the Oregon Health Authority, presented its *Action Plan for Health*, which laid out strategies that would help Oregon achieve the Triple Aim, envisioning a health care system in which "communities and health systems work together to find innovative solutions to reduce overall spending, increase access to care and improve health" (https://www.oregon.gov/oha/Pages/Action-Plan-Health.aspx). In a number of ways, Oregon's redesigned health care system fosters relationships between health care providers and the community, thereby promoting the ability to achieve health. Specifically, the entities that form Oregon's Medicaid delivery system, Coordinated Care Organizations (CCOs):

- are required to connect with the community through Community Advisory Councils (CACs);
- have increased latitude to pay for community-based supports through the provision of health-related services; and
- are financially incentivized to engage with the community through outcome metrics.

All of these factors have contributed to perhaps the most important feature of Oregon's health system transformation: a health care system culture shift, resulting from the acceptance that the system needs to expand beyond the clinic walls into the community to promote health.

## COMMUNITY ADVISORY COUNCILS

When Oregon's CCOs were being designed, CACs were seen as an integral component. The concept originated with a few Oregon Health Policy Board members who were passionate about consumer engagement and public health. They modeled CACs on the governance board requirement for community health centers (otherwise known as federally qualified health centers) that stipulates a majority of board members must be served by the center. The effectiveness of this requirement had been demonstrated by over four decades of community health center experience, and was seen a proven means of holding the health care system accountable. Connecting the community health center consumer-board requirement with the goals laid out in Oregon's *Action Plan for Health* resulted in consumer-majority CACs that were focused on the health of all Oregonians, not just the CCOs' membership.

Once the Oregon Health Policy Board recommended the CAC concept be a key part of CCOs, Oregon Health Authority leadership managed the process of obtaining buy-in from key legislators, which resulted in the CAC concept being codified in Oregon statute (http://www.oregonlaws.org/ors/414.627). Specifically, the legislation dictates that CCOs must have a CAC "to ensure that the health care needs of the consumers and the community are being addressed." It is important to note that the legislation calls out not just the CCOs' members, but their entire community. Making CCOs responsible for the health of everyone living in their geographic area is unique in its own right. In addition, CACs must include representatives "of the community and of each county government served by the coordinated care organization, but consumer representatives must constitute a majority of the membership." Requiring majority Oregon Health Plan membership on CACs is another groundbreaking piece of the legislation. No other state's Medicaid delivery system structure includes such an intentional inclusion of Medicaid members.

### The Community Health Improvement Plan

Overseeing CCOs' Community Health Improvement Plans (CHIPs)—which outline strategies to support improved health of individuals and the community—is a core component of the CACs' responsibilities. According to Oregon legislation, the duties of the CAC include, but are not limited to: "Identifying and advocating for preventive care practices to be utilized by the CCO"; "overseeing a community health assessment and adopting a community health improvement plan to serve as a strategic population health and health care system service plan for the community served by the CCO"; and "annually publishing a report on the progress of the community health improvement plan." CCOs' CHIPs must provide a plan for both population health and the health care system, and describe the scope of the activities, services, and responsibilities for the CCO to consider upon implementation of the plan (http://www.oregonlaws.org/ors/414.627).

### *The Community Health Improvement Plan Development Process*

For most CCOs' CACs, creating their CHIPs involved a process whereby the health care system engaged community partners to collectively identify community needs through the community needs assessment, and identified strategies for addressing these needs through the CHIPs. Taking health care system/community collaboration to a new level, the three CCOs that serve the southern Oregon region demonstrated collaboration across their CCOs as well as with the community. The process began with CAC members from all three CCOs reviewing the data from their two counties highlighted in their 2013 community health assessment (on which they also collaborated) and collectively identifying three major topic areas for all three CCO CHIPs to focus on: healthy beginnings, healthy living, and health equity. The next collaborative step involved gathering extensive community input about possible strategies to address the three health priority areas. Surveys and public meetings captured over 1000 comments and survey data from over 600 community members and individuals that deliver health and social services in the two counties. All three CCOs provided resources and data analysis support for this community input process. Combining this community input with each CCO's CAC's input, guiding philosophies and organizational resources, each CCO chose their own strategies and drafted their own CHIP. However, each CHIP maintains the shared health priority focus areas, format, and design, and includes strategies for members of their CCO as well as strategies for the community at large.

Some CCOs faced challenges in developing their CHIPs, especially with regard to developing meaningful baseline data on health disparities, which was a CHIP requirement. Unfortunately, in most cases, countywide or CCO data were not sufficient to identify health disparities within the CCOs' service areas. This made it very difficult for CCOs to assess the current gaps and needs and identify strategies for improvement. However, in some cases, CACs overcame this challenge by convening focus groups with communities experiencing disparities to better understand their experiences and identify disparities in order to develop strategies their CCO could consider. For example, Eastern Oregon CCO conducted one-on-one interviews, community visioning meetings, and focus groups to assess the specific communities' specific needs.

## *Community Health Improvement Plan Priorities*

CCOs submitted their CHIPs to the Oregon Health Authority in June 2014. The majority of the priorities identified in the CHIPs focused on strategies that required moving upstream from the traditional medical care system to address social determinants of health, public health, and health equity priorities. Specifically, almost 60% of CHIP priorities included topics such as access to healthy food and opportunities for physical activity; availability of affordable housing and transportation; fostering connections with the early childhood system; chronic disease prevention by decreasing tobacco use and obesity; and improving health equity. The remaining 40% of priorities were focused on addressing more clinical issues that still fell outside of the delivery of physical health services, including addressing mental health, substance abuse, oral health, and access to care (Fig. 9.1).

While CCOs' CHIPs are an important guidance document for CCOs' work, the degree to which they are used by each CCO to guide their work is

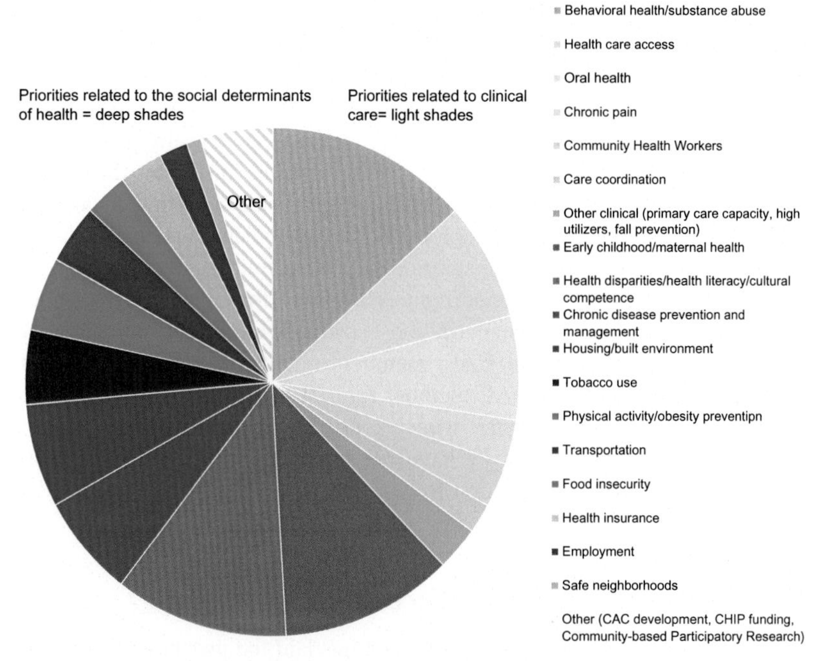


**FIGURE 9.1** CCOs' Community Health Improvement Plan priority areas.

the true test of their value. On this front, there are wide differences between CCOs. Since CCOs are not required to implement the priorities identified in their CHIP and are not accountable for reaching its goals and outcomes, CHIPs are sometimes a lower priority for their CCOs than attaining the goals of their required Transformation Plans or their incentive metrics, for which CCOs can receive a financial reward (see later). CCOs that report making progress implementing their CHIPs indicate that this progress is strongly connected to aligning the CHIP with their work focused on other CCO requirements, the incentive measures, and existing funding. This alignment—and thus prioritization of CHIP implementation—can be further fostered in cases where there is a solid relationship between CCOs and CACs, as discussed later.

## Value of the Community Advisory Council: New Partnerships

While the CHIP-development process required CCOs to engage their community, the CACs provide a new table whereby the community is engaged with the CCO on an ongoing basis. A frequently expressed sentiment about the value of the CAC table is that it provides a vehicle for relationship-building and new partnerships that would not happen otherwise. Depending on the CCO, CACs include people with work experience in areas such as housing, public health, early childhood, and mental health who are able to make instrumental connections between the CCO's CAC and other organizations focused on delivering social services. For example, in one CCO, a former CAC member had work experience as a housing provider, which led to increased CAC interest in housing and the inclusion of housing as one of its CHIP priorities. In addition, because of connections made by this former CAC member, the CAC is having ongoing conversations with a local housing workgroup about how the region invests in housing via the CCO and how to use the CCO to leverage additional housing funding.

The partnership-building opportunities afforded through the CAC model may well be one of the model's most lasting, impactful outcomes. In response to a survey of the CAC members about what helps them stay engaged in their CAC, one CAC member said, "I love that we have had conversations between social service agencies, primary care, CCO members, and schools." Another responded, "I have never experienced such community involvement and such a dedication to working with all community providers for a common goal."

## Community Advisory Council Successes

Oregon's health care system transformation has laid the groundwork, through CACs, for fostering community/health care system connections and moving the health care system upstream to address social determinants of health.

A key factor of CACs' initial success in bridging the community with health care is that they give Oregon's Medicaid members a voice in setting their CCOs' priorities. Many CCOs' CACs have begun to successfully connect the CCO with community partners to address social factors that influence health, thereby beginning to break down long-standing silos between entities that deliver health care and promote health. They are driving dozens of partnerships between CCOs and the community around projects as diverse as a health program for Latino adolescents in a local high school, parenting and exercise classes for new mothers, purchasing books for young children that the CAC members hand out at community events, and a food-delivery program for patients recently discharged from the hospital.

For example, one CCO's community health assessment indicated cultural and linguistic barriers to accessing traditional health services, including adolescent well-care visits, for the Latino population. To address these issues, the CCO funded a community health worker to support pregnant or parenting adolescents—many of whom are Latino—in a local high school. The community health worker not only links these teens to health services, but also provides support to address the social determinants of health, such as housing, food security, and income stability. She recently helped a parenting teen apply for safe and affordable housing through the local community housing agency. Over the last few years, the community health worker has made 130 contacts, made 38 referrals, and had 16 intensive contacts with school-age children in the county.

## Characteristics of Effective Community Advisory Councils

The degree to which CACs are effective vehicles for promoting health care system/community partnerships varies widely across CCOs. After almost 5 years, Oregon has ample anecdotal data from which to make conclusions about the characteristics inherent within successful CACs.

First, it is important that all CAC members, including their Oregon Health Plan members, feel engaged, valued, and able to be full contributors to the CAC. Many CAC members, especially those that represent the Oregon Health Plan population, are new to serving on advisory committees, and the most successful CACs offer orientation and training to their new CAC members on how the CAC functions and—given the technical nature of the health care and community health fields—the subject matters discussed at CAC meetings. Some CACs have a mentor or buddy system between seasoned and new CAC members to ensure the new members become comfortable with their new role. Retention strategies include providing meals and childcare to CAC members to make meetings more accessible, and holding meetings during evening hours so as to not conflict with daytime work schedules. The Oregon Health Authority Transformation Center has focused on providing targeted supports for CACs' recruitment and retention strategies, including developing a CAC recruitment video and a CAC 101 orientation presentation for use by all CACS, and making consultants available to help CACs meet their specific recruitment and retention goals.

CAC and CCO relationships vary across CCOs. When CACs communicate regularly with their CCOs, when CCO leaders engage in CAC activities, and when CAC members are invited to participate in CCO activities, CAC members are more engaged and empowered to participate in health care system transformation efforts. For example, effective CACs have built-in connections between the CAC and CCO leadership. While all CCO boards are required to include one representative from the CAC, innovative CCOs ensure that CCO leaders attend all CAC meetings, or that the CCO board has a standing update from the CAC on its agenda. One CCO with an effective CAC relationship has two CAC members on its governance board—the CAC Chair (a local health department representative) and an Oregon Health Plan member—and the CAC is currently involved in the CCO's strategic planning process.

In addition to connections with CCO leadership, effective CACs need to have a defined purpose beyond the legislatively mandated responsibility of developing the CHIP. Successful CACs influence CCO programs and initiatives, and some CACs even receive financial support to implement CHIP priorities. For example, one CCO has delegated authority to the CAC for managing the dollars it sets aside from its Oregon Health Authority funding, or global budget, for prevention work. The CCO invests $1.33 per member per month in prevention programs intended to reduce the demand for chronic care and improve the health of community members. In partnership with the local health department, the CAC is charged with recommending effective evidence-based prevention programs that can help move the needle on the CCO's CHIP goals, which include tobacco and obesity prevention and mental health promotion. A subgroup of the CAC, which includes public health staff, physicians, and representatives from community stakeholders, meets monthly to identify objectives and potential local projects. The proposals are then reviewed and approved by the CCO's finance committee, CCO management staff, the CCO's clinical advisory panel, and ultimately the CCO Board. As a result of this process, funded projects have included the purchase of summer pool passes for youth members, and the implementation of the evidence-based "Good Behavior Game" (Poduska et al., 2008) in selected local schools. This program—which impacts all children in the schools, not just the Medicaid members—has been shown to have a positive impact on kids' futures in a variety of ways, including reducing the incidence of mental health issues and substance abuse.

Another CCO with a successful relationship with its CAC established a "Community Health Investment Fund" with dollars contributed by the two entities that formed the CCO. CACs in each of the CCO's three counties collaborate on the application process with community partners, reviewing each application that comes in, and making funding recommendations to the CCO Finance Committee and CCO Board. One criterion for all funded projects is they must connect with the CCO's CHIP priorities. Each project touches community residents beyond the CCO's Medicaid members. The money funds county-specific projects as diverse as food for all low-income residents; trauma-informed care trainings for school personnel, community partners, parents, and students; parenting education; temporary emergency relief services, including safe/sober housing and food for homeless residents; and increased screening and referral capacity for domestic violence organizations in all three counties. Each local CAC is regularly updated on their county's projects, and receives a final in-person presentation on the project at the conclusion of the funding cycle.

## PAYMENT INCENTIVES FOR COMMUNITY ENGAGEMENT

The incentives inherent within the health care system's payment methodology can provide either substantial barriers or formidable enticements to the system engaging with the community to address the social determinants of health. Through its delivery system reform efforts, Oregon has taken a different path that addresses the barriers that have historically resulted from the health care payment system.

### Health-Related Services

The traditional fee-for-service payment system is rife with barriers to promoting the health system engaging with the community to address the social determinants of health. Under Medicaid fee-for-service, the state can only pay for medical benefits covered through its Medicaid State Plan. However, per its 1115 waiver authority from the Centers for Medicare and Medicaid Services (http://www.oregon.gov/oha/HPA/HP-Medicaid-1115-Waiver/Pages/2012-2017-Demonstration.aspx), Oregon has incorporated an innovative feature within its delivery system design, called health-related services, to address these payment barriers. Specifically, instead of being reimbursed for each service delivered, as is the case with fee-for-service, CCOs receive a global budget that combines the funds for their physical health, mental health, and oral health services into one budget that has very few restrictions when compared with traditional Medicaid. In particular, through the provision of health-related services, CCOs have additional latitude with regard to the services they provide, allowing them to engage community partners to provide health-related supports aligned with their members' needs that fall outside of the traditional health care system. Health-related services need to be justified by the health care provider as helping manage a health-related condition in a cost-effective manner, essentially resulting in better health for the Medicaid members at a lower cost. In addition, since they are nonmedical services, health-related services cannot be a covered benefit under Oregon's State Plan or have billing or encounter codes.

In general, CCOs have used health-related services that fall into three categories: housing-related supports, wellness supports, and mental health and counseling supports. In most of these cases, the implementation of health-related services required the CCO to collaborate with community partners, such as housing providers, schools, exercise or nutrition programs, and transportation providers.

Some CCOs use health-related services to provide members with housing-related supports. CCOs report occasional use of health-related services to work with service providers on the ground to coordinate housing services, helping members with no other place to go find stable, transitional housing after a hospital discharge to help facilitate the healing process. In addition, CCOs have used health-related services to cover home improvements, such as a vacuum to help an asthma patient manage their condition, and help members with rent, deposits, and application fees.

The second broad category of health-related services includes wellness supports, including exercise shoes, gym memberships, and healthy cooking and exercise classes. For example, one CCO with high obesity and diabetes rates has developed a "Community Health Improvement Program," which emphasizes exercise and lifestyle modification, including tips on grocery shopping.

The third category encompasses mental health and counseling supports for community members beyond the Medicaid population, which include programs as diverse as mental health courts—which link offenders who would ordinarily be prison-bound to long-term community-based treatment—and employment counseling to support job searches. Another CCO has used health-related service dollars to fill a gap in the availability of mental health counseling available for students in an eastern Oregon county. The CCO funded the local community mental health agency to embed mental health specialists in schools; the program has recently expanded to all schools in the county. The program has realized impressive results: over three-quarters of the students served did not require referral into more intensive outpatient services, resulting in significant cost savings from diverted outpatient treatment. Other examples of how CCOs have used health-related services include transportation for nonmedical services, cell phones, and gift bags to incentivize adolescent well-care visits.

### *Barriers to Using Health-related Services*

Despite the successes identified earlier, CCOs are not using health-related services extensively. Approximately 7% of the CCOs' 2015 budgets were applied to health-related services, with 5 of the 16 CCOs reporting no use of health-related service dollars. CCOs have reported a number of barriers to their use of health-related services. For example, some CCOs are uncertain how to communicate the availability of health-related services with members, citing concern about their capacity to handle demand if health-related services are widely advertised. Some CCOs also report that providers lack understanding about when to recommend health-related services. In addition, CCOs report administrative challenges such as the time required to process gym memberships; balancing the need for timely decisions with the requirement to obtain provider authorization; and uncertainty how to identify health-related services' return on investment. Finally, while some CCOs would like more direction and guidance from the state on what counts as a health-related service and how to adequately report the number of members impacted and outcomes, other CCOs appreciate the lack of guidance, thus the ability to implement the program to meet their specific needs.

Despite these challenges, CCOs are generally optimistic about the opportunities afforded by health-related services. One CCO representative has noted that it does not cost a lot to improve someone's living situation or pay for exercise classes, so the fact CCOs are not spending significant dollars on health-related services is not indicative of their value. Another CCO staff member said that CCOs have only been managing health-related services for a few years, and need more time to prove they work.

## Incentive Metrics

Another feature of Oregon's Medicaid model that fosters community engagement to support the Triple Aim of better health, better care, and lower costs is an annual financial pool or "quality pool," established through Oregon's 1115 demonstration waiver that rewards CCOs for the quality of care provided to Oregon Health Plan members. This model rewards CCOs for their outcomes, rather than their members' utilization of services. In essence, the quality pool is a withhold from the CCOs' global budgets (currently at 4.25% of CCOs' total funding, having increased gradually from an original withhold of 2% when CCOs were created) that each CCO receives if they achieve the statewide benchmark or CCO-specific improvement target on performance on each of the 17 incentive measures (http://www.oregon.gov/oha/HPA/ANALYTICS/Pages/CCO-Baseline-Data.aspx).

There are two ways in which the quality pool fosters CCO engagement with the community to support health. First, a few of the CCO incentive measures now focus on population health, and thus promote community partnerships. Whereas most incentive measures were clinical in nature when they were first implemented in 2013, during the last few years the state's Metrics and Scoring Committee has added population-health measures such as tobacco use prevalence, effective contraceptive use among women at risk of unintended pregnancy, and dental sealants on permanent molars for children—all of which fall outside the traditional medical health model. These newer metrics incentivize CCOs to work with community partners so they can achieve their metrics goals. For example, to address tobacco use, one CCO implemented programs—with guidance from its CAC—to help its members stop using tobacco, including partnering with the Women, Infants, and Children (WIC) program to implement a tobacco cessation program for pregnant women.

The conversations at the Metrics and Scoring Committee indicate that more population-health-related metrics may be adopted in the future. For example, the committee has engaged in numerous conversations on metrics related to kindergarten readiness and obesity prevalence, and is considering adding metrics for both topics in the near future. If this happens, CCO engagement of their CACs and community partners will become more important as they become increasingly responsible for—and have greater financial incentives to meet metric goals related to—population health.

The quality pool also indirectly supports community engagement for the few CCOs that allocate funds secured through the pool to their CACs, giving these CACs oversight as to how to use the funds to support health of both CCO's members and the community at large. For example, one CCO manages a grant program for its CACs to manage the money obtained through the quality pool, and many of the funded projects entail partnerships between the CCO and entities in the community. Grant-funded programs include supporting services in schools such as oral health treatment, mental health counseling, and an empowerment program for female students. Another project entails a collaboration between the CCO and the local jail, supporting a mentor who helps offenders who are due to be released soon access mental health and primary care treatment post release.

## LOOKING TOWARD THE FUTURE

After almost 5 years of health system transformation, Oregon's Medicaid delivery system continues to evolve. The importance of the state's role in fostering connections between CCOs and their community-based partners to promote health for both CCO members and the broader community is clear. To that end, in addition to the possibility of new CCO incentive metrics related to population health, Oregon's recently renewed 1115 waiver from the Centers for Medicare and Medicaid Services will allow the state to play an even stronger role in promoting these activities.

The waiver states that the Oregon Health Authority will transition to a payment system that rewards improvement in health outcomes and not volume of services and that, as part of this transition, the state will work with CCOs and network providers to develop a value-based payment plan that describes how they will collectively achieve a set target of value-based payments by the end of the waiver demonstration period. The state envisions that health-related services will be a key component of CCOs' value-based payment approaches. The waiver also states that the Oregon Health Authority will update the CCO contract language to require CCOs to consider using alternative services including health-related services. In addition, Oregon's waiver notes that, as CCOs provide health-related services that are more cost-effective than Medicaid state plan services, the per-capita growth rate for capitation rates should gradually decrease over the five-year waiver period. The waiver allows the state to offset these decreases with changes in the rate development methodology. Specifically, the Oregon Health Authority plans to develop capitation rates with a profit margin that varies by CCO, as opposed to a fixed percentage of premium for each CCO. The capitation rates for CCOs identified as high performing (i.e., those showing quality improvement and cost reduction in the previous years) will have a higher percentage of profit margin built into their capitation rates than lower-performing CCOs. If implemented, this approach would ensure CCOs are not penalized for providing cost-effective health-related services.

## CONCLUSION

The model CCOs are implementing is showing early signs of success in bridging the gap between the community and the health care system. While Oregon's Medicaid health care reform efforts are broadly serving as an inspiration for other states, the ways in which the reformed system encourages community engagement in support of achieving health has been particularly resonant for policymakers, public health, and social service leaders across the country. The tools Oregon has used to promote community engagement and address the social determinants of health are available to other state and local health systems. Giving Medicaid members a voice in their health care through models such as CACs influences the health care systems' perspective on how to promote health and encourages them to partner with community-based organizations. Modifying the health care payment system to allow for payment of activities and supports that promote health outside of the traditional medical-care model can also foster community-based partnerships. Finally, integrating financial incentives that encourage the health care system to focus on the social determinants of health promotes community-based partnerships beyond the health care system walls. If all health care delivery systems in the United States modeled their design on these three components of Oregon's model, the Triple Aim and the achievement of health for all would be less of a vision and more of a reality.

## REFERENCES

Centers for Disease Control and Prevention. *Social determinants of health*. Available from: www.cdc.gov/socialdeterminants/FAQ.html.

DeCubellis, J., & Evans, L. (2014). Investing in the social safety net: Health care's next frontier. *Health Affairs Blog*. Available from: http://healthaffairs.org/blog/2014/07/07/investing-in-the-social-safety-net-health-cares-next-frontier/.

Kindig, D. (2015). *What are we talking about when we talk about population health?* Health Affairs Blog, April 6, 2015. Available from: http://healthaffairs.org/blog/2015/04/06/what-are-we-talking-about-when-we-talk-about-population-health/.

Magnan, S., Fisher, E., & Kindig, D. (2012). Achieving accountability for health and health care. *Minnesota Medicine*, *97*(11). Available from: http://www.minnesotamedicine.com/Past-Issues/November-2012/Achieving-Accountability-for-Health.

Poduska, J. M., et al. (2008). Impact of the Good Behavior Game, a universal classroom-based behavior intervention, on young adult service use for problems with emotions, behavior, or drugs or alcohol. *Drug and Alcohol Dependence 2008*, *95*(1), S29–S44.

Shortell, S. (2013). Bridging the divide between health and health care. *JAMA*, *309*(11), 1121–1122.

Chapter 10

# Promoting Health Equity

**Latricia Tillman**[1,2]
[1]*Multnomah County Health Department, Portland, OR, United States,* [2]*Oregon Health Authority, Portland, OR, United States*

## INTRODUCTION

Like the rest of the nation, Oregon has an inconsistent history of mitigating the drivers and consequences of health inequity. Yet, work within the last decade deepened awareness of the diverse communities across the state and brought a renewed focus on their health concerns, increased those communities' access to the people and places where health policy is changed, and led to health equity being a key part of Oregon's Health System Transformation. This chapter details not only the work to make health equity a part of Oregon's Coordinated Care Organizations (CCOs), but also broader system-wide operational and policy changes that promote health equity. Given the rapid national political pivot away from values of health equity, diversity, and inclusion, these policies and practices are even more essential in order to inoculate communities made vulnerable from a sociopolitical environment that is increasingly likely to experience health inequities.

## BACKGROUND

### Defining Health Equity

Prior to 2000, when the US Department of Health and Human Services established Healthy People goals for the nation, it was acceptable to have different expectations for population health—one for white Americans and one for Black Americans and other people of color. This was intended to recognize and shine a spotlight on the differences in health status that have existed along racial lines for hundreds of years. Since then, awareness has shifted from tolerance of health disparities to the understanding that these differences are the result of racism, including interpersonal racial bias and institutional practices, and the expectation that health care professionals actively work for the elimination of these differences and the achievement of

**Health Reform Policy to Practice. DOI: http://dx.doi.org/10.1016/B978-0-12-809827-1.00010-0**

health equity—which is the ability for all to achieve their full health potential.

## Historical Considerations

Oregon's largest city, Portland, has the dubious distinction of being the whitest major city in the United States. And while Oregon is not the whitest state (20 states have higher proportions of their populations who identify as non-Hispanic White), it has a particular history of exclusion, termination, internment, exploitation, and marginalization of diversity that established a dynamic that has made work around equity uniquely challenging. Attempts to avoid addressing equity include characterizing issues of equity, diversity, and inclusion as "urban" issues that are not relevant to the entire state. Interestingly, this is an erroneous and antiquated impression of the state, as some of the most proportionally diverse counties in Oregon are rural. Immigrants and refugees who are priced out of urban housing markets are increasingly seeking to restart their lives in rural and even frontier communities that are ill-equipped to welcome them. Furthermore, discussions of urban and rural inequities often fail to determine whether people of color, immigrants, refugees, and Native Americans living in rural Oregon have less access to public health services, poorer quality health care, or worse health outcomes than their White neighbors.

In addition, while it is critical to note that Oregon has significantly less diversity that many states (approximately 20% of the population of Oregon was nonwhite in 2010, according to the US Census Bureau), 37% of Oregon's Medicaid population in fiscal year 2011 was nonwhite (http://kff.org/medicaid/state-indicator/medicaid-enrollment-by-raceethnicity/?currentTimeframe = 0). Prior to Oregon's Health Systems Transformation efforts, the reality of a disproportionate representation of diverse Oregonians in the Medicaid population had not led to an elevated sense of urgency related to any discussions of health inequities in past state health policy conversations. In Oregon, leaders from communities of color, health care, and public health had been actively working to achieve health equity for generations. However, due to Oregon's history, those efforts and consistent recommendations for change resulted in minimal funding or implementation beyond acknowledgment of the challenges of inequality. Nevertheless, the work of practitioners and activists, most notably from the civil rights era onward, prepared Oregon to focus on health equity as a significant part of Health Systems Transformation.

For example, during the late 1990s, Governor John Kitzhaber, at the urging of Senator Avel Gordly, Oregon's first African American senator, established the Racial and Ethnic Health Task Force. This body worked from 1999 to 2000, to identify the cross-cutting barriers to health equity in Oregon as well as several pressing health-specific equity issues, and made several recommendations for institutional changes that would position the state of Oregon to address fundamental barriers to equity. These recommendations included more granular race and ethnicity data, a more diverse and culturally competent health care workforce, and accountability for eliminating health inequities. Unfortunately, the recommendations from this effort went largely ignored by Oregon's Department of Human Services, which at the time included Medicaid and Public Health services.

In 2007 when efforts to plan for future Oregon health reforms established the Oregon Health Fund Board, the Board's only African American member and the president of the Urban League of Portland, recommended the creation of a Health Equities Committee of the Board. The Committee's charge was to help develop strategies to assure that health reforms would include a strong focus on specific strategies necessary for assuring the optimal health of people of color, immigrants, and refugees, and people with disabilities who have traditionally been underserved in the health care system.

In other settings, governmental efforts had identified other strategies for advancing equity in health care and public health on a national scale. Consistently, recommendations for advancing health equity in Oregon and nationally have focused on the collection and use of more granular racial and ethnic health status data; assuring the provision of more culturally competent health care (including better training for providers and a more diverse health care workforce); and the integration of community health workers, trusted community members trained in health promotion and disease prevention. These recommendations also included a more active enforcement of the Civil Rights Act, which was the legal force responsible for desegregating health care, addressing discriminatory health practices, and assuring access to care for individuals with limited English proficiency.

When Oregon began its journey in 2009 toward the establishment of CCOs as the flagships of Health Systems Transformation, numerous bodies had been recommending a consistent set of changes to improve the health of communities of color, immigrants, and refugees. The state was encouraged to look to these historical efforts to advance health equity and take swift action on these languishing strategies, rather than continue to engage in the exercise of further assessment activities. These foundational building blocks effectively created the foundation of the health equity work that was embedded in Oregon's Health Systems Transformation.

## ADDRESSING EQUITY IN HEALTH SYSTEM CHANGE

### Structural Support for Health Equity

It could be appealing to consider the policy and program efforts of embedding equity in Oregon's Health Systems Transformation separately from the way the work was situated in a broader department. However, the structural support for health equity was essential to the successful programmatic and policy work undertaken by the Oregon Health Authority. The work of embedding health equity into Health Systems Transformation efforts was led by the Office of Equity and Inclusion (OEI). Nationally, over 30 states have similar offices. However, this office was structured to influence significant change in both intent and action related to health care reform.

Initially, OEI was established in 1993 to bring attention and resources to efforts to eliminate health inequities in communities of color across Oregon. For its first 10 years of existence, it was called the Office of Multicultural Health and existed within the Public Health Division without an agency-wide focus. Like many similar offices across the country, it did not have a solid budgetary foundation or a strong role in broad system change initiatives.

In 2009 OEI was moved into the Director's Office of the Department of Human Services and moved to the Oregon Health Authority when it was created in 2011. Its charge was to bring attention and resources to efforts to promote health equity, as well as diversity and inclusion across the entire agency. Existing staff was joined by staff members who were reassigned from Human Resources, which allowed the office to develop both an internal workforce development focus and an external community-driven public policy focus.

Budget is often cited as a significant constraint to work on health equity. Prior directors of the OEI identified limitations in funding, in addition to lack of control over their budgets as significant challenges in advancing the work of the office. Without budgetary control, office administrators cannot plan and make investments in strategies that advance equity. A significant part of the OEI budget came from stable funding sources. As such, staff could concentrate on policy issues rather than on grant writing. The structural location of OEI with a department-wide perspective from the Director's Office allowed for the use of Medicaid Administrative Match funds. The work of OEI was focused on system-wide improvements, including the Medicaid program, which met the needed criteria to justify the use of these funds. In addition, OEI was recognized as contributing to the overall function and improvement of the agency, which allowed the office to be funded similarly to Human Resources, Finance, and Communications—through a cost-allocation formula. This reinforced and emphasized the mandate for OEI to work across the department using a strategic organizational development and policy approach, rather than a disease or community focus.

In addition to organizational structure and foundational funding, supportive executive leadership was preeminent to the success of Oregon's health equity work. Similar offices or positions often find their work hampered due to low expectations for equity and diversity work, unwillingness to provide access to critical decision-making tables, lack of transparency and mentorship to support staff in learning both formal and informal processes used in making policy and navigating political processes.

It is hard to overstate the amount of executive support enjoyed by the OEI under the OHA, as well as how that support was maintained and deepened throughout the transformation effort. Beyond the willingness to move the office into the Director's Office and to assure a reasonable budget to accomplish the office's mandate, the Office enjoyed support from the agency director and the deputy director, which made it possible to take on both political and administrative challenges; to take risks, develop approaches over time, learn from failure; and to create a culture of learning and accountability for equity, diversity, and inclusion at executive levels of leadership across the agency.

## Community Engagement

As mentioned in the "Introduction," health equity advocacy was not new to Oregon. For decades community leaders had been making recommendations for policies that would advance equity. What was new was that the state agency leading the health system reform effort had a sustained organizational commitment to engage communities in leadership roles to advance those policy recommendations.

Consistent health equity recommendations from communities of color, immigrants, and refugees across the state and the nation included culturally and linguistically competent health care, so that people could expect to feel that their cultural health practices would be at least respected, if not also understood; a diverse workforce to help bridge the divide and build trust between providers and patients; the use of Community Health Workers, trusted community members who are trained to share health information and health promoting practices with communities; and better collection of race, ethnicity, and language data—and the consistent reporting of data by increasingly granular demographic factors to better identify where poor health outcomes are burdening specific communities.

Community engagement remains core to all of the work of the OEI. The OEI produced an annual report, in part, to be accountable to divisions across the agency who contributed to the office's funding, but primarily to document to community partners how the Oregon Health Authority and the Office were responding to their community priorities. An annual in-person meeting was the companion to the annual report. Both efforts helped to demystify government to communities historically excluded from public processes, gave diverse communities access to agency leadership in a less intimidating, more relational setting, provided opportunities for engagement of both agency leadership and community leadership in celebrating accomplishments, as well as built shared expectations of accountability for an action-oriented, sustainable office. These opportunities were the first time many culturally diverse community leaders interacted with state agency leadership. It was an opportunity to appreciate community members who served and to recruit others to participate on more formal state policy and rulemaking bodies, including several that were run out of the OEI.

Community members were engaged in a number of committees that supported the governance of the OEI and the development and implementation of state policies. The primary oversight body was the Office's Community Advisory Board, which was established informally (no bylaws or formal charter was established to maintain an advisory rather than a governing role) to deepen the community ownership of, knowledge of, and expectations for the OEI. The Board was appraised of and provided input into the budget, advised on program investments, participated in strategic planning, and assisted with position recruitments and funding processes.

The Health Equity Review Committee was established for a finite time as a committee of the Oregon Health Policy Board (OHPB) to help assure that health equity would be considered throughout the process of health systems transformation. The Committee was established to provide a review as OHPB Committees brought forward policy recommendations that would be incorporated into Oregon's Action Plan for Health. The effectiveness of this committee was limited by the timing of their engagement in the process of developing recommendations. Rather than being engaged earlier in the developmental process (which happened in a very compressed time frame), the committee was often engaged after policies were well developed. Staff members who brought policy recommendations to this committee were often surprised by the depth of feedback they received, which highlighted limited internal agency knowledge, skill, and capacity to consider the unique strengths, needs, and circumstances faced by people of color, immigrants, and refugees in Oregon when developing policy that would impact an increasingly diverse Oregon population.

At the end of the formal charter period for the Health Equity Review Committee, members of the committee and other community members expressed appreciation for the opportunities to learn about the agency and Oregon's health policy priorities. They requested that the committee continue through the 2011 legislative session, and at that point, requested that it continue indefinitely. The committee morphed into the Health Equity Policy Committee to provide opportunities to develop community priorities and for agency leaders to vet policies in an earlier stage of development for deeper discussion and more meaningful real-time input.

### *Health Care Interpreter Council*

The Health Care Interpreter Council was established in 2001 to implement legislation that required the state of Oregon to create a mechanism to certify and qualify health care interpreters. Certification requires training, practice hours, and the passage of a test on medical terminology and professional and ethical standards, which demonstrates bilingual proficiency in lay and health care terminology. Certification is required for health care interpreters in the seven most common languages spoken in Oregon. Qualification requires demonstration of bilingual proficiency, training, and practice hours, recognizing that establishing a certification test in all potential languages is not practical. The Health Care Interpreter Council has served as a body that has advised on the administrative rules and rule changes for implementation of legislation. It informed the implementation of the law, which began in earnest in 2010, a decade after the legislation passed.

### *Community Health Worker Advisory Committee*

The OEI established a Community Health Worker advisory committee to assess the needs of community health workers in terms of defining scope of practice, identify training gaps, exploring the benefits and pitfalls of certification, and pursuing funding sustainability. This committee was formed to respond to recommendations for culturally competent health care and integration of community health workers in public health and health care interventions. The office recognized the long history of lay health advisors, promotoras de salud, and other community health workers indigenous to communities of color, immigrants, and refugees in Oregon and the nation, as well as internationally. The purpose of this committee was initially exploratory, with the expectation that strategies for implementation would emerge from an assessment of the current needs of community health workers.

The work of this committee was fast-tracked and modified when references to community health workers were inserted in health systems transformation legislation in 2011. Subsequently, Governor Kitzhaber made a commitment to train and certify 200 community health workers and established a funding stream for community health worker training. This funding was limited in its impact because it was directed to community colleges without requiring partnerships with community and health organizations that had hired, trained, and worked with community health workers for decades.

The context in which the OEI operated, including the state's history of inaction, the agency leadership's commitment to advance equity as part of health system transformation, and deep community support, facilitated the policy and organizational change that advanced health equity significantly over a short period of time.

### *Regional Health Equity Coalitions*

The Office of Equity and Inclusion established regional coalitions focused on racial and ethnic health equity to address diverse priorities and concerns of culturally diverse communities across the state (http://www.oregon.gov/oha/oei/Pages/rhec.aspx). These primarily rural coalitions were funded for a planning year in which they developed geographically, culturally, and professionally diverse membership, identified local opportunities to advance health equity, and developed plans for action. Subsequent years focused on moving planning priorities into action, including working with local Coordinated Care Organizations on their community-health improvement and transformation plans, providing cultural competence training for health-care providers and community members, and advancing local and statewide health equity priorities. These coalitions expanded, diversified, and deepened local health leadership, and supported understanding of and capacity to advance health equity using a variety of tailored strategies relevant to their communities.

## Health Equity Policy Changes

Prior to the 2010 legislative session, OEI held listening sessions with community partners to identify policy priorities. In these sessions, community organizations were invited to present their top three public policy priorities. Through these presentations, the priority of a culturally competent health care workforce emerged—both in terms of the use of community health workers as well as the training of health care providers. While OEI launched a committee focused on community health workers, it also developed a legislative policy concept that would require continuing education training for health care providers. At that time, only seven states in the nation had passed similar legislation, most of which authorized further exploration of cultural competence training standards but did not mandate training.

The process for developing this concept involved engaging community organizations, health care reform coalitions and health systems, and securing their endorsement. With community input, the language of the concept expanded beyond cultural competence focused on race and ethnicity to include sexual orientation and identity, disability, and veterans. Community partners and OEI staff made presentations to health care licensing boards and health care professional organizations, which for the most part did not favor the concept because of the additional work it would create for their limited operations. Uniquely, the Board of Nursing did not favor any training requirements because it had no existing training requirements for licensure and did not want to establish that precedent.

These efforts led to the introduction of Senate Bill 97 (SB 97). When introduced, SB 97 included a requirement for cultural competence continuing education for all licensed health care professionals. Testimony to the Senate Health Care Committee was largely supportive; however, several professional associations quietly, but powerfully, opposed the mandate requirement in the legislation. Ultimately, requirements for continuing education were removed, and the final legislative language called for three activities to be conducted and reported back to the legislature:

- develop standards and list of opportunities for continuing education in cultural competence,
- develop and implement such education for licensed health care providers, and
- study cost to operate and opportunities to provide cultural competence continuing education for health care professionals, in various modalities.

Support and opposition to the legislation fell largely along party lines. While supportive, some Democratic legislators were wary of the effectiveness of professional continuing education. Others who supported the concept felt that much of the intent could be addressed through employee training of health systems, which should extend beyond providers to include all staff. It is notable that the Joint Commission, which accredits many health systems in Oregon and nationally, requires staff orientation and ongoing education and training related to cultural diversity and the specific needs of populations being served (The Joint Commission, 2014).

Ultimately, and uncharacteristically, the legislation passed the Senate but the bill was defeated on the House floor after legislative champions brought it forward twice for a floor vote. However, this failure led to greater success in four ways:

1. While efforts to advocate for the bill on the Senate side were led by the OEI, leadership for advocacy was transitioned to the community partners when the bill moved to the House. This created an excellent opportunity for community organizations and their constituents to engage directly and repeatedly in advocacy and education with legislators.
2. Advocates for equity learned a valuable lesson in the impact of electoral politics on policy change. With SB 97 falling one vote short of passage, the impact of changing one legislative district to be more favorable to issues surrounding health equity was not insignificant and could impact the possibility of passing other policies focused on racial justice. Numerous bills related equity and racial justice also failed that year, compelling community partners to publish *Facing Race: Legislative Report Card on Racial Equity* to evaluate the voting record of the Oregon legislature and create a platform that advocacy organizations used to advance significant legislation in future sessions (http://www.westernstatescenter.org/tools-and-resources/Tools/2013-facing-race).
3. In his floor speech, a prominent legislator who opposed the bill noted that the legislation was symbolic and the agency could implement everything the bill without legislative approval. This statement led the Office to establish the Cultural Competence Continuing Education Committee, which met between September 2010 and December 2011 to address the elements of the legislation. This work led to the ultimate passage of cultural competence continuing education legislation (HB 3100) in 2012.
4. By far the biggest champion of cultural competence continuing education legislation was Representative Tina Kotek, one of the legislators involved

in negotiating the health system transformation bill. Her commitment to health equity translated into an opportunity for OEI to engage in the much more significant and substantive legislative conversation happening that year, which focused on health systems transformation. Representative Kotek requested that OEI provide her with a health equity analysis of HB 3650, the bill that framed Oregon's Health Systems Transformation efforts. As a result of this analysis, language was added to the legislation that included numerous significant policy priorities that community leaders had been advocating for both locally and nationally. These included:

**a.** *Inclusion of underserved communities in policy and program development*: Sec 2(2). The Oregon Health Authority will seek input from groups and individuals who are part of underserved communities, including ethnically diverse populations, geographically isolated groups, seniors, people with disabilities, and people using mental health services, and will also seek input from providers, CCOs and communities, in the development of strategies that promote person centered care and encourage healthy behaviors, healthy lifestyles and prevention and wellness activities, and promote the development of patients' skills in self-management and illness management.

**b.** *Culturally appropriate health care delivery*: Sec 4 (1)(k). Members have a choice of providers within the CCO's network and providers participating in a CCO: (G) Work together to develop best practices for culturally appropriate care and service delivery to reduce waste, reduce health disparities, and improve the health and well-being of members.

**c.** *Accountability to executive leadership on efforts to eliminate health inequities*: Sec 26 (7). Each CCO will work to provide assistance that is culturally and linguistically appropriate to the needs of the member to access appropriate services and participate in processes affecting the member's care and services.

Sec 2 (3)(b). The authority will regularly report to the OHPB, the Governor and the Legislative Assembly on the progress of payment reform and delivery system change including progress toward eliminating health disparities.

**d.** *Culturally relevant community-based care settings*: Sec 4(1)(f). Services and supports are geographically located as close to where members reside as possible and are, if available, offered in nontraditional settings that are accessible to families, diverse communities, and underserved populations.

**e.** *Culturally appropriate care and service delivery*: Sec 4(1)(k)(G). Providers participating in a CCO work together to develop best practices for culturally appropriate care and service delivery to reduce waste, reduce health disparities, and improve the health and well-being of members.

f. *Culturally diverse workforce*: Sec 19(1)(L). The authority will: Implement policies and programs to expand the skilled, diverse workforce as described in ORS 414.018 (4). Sec 30(1)(a). Workforce data collection. Using data collected from all health care professional licensing boards, including but not limited to boards that license or certify chemical dependency and mental health treatment providers and other sources, the Office for Oregon Health Policy and Research will create and maintain a health care workforce database that will provide information upon request to state agencies and to the Legislative Assembly about Oregon's health care workforce, including:
   i. Demographics, including race and ethnicity.
   ii. Incentives to attract qualified individuals, especially those from underrepresented minority groups, to health care education.
g. *Granular race and ethnicity data collection*: Sec 10(2) (2). The authority will evaluate on a regular and ongoing basis key quality measures, including health status, experience of care and patient activation, along with key demographic variables including race and ethnicity, for members in each CCO and for members statewide.

   Quality measures identified by CCOs are expected to collect or maintain race, ethnicity, and primary language for all members on an ongoing basis in accordance with standards jointly established by OHA and the Oregon Department of Human Services. CCOs can then track and report on any quality measure by these demographic factors and will be expected to develop, implement, and evaluate strategies to improve health equity among members (Coordinated Care Organizations Implementation Proposal; House Bill 3650 Health System Transformation, Oregon Health Policy Board, January 24, 2012).
h. *Community health workers*: (e) Members receive assistance in navigating the health care delivery system and in accessing community and social support services and statewide resources, including through the use of certified health care interpreters, as defined in ORS 409.615, community health workers and personal health navigators who meet competency standards established by the authority under Section 11 of this 2011 Act or who are certified by the Home Care Commission under ORS 410.604.
i. *Focus on eliminating differences in health outcomes between racial and ethnic groups*: Sec 1 (3)(b). Health care services, other than Medicaid-funded long-term care services, are delivered through coordinated care contracts that use alternative payment methodologies to focus on prevention, improving health equity and reducing health disparities, utilizing patient-centered primary care homes, evidence-based practices, and health information technology to improve health and health care.

i. Sec 2 (3)(b). The authority will regularly report to the OHPB, the Governor, and the Legislative Assembly on the progress of payment reform and delivery system change including: Progress toward eliminating health disparities.
ii. Sec 4 (1)(k)(G). Members have a choice of providers within the CCO's network and that providers participating in a CCO: Work together to develop best practices for culturally appropriate care and service delivery to reduce waste, reduce health disparities, and improve the health and well-being of members.

j. *Culturally and linguistically appropriate health care*: Sec 8 (1)(c). Consumers must have access to advocates, including qualified peer wellness specialists where appropriate, personal health navigators, and qualified community health workers who are part of the member's care team to provide assistance that is culturally and linguistically appropriate to the member's need to access appropriate services and participate in processes affecting the member's care and services.

Without Representative Kotek's advocacy for health equity, emboldened by the blazing defeat of a symbolic piece of cultural competency legislation, Oregon's reform efforts would be more typical of historical health policy—color blind and race neutral, thus upholding status quo for communities of color shouldering the burdens of poor health. Ultimately, the legislation that established Oregon's Health Systems Transformation effort had numerous explicit references to the improvement of racial and ethnic health outcomes that codified significant health equity strategies.

## Health Equity in Coordinated Care Organizations

The legislation established a clear mandate for health equity as part of the establishment of CCOs. The strength of this mandate was essential in order to extend the commitment to health equity throughout each of the subsequent steps of policy implementation. The administrative rules request for proposals for CCOs and the review process for those proposals, the contracts for CCOs, and their quality assurance processes all include a focus on health equity.

In each of these administrative steps, the state maintained the expectation that CCOs develop systems and practices to improve health equity in the Medicaid population and to expand work toward health equity as the CCO model expanded into the commercial market. At every step in implementation of health systems transformation, the state communicated the importance of culturally and linguistically appropriate health care, the use of community health workers, the collection of more granular race, ethnicity and language data, and the engagement of CCOs in community health assessment and improvement planning that included a focus on health equity. In order to track progress toward health equity, the Oregon Health Authority committed to publishing CCO metric data by race and ethnicity.

### CCO Transformation Plans

A unique element of the CCO Contract is the expectation that organizations develop and implement transformation plans. These plans included eight elements, including increased use of electronic health records and transitioning to patient-centered primary care homes. While CCOs have the flexibility latitude to identify their own objectives and set benchmarks, these require approval by OHA. Notably, four of the eight elements of the transformation plan aim to directly or indirectly support CCOs to address health inequities both in the Medicaid population and in the broader community:

1. Preparing a strategy for developing Contractor's Community Health Assessment and adopting an annual Community Health Improvement Plan consistent with ORS 414.627. (While not explicitly focused on health equity, guidance for community health improvement planning included looking at health assessment data by race, ethnicity, and language, engaging culturally diverse community partners in planning, and establishing community-based health equity outcomes as part of the plan.)
2. Assuring communications, outreach, member engagement, and services are tailored to cultural, health literacy, and linguistic needs.
3. Assuring that the culturally diverse needs of members are met (cultural competence training, provider composition reflects member diversity, Certified Traditional Health Workers, and Traditional Health Workers composition reflects member diversity).
4. Developing a quality improvement plan focused on eliminating racial, ethnic, and linguistic disparities in access, quality of care, experience of care, and outcomes.

According to the Oregon Health Authority's legislative report on Oregon's Health System Transformation (http://www.oregon.gov/oha/hpa/analytics/pages/index.aspx), "each of the following activities is reflected in the transformation plans for at least half of the CCOs:

- Providing cultural competency, cultural diversity, and/or health equity trainings for their clinic and provider networks;
- Reviewing and revising member materials for appropriate health literacy; and
- Analyzing, reporting, and disseminating their quality performance measure data stratified by member demographics, such as race, ethnicity, language, age, gender, and disability."

### *Health Literacy: Equity in Communication*

CCOs invested a significant amount of effort in basic foundational work of assuring equitable and accessible access to information for Medicaid members. While the legislation that established CCOs does not specifically mention health literacy, this emerged as a key strategy and a basic tenet of health equity. Given the amount of significant change to Medicaid through Oregon's transformation efforts along with the passage of the Affordable Care Act, it was essential to assure that existing and new Medicaid members understood their benefits and rights. For example, access to community health workers were new benefits. Access to an interpreter was not new, but the fact that those members of the health care workforce now needed to be certified or qualified, was a new element to an existing benefit. Clarity about client civil rights was new, in spite of the fact that the Civil Rights Act was approaching its 50 year anniversary at the time that transformation was happening in Oregon's Medicaid system.

It was essential that the Membership Handbook was written in a way that was clear to members of the community who might have limited literacy in general, and limitations in literacy related to complicated health and insurance terminology. In addition, with an emerging understanding of the Civil Rights Act, it was important that the Oregon Health Authority and CCOs provide the handbook in more languages than providers and state agencies had done in the past. Federal guidance on language access provides clarity about when documents must be available in languages other than English and clarifies that all languages must be provided in languages other than English upon request (https://www.lep.gov/interp_translation/trans_interpret.html).

Furthermore, OHA and CCO staffs needed to assure that the translated handbooks were also accessible to Medicaid members with limited literacy in their native languages. A significant amount of work was invested by the Oregon Health Authority and the OEI to refine complicated handbook jargon and include information about the civil rights of Medicaid members. While it would be useful to engage actual Medicaid members to review written materials to assure that they are conveying the information as intended, a general OHA membership handbook for all Medicaid members served as a template that CCOs could use, tailor, and refine for their members.

## Supplementary Policy Development

In addition to the policies developed as a result of Health Systems Transformation legislation, a number of health equity policies were developed in a way that was complementary to the legislation. In some cases, the legislation was integrated into health systems transformation, and in others, the policy remains separate, but change the environment in which health care reform continues to advance and evolve.

## *Doulas*

In 2011 at the same time as the state was developing CCOs, HB 3311 was introduced that compelled the Oregon Health Authority to explore the use of doulas as a strategy for addressing poor birth outcomes among women of color on Medicaid. The legislation required analysis of birth outcome data by race and ethnicity, a literature review of best practices for using doulas to address health inequities, a recommendation of an appropriate scope of practice for doulas, recommended training components, and an analysis of reimbursement mechanisms. This work was assigned to the OEI and completed by the HB 3311 Implementation Committee, made up of culturally and professionally diverse maternal and child health experts. This Committee determined that doulas were very similar to community health workers with a focus on early childhood and family health. As such, doulas were integrated into the work to establish Community Health Workers as part of a multidisciplinary health care team. Oregon is one of two states who will reimburse doulas who are certified in Oregon for labor and delivery services, as well as prenatal and postpartum care, for Medicaid members in Oregon (Oregon Administrative Rules 410-130-0015 http://arcweb.sos.state.or.us/pages/rules/oars_400/oar_410/410_130.html).

## *Race and Ethnicity Data*

In response to the consistent calls from various advocacy groups and advisory bodies for more granular analyses of health status by race/ethnicity, the OEI developed the State of Equity Report to report on key health indicators with a focus on health equity. Prior to 2009, the only health equity reporting that had been done was focused on Medicaid outcomes. The State of Equity report built on those efforts and expanded to include indicators across health and human services sections of the Department of Human Services and the Oregon Health Authority. Of note, 2011 and 2013 reports illuminated significant differences in outcomes and performance that had been hidden by reporting only aggregate data, most of which was accumulated among African American and Native American service recipients and community members (OHA, 2011, 2013). Furthermore, the report found little consistency across programs and divisions in the way data were reported by race and ethnicity and significant amounts of missing race/ethnicity data.

Using process and outcome findings, the Office of Health Equity established three agency-wide responses:

- It convened and supported an internal committee of data analysts to develop and recommend standards for the collection, analysis, and reporting of data by race and ethnicity across all programs within the

Department of Human Services and the Oregon Health Authority. Standards included separating Hawaiian/Pacific Islander ethnicity from Asian ethnicities and asking Hispanic ethnicity separate from racial identity questions. These standards were adopted by both agencies in July 2012. From the adoption of the standards, the agencies began exploring ways to bring data systems and program eligibility applications for Medicaid into closer compliance with the standard.

- It established an annual meeting with the African American community to bring agency executives into dialog with African American community leaders, members, service recipients, and professionals in physical and behavioral health care, human services, and public health to identify strategies to address the poor outcomes. Executives and their leadership were expected to maintain connection with community members throughout the year to advance community recommendations to improve performance and outcomes. Similar efforts to bring agency executives into dialog with tribes were much more challenging due to geographical and political dynamics.
- Health Systems Transformation metrics were reported by race and ethnicity to create public, although not financial, accountability for assuring that all populations would benefit from Oregon's health reform efforts. The baseline metrics report for Oregon's CCOs did not report by race and ethnicity, but subsequent reports have consistently shown differences in outcomes by racial and ethnic groups.

Community partners, led by the Oregon Health Equity Alliance, wanted to further respond to long-standing recommendations for better data to understand health inequities, build upon agency-wide actions, codify these commitments in law, and assure that state agencies would continue to make significant investments in standardizing race and ethnicity data collection. In 2013, HB 2134 was introduced to address that long-standing need. In 2014 the Oregon Health Authority established a standard (OHA, 2014), based on work in Massachusetts and local community and census demographic data, for the collection of race, ethnicity, and disability demographic data. The OHA in collaboration with the Oregon Department of Human Services and the advisory committee are required to review the standards at least once every 2 years to ensure that the standards are up to date and consistent with current health equity practices.

### *Civil Rights Implementation*

In implementing changes in Medicaid, it became clear that with the establishment of the Oregon Health Authority, the agency also needed to establish its own policies and procedures to protect the civil rights of Medicaid members and other clients served by other programs in the agency. In May 2014, in the 50th anniversary year of the passage of the Civil Rights Act of 1964, the Oregon Health Authority established a nondiscrimination policy and procedure for reporting and investigating discrimination complaints, as well as expectations for contractors to adhere to the same standards (http://arcweb.sos.state.or.us/pages/rules/oars_900/oar_943/943_005.html). The establishment of training, procedures for investigation, and data collection related to civil rights complaints helps the Authority and its contractors, including CCOs, actively prevent and remedy practices that contribute to health inequities.

## LESSONS LEARNED

The work of advancing health equity between 2009 and 2015 would not have been possible without trusting relationships between a number of partners. Building on relationships with community-based organizational leadership and creating opportunities for those leaders to engage and build their own relationships with agency and legislative leaders were critical to the work moving forward. In many cases, community leaders were more effective than agency staff in advancing the health equity agenda assertively and urgently. It is possible for government, community, and legislators to work together effectively to advance health equity policy and to assure effective implementation of that policy; transparent communication, mutual accountability, and value alignment are elements of the relationships that led to effective, outcomes-focused collaboration.

For many health equity advocates, moving policy to support change can be mysterious and ephemeral. Legislators, lobbyists, and policy analysts can help demystify the process and help communities who have been historically kept outside of the policy-making process navigate the complexities of policy change. In some cases, legislators requested a health equity analysis of policies, which opened opportunities to embed the objectives of communities of color in legislation. In other cases, legislators engaged leaders from communities of color in negotiations to assure that lobbyists for dominant culture organizations had to make a formal compromise rather than dismiss a health equity agenda. Some lobbyists helped health equity advocates consider the value of incremental change as compared to taking a more ambitious, but likely philosophical stand. Others helped advocate for legislative change for health equity policies as part of their overall organizational platform.

However, some legislative champions and policy analysts also created vulnerability in health equity policy by getting ahead of community-based leadership. The adoption of Community Health Worker language without a clear understanding of what this worker is, and is not, created an opportunity for entities with no experience with the discipline to co-opt it. Funding was allocated to institutions to train Community Health Workers, even though they had only a superficial understanding of the work they do and their contribution to community and clinical health outcomes and health equity. The pace of policy making requires either preparation or a significant stretch on the part of advocates for health equity who have not been engaged in the policy process; however, advancing health equity policy requires legislators, lobbyists, and large organizations to work collaboratively with marginalized groups, which may mean a shift in pacing and increased transparency about policy strategy.

While the Oregon Health Authority was very successful embedding health equity into policy, health equity efforts would have been much stronger if they had been tied to financial incentives for CCOs and health care practices and providers. Other important transformation elements like electronic health records, patient-centered primary care, and various quality metrics were tied to significant funding as a way to incentivize health care providers to develop plans and strategies to manage their own operational and organizational changes. While the mandate for health equity was clear, a significant limitation of the effort to embed equity into CCOs is that none of the elements of health equity written into implementation carried sufficient financial incentives or relative weight to create the conditions for rapid change. Given the relative weight of health equity compared to incentivized transformational elements, the lack of meaningful responses to health equity elements of transformation did not create the need for significant negotiation with CCOs or a reconsideration or denial of any applicants. Consequently, achievement of health equity carries no financial incentive, and conversely, failure to advance health equity by a CCO has not to date caused any penalties or other repercussions.

## CONCLUSION

Decades of work to advance civil rights and health equity has become combined with a moment in time rich with opportunity to transform health care in Oregon and nationally. Tireless community calls for change in areas of cultural competency, better data, community health workers, and an opportunity for communities of color to enjoy optimal health in their communities was not new. Those years of advocacy, advances in practice, leadership by national health care organizations, and local legislative and community partnerships has culminated in alignment with the newly formed Oregon Health Authority and the OEI to advance health equity policy and practice in numerous tangible ways.

As the national political landscape rapidly retreats from values of equity, inclusion, and diversity, the challenge will be to protect and advance these gains at the same time as inequity is likely to increase because of the renewed and extreme marginalization of immigrants, refugees, Native Americans, communities of color, and sexual minorities. Documentation of processes and outcomes, strategies, and policies may help preserve the institutional and community memory of recent health equity and civil rights victories, along with the organizational conditions that supported those successes. Community capacity to navigate the policy landscape and the halls of the legislature must boldly persevere. Policy makers and community advocates in Oregon should demand the ongoing implementation of these policies—and health equity practitioners in other states must replicate, modify, or leverage them to support emerging local policy and practice solutions.

While no time period in the United States or Oregon history can claim success in eliminating avoidable gaps in health outcomes, the past decade has seen an indelible increase in the public commitment of the Oregon legislature to equity in health, education, and other aspects of policy. The approaches and lessons learned in Oregon and codified in statue should continue to be protected to the greatest extent possible and contribute to the state's deepening foundation of health equity work. Like other efforts to promote health and social justice, political opposition is expected and should not be a deterrent. Health equity work must advance in spite of current political realities—until such time as they can regain momentum and permanently shift the practice of public health and health care.

## REFERENCES

The Joint Commission. (2014). *A crosswalk of the national standards for Culturally and Linguistically Appropriate Services (CLAS) in health and health care to the joint commission ambulatory health care accreditation standards.* Available from: https://www.jointcommission.org/assets/1/6/Crosswalk_CLAS_AHC_20141110.pdf.

OHA (Oregon Health Authority). (2011). *State of equity report: Phase 1.* Department of Human Services and Oregon Health Authority. Available from: http://www.oregon.gov/OHA/OEI/Reports/State%20of%20Equity%20Report%20-%20June%202011.pdf.

OHA (Oregon Health Authority). (2013). *State of equity report: Phase 2.* Department of Human Services and Oregon Health Authority. Available from: http://www.oregon.gov/OHA/OEI/Reports/State%20of%20Equity%20Report%20-%20Phase%20II%20-%20September%202013.pdf.

OHA (Oregon Health Authority). (2014). *Race, ethnicity, language, and disability demographic data collection standards.* Oregon Health Authority, Office of Equity and Inclusion. Available from: http://www.oregon.gov/oha/OEI/Policies/Race-Ethnicity-Language-Disability-Data-Collection-Standards.pdf.

Chapter 11

# Measuring Success: Metrics and Incentive Payments

**Tina Edlund[1] and Lori Coyner[2]**
*[1]Health Management Associates, Portland, OR, United States, [2]Oregon Health Authority, Portland, OR, United States*

## INTRODUCTION

Accountability and transparency took center stage in Oregon as we began the work of redesigning the health care delivery system from a system that rewards volume to one that rewards value. In 2009 the state combined its health care purchasing into a single state agency, the Oregon Health Authority (OHA), which now had consolidated responsibility for purchasing health care for approximately 25% of the state's residents. As state policy makers, we recognized that with the right tools, we could lead the market as a purchaser, but becoming a smarter purchaser required developing a system of accountability that included improved measurement systems, the right metrics, and transparency in reporting costs and outcomes. Equally important was the development of a meaningful incentive structure aimed at pushing the delivery system toward the kinds of outcomes the state was looking to achieve. A system of measurement would need to answer the question at the center of Oregon's reform efforts, "Are we making progress toward the three-part aim of better care, better health and lower per capita costs?" And finally, if the transformation was going to be continuous and dynamic, we needed data that would allow us to make course corrections along the way.

Oregon holds its Medicaid Coordinated Care Organizations (CCOs) accountable for performance and measures progress toward success in multiple ways, but central to these efforts is a system of 17 quality and access metrics with established improvement goals and benchmarks that are directly tied to a percentage of the CCO global budget that started at 2% and will increase annually to 5%. It is a system developed on the premise of aligning incentives with desired outcomes, but also on simplicity, flexibility, and a recognition that not every CCO was in the same place on a continuum toward the goals of health care transformation.

**Health Reform Policy to Practice. DOI: http://dx.doi.org/10.1016/B978-0-12-809827-1.00011-2**

These metrics not only provide an objective measure of progress, but also prove to be a key part of engagement as CCOs and providers join together to identify successful approaches to improving outcomes. Oregon has seen some dramatic improvements as reflected in these metrics. CCOs continue to reduce avoidable emergency department visits and preventable hospital inpatient admissions. Patient-centered primary care home (PCPCH) enrollment has improved or remained stable even with the large influx of members during the Affordable Care Act (ACA) Medicaid expansion.

The program has not seen across the board improvement, however. Adolescent well-child visits and self-reported access to care have remained relatively flat. But because metrics are used to monitor where CCO performance is strong as well as where performance is weak, quality improvement efforts can be targeted. And because it is an open and transparent process with a system of supports and learning collaboratives built in as part of the overall transformation model, CCOs can learn from each other.

This chapter is intended to describe how we developed and implemented a system of metrics and incentives that hold CCOs accountable for improved outcomes and to offer some lessons learned along the way.

## HISTORY

The status of measurement and reporting in Oregon's publicly financed health care programs today is the synthesis of efforts that began in the early 1990s. Oregon, like other states, has a long history of experimenting with the idea of measurement and transparency in health care. The integration of data collection, metrics development, and public reporting as part of health reform in Oregon is the result of focused and evolutionary attention to health system transformation from the 1990s through today. Early efforts at measurement included the Oregon Consumer Scorecard Consortium, an effort in the early 1990s funded by the federal Agency for Healthcare Policy and Research (AHCPR), to create a consumer-facing "scorecard," or health plan comparison. The Oregon Health Policy Commission (2003–07) drove an early effort to publicly report hospital inpatient quality indicators. But the operational capacity to collect and report data, the meaning of health care metrics, and the idea of paying for performance was in its infancy, and none of these efforts was part of an overall financial realignment or model of delivery system redesign.

A major shift came in 2007 with the creation of the Oregon Health Fund Board. Legislative leadership in Oregon recognized that the proportion of state budget being expended on health care was growing at a rate that outpaced state revenues, household income growth, and all other economic indicators. Rapidly increasing health care costs were clearly not sustainable and beginning to threaten the state budget share for other vital public services, such as education and public safety. The Board was chartered with developing a comprehensive plan to ensure access to health care for Oregonians, contain health care costs, and address issues of quality in health care. In November 2008, the Board submitted a comprehensive action plan, "Aim High: Building a Healthy Oregon," to Governor Kulongoski and the Oregon Legislature, providing a blueprint for reforming Oregon's health care system. The Fund Board's work established measuring, reporting, and transparency as one of the seven essential building blocks to health reform in Oregon, stating that "ensuring transparency of costs and health outcomes throughout the system will create competitive pressure between providers to continuously improve."

In June 2009, the Oregon legislature passed historic health reform legislation based on the recommendations of the Board, including creation of the OHA and the Oregon Health Policy Board (OHPB). OHA is responsible for streamlining and aligning state health purchasing to maximize efficiency, organize state health policy and health services, and for implementing health reform policies and programs. The OHPB was established as the policy-making and oversight body for OHA and was directed, as one of its primary duties to, "establish and continuously refine uniform, statewide health care quality standards for use by all purchasers of health care, third-party payers and health care providers as quality performance benchmarks." The work completed by OHPB over the following year established a framework for metrics development in the state.

## Foundational Work

In 2010, the Health Policy Board established the Incentives and Outcomes Committee and chartered it to start Oregon on the path toward a transformed delivery system that:

- Fosters provider accountability through a mature measurement infrastructure that provides meaningful, accurate, and actionable data on performance at the provider, practice, and institutional levels.
- Measures health outcomes and cost metrics relative to historical performance, peer performance, and explicit benchmarks.
- Pays for care in a way that initially rewards performance and ultimately is tied to a budgeted cost for efficient provision of necessary care.

The Incentives and Outcomes Committee was seen as an early stage of the work to establish a structure of accountability. The committee laid out broad principles and an array of potential metrics and benchmarks, directing that the next stage should focus on "measurement and payment efforts in areas of significant cost impact or significant defects in the quality of care, where the potential for improvement is the greatest." This set the stage for the next phase of the work, which created the system of metrics Oregon uses today, a system that is not separate, but part of an overall quality strategy for a transformed delivery system that would begin with that largest share of the state's health care budget: the Medicaid program.

In 2011, the Oregon Legislature and Governor Kitzhaber created the framework for a transformed Medicaid delivery system through CCOs. Essential elements of that transformation were: integration and coordination of benefits and services; local accountability for health and resource allocation; standards for safe and effective care; and a global Medicaid budget tied to a sustainable rate of growth. The legislation directed the OHPB to develop a business plan and to bring the business plan back to the legislature in the 2012 session for approval. As part of developing the business plan, the OHPB established an Outcomes, Quality, and Efficiency Work Group that further refined principles for measurement and accountability for CCOs. The measurement principles were used as the basis for the work to come:

- *Transformative potential*: Measure should help drive system change rather than reinforcing the status quo.
- *Consumer engagement*: Measure should successfully communicate to consumers what is expected of CCOs.
- *Relevance*: The condition or practice being measured should have significant impact on issues of concern or focus.
- *Consistency with existing state and national quality measures*, with room for innovation if needed.
- *Attainability*: It is reasonable to expect improved performance on this measure (can move the meter).
- *Accuracy*: Changes in CCO performance will be visible in the measure and measure distinguishes between different levels of CCO performance.
- *Feasibility of measurement*: Measure allows CCOs and OHA to capitalize on existing data flows and data collection will be supported by developing Health Information Exchange (HIE) and Health Information Technology (HIT) infrastructure.
- *Reasonable accountability*: CCO has some degree of control over the health practice or outcome captured in the measure.
- *Range/diversity of measures*: Collectively, the set of CCO performance measures covers the range of topics, health services, operation and outcomes, and populations of interest.

The work plan was approved and implemented through HB 1580, which went into effect March 2, 2012. As part of the implementation plan, HB 1580 created the Metrics and Scoring Committee, a nine-member committee (made up of three CCO representatives, three members with expertise in health care outcomes measurement, and three at-large members). Statutory direction reflected the work of the Health Policy Board's Outcomes, Quality and Efficiency Work Group; the committee was directed to "use a public process to identify objective outcome and quality measures, including measures of outcome and quality for ambulatory care, inpatient care, chemical dependency and mental health treatment, oral health care and all other health services provided by coordinated care organizations." The statute further directed the Metrics and Scoring Committee to use metrics that are consistent with state and national quality measures.

## ESTABLISHING ACCOUNTABILITY

At the same time Oregon was establishing new statutory authority and a policy framework for a transformed Medicaid delivery system, state leadership and staff were working with the Centers for Medicare and Medicaid Services (CMS) to ensure that the appropriate federal authorities were in place. Medicaid is authorized under the Social Security Act and operates as a joint federal-state program funded with both federal and state dollars. As such, changes in a state Medicaid program require certain federal authorities and permissions. We had submitted an original concept paper to CMS at the same time our original legislation was moving through the legislative process to allow time for CMS to ask some broad questions about our goals and to determine what specific authorities we would need in order to move forward. Our goal was to be ready to submit a formal 1115 waiver request on the same day the legislature passed the enabling language into law ("1115" refers to a specific section of the Social Security Act which gives states additional flexibility to demonstrate and evaluate new policy approaches).

Oregon's request to CMS centered on increasing value in the Medicaid delivery system: decreasing costs while improving outcomes. It encompassed a request for additional flexibilities to integrate behavioral, physical, and oral health care into a single global budget for each of the Medicaid CCOs in the state, as well as providing additional federal funding through the Designated State Health Program (DSHP) mechanism. In return, Governor Kitzhaber committed to reducing Medicaid cost trend in the state by 2 percentage points, from 5.4% to 3.4% per capita over the 5-year life of the demonstration (2012–17). The agreement included a noteworthy commitment to improving health outcomes; discussions with CMS throughout the spring, summer, and fall of 2012 focused on ensuring that reducing the cost trend would not only "do no harm" to quality or access to care, but improve performance on specific quality and access metrics over the life of the demonstration.

### State Accountability: Quality and Access Test

As part of the state's agreement with CMS, we developed an accountability plan that included a state "test" for quality and access, a measurement strategy, development of a quality incentive pool, and ongoing expenditure review, among other requirements. The state-level "test" would be an assessment of whether or not the state was meeting its obligation to improve health care quality and access. There were significant financial penalties if the state failed the test as measured through a broad set of metrics. Working from the framework created by the Health Policy Board's Outcomes, Quality and Efficiency Work Group, CMS, a transitional Metrics and Scoring Committee (the permanent committee was chartered in August 2012), and OHA, staff identified a set of metrics across seven quality improvement focus areas as well as specific metrics related to access and Electronic Health Record (EHR) adoption that would be used for the test. In order to avoid federal financial penalties during the first 2 years of the demonstration, state performance could not decline across the 33 metrics at an aggregate level and for the remaining 3 years, performance was to improve. Quality improvement focus areas were identified by looking at areas where historical performance showed significant room for improvement, and/or areas that were major drivers of cost. The seven focus areas were:

- Improving behavioral and physical health coordination,
- Improving perinatal and maternity care,
- Reducing preventable rehospitalizations,
- Ensuring appropriate care is delivered in appropriate settings,
- Improving primary care for all populations,
- Reducing preventable and unnecessarily costly utilization by beneficiaries with complex needs,
- Addressing discrete health issues (such as asthma, diabetes, hypertension).

Besides these seven quality improvement focus areas, the 33 measures also included a metric for EHR adoption and five metrics that address improving access to effective and timely care.

The state-level quality and access test consists of two parts: the first is a relatively simple comparison of annual performance on the 33 metrics against a baseline; the second is a more complex analysis of the association between transformation activities and performance on access and quality. The second part of the test is conducted only if the state fails part 1 of the test. In order to simplify the methodology, the 33 metrics are rolled up into a single, aggregate indicator, or composite score. To date, the state has not failed the first test.

## Making Metrics Matter: Putting CCOs at Risk for Quality

An important element of Oregon's overall agreement with CMS was how best to hold the CCOs accountable for quality. Transparency and public reporting had always been a part of our overall quality strategy, but without financial incentives, we felt that the effectiveness of collecting data and reporting progress would be severely limited. One important goal of health system transformation in Oregon was to shift from pure capitation to paying for outcomes, but we needed a bridge strategy that would incrementally move the payment system in that direction. There was interest from both CMS and the state in developing a quality-based incentive program, driven by a subset of the 33 metrics for which the state was to be held accountable. But there was some internal disagreement at CMS about the effectiveness of such programs. The literature to date was mixed on the topic, mostly focused on physician-level programs where there was evidence that incentives as high as 10% of payment would be required to drive any change in clinical effectiveness. We were ultimately successful in our argument that while 10% may be what is required at the physician level, we were addressing Medicaid managed care organizations with 1%–5% margins; in this circumstance a 5% incentive structure is meaningful. CMS ultimately agreed that putting an amount at risk for quality that is essentially equal to an organization's margin could be a compelling driver of change. The waiver agreement directed the Metrics and Scoring Committee to develop parameters for the CCO incentive program, including identifying a subset of the 33 measures to be included, establishing the performance benchmarks and targets, and endorsing measure specifications.

The state agreed to put an increasing proportion of the CCO global budget at risk for quality (in 1 percentage point increments over the life of the waiver) and was given 120 days to develop an incentive pool structure. CMS requirements were that the baseline year would be 2011, to clearly separate the "before" to a time period before legislation and before CMS negotiations; the first measurement year would be 2013, recognizing that there would likely be significant "noise" in the system as Medicaid transitioned from one delivery system structure to another; we would allow the dust to settle during 2012; all quality pool dollars would be distributed for a given measurement year (there would be no carry over) and, finally, measures would remain the same for the first 2 years of the program.

In the first year of the Quality Pool program, 2% of the global budget, or approximately $46 million was distributed to CCOs based on performance on 17 metrics. In 2014 the pool grew substantially as Oregon chose to expand Medicaid under provisions of the ACA and Medicaid enrollment increased from about 650,000 to almost a million people. Four percent of the global budget was included in the quality pool in 2015, amounting to $168 million.

## DEVELOPING A QUALITY POOL METHODOLOGY

Even with the parameters of a quality-based incentive program defined, the Metrics and Scoring Committee had a short time frame in which to select the subset of metrics that would be included and to develop a methodology for distribution of funds. We brought in consulting expertise to assist with the incentive program design and implementation planning. The first step was to develop a framework to guide program design; beyond the principles established earlier and described above (e.g., measures from nationally vetted measure sets), the committee agreed that:

- The program would reward meeting either performance or improvement benchmarks.
- The methodology for measurement and for payment would be clear and transparent in order that providers as well as the consumer can understand what is being measured.
- Metrics should be chosen:
  - with awareness that they will drive CCOs to prioritize the underlying services that are being measured;
  - that largely rely on available data (e.g., claims), but with selected measures of outcomes that will build on the developing HIE, HIT infrastructure in the state;
  - that are not significantly skewed by patient mix.
- The methodology should recognize the impact of small denominators.
- Essential to ongoing performance improvement, OHA would develop a process so that CCOs could review numerators and denominators at the member-level on a regular basis (preferably monthly) so that they could review, correct errors, and monitor progress during the year.
- Potential pool award would be determined by plan size (per member per month) with a minimum amount established as a floor for all CCOs.
- After the first 2 years, the Metrics and Scoring Committee would review all metrics annually to assess them for effectiveness and for any unintended consequences.

## The Metrics

The Metrics and Scoring Committee, after reviewing the 33 metrics against their principles and framework for measurement, adopted 17 measures as CCO incentive metrics (Table 11.1).

The selection of these metrics was not without intense discussion within the committee. There was a healthy tension between selecting measures meeting the principles that were "good enough" and could be collected without added data collection burden versus measures that were considered "transformational" but may require the creation of special registries or changes in work flow at the clinic level. Screening, Brief Intervention, and Referral to Treatment (SBIRT) is perhaps the best example of the conundrum faced by the committee. SBIRT is an evidence-based, effective method to treat alcohol and drug misuse, and the literature indicated that it reduced alcohol consumption and heavy drinking episodes in adults. There was a federally supported SBIRT training program being established at the

**TABLE 11.1 Oregon CCO Quality Incentive Measures**

| Measure | Measure Source |
|---|---|
| Alcohol or other substance misuse/ SBIRT | OHA |
| Follow-up after hospitalization for mental illness | National Quality Forum (NQF) 0576 |
| Screening for clinical depression and follow-up plan | NQF 0418 |
| Mental and physical health assessment within 60 days for children in Department of Human Services (DHS) custody | OHA |
| Follow-up care for children prescribed ADHD medication | NQF 0108 |
| Timeliness of prenatal care | NQF 1517 |
| PC-01: Elective delivery before 39 weeks | NQF 0469 |
| Ambulatory care: Outpatient and ED Utilization | HEDIS, California Department of Health Care Services Medi-Cal Managed Care Division |
| Colorectal cancer screening | HEDIS, 2016 |
| PCPCH enrollment | OHA |
| Developmental screening in the first 36 months of life | NQF 1448 |
| Well-child visits in the first 15 months of life | NQF 1392 |
| Adolescent well-care visits | HEDIS, 2016 |
| Controlling high blood pressure | NQF 0018 |
| Diabetes HbA1c Poor Control | NQF 0059 |
| EHR adoption | OHA |
| CAHPS adult and child composite<br>• access to care<br>• satisfaction with care | HEDIS, 2016 |

Complete technical specifications are available from: http://www.oregon.gov/oha/HPA/ANALYTICS/Pages/CCO-Baseline-Data.aspx.

academic medical center, but it was nascent and there was little evidence that the screening was being done in any kind of numbers in the state. Effectively implementing this metric would require establishing new workflows and educating plans and providers about the screening as well as resolving billing and payment issues—baseline was essentially at zero. But the committee felt strongly that this measure represented the heart and goals of transformation and so included it. OHA would need to partner with other efforts and provide technical assistance through its Transformation Center in order to see success.

The adolescent well-care metric was also a particular focus in that some committee members felt the measure did not rise to a level of import that justified inclusion in what needed to be a relatively small set of metrics. Others were ultimately successful in arguing that these visits represented an important opportunity to discuss detrimental health behaviors often beginning in adolescence, such as unsafe sex or the use of drugs, alcohol, and tobacco. Because the metric does not assess that those conversations actually happen in the visit, it was another area identified as needing strong provider education and CCO and OHA technical support.

## Benchmarks and Performance Targets

CCO performance on each of the metrics is judged on a pass/fail basis, but in line with one of the key principles for the program, performance is rewarded on the basis of reaching an improvement target or on reaching an established benchmark. The Committee generally chose benchmarks of either the 75th or 90th percentile score for Medicaid managed care plans nationally based on the reasoning that the state should aim for being a top performer as well as the notion that if other Medicaid programs had attained a certain level of performance, it was by definition achievable. For those metrics where there were no national Medicaid data available, the Committee chose a benchmark based on what the state's baseline score was and what they felt was reasonably achievable in a 10-year period. The benchmarks are the same for all CCOs, regardless of geographic region and patient mix.

One of the overarching principles of the Quality Pool Program was to keep it as simple and understandable as possible, so the Committee did not want to set performance improvement goals based on a set of algorithms that would be indiscernible to plans and providers. They chose to base improvement targets on the Minnesota Department of Health's Quality Incentive Payment System referred to as the "Minnesota method" or "basic formula." (Minnesota Department of Health, 2016). This method requires at least a 10% reduction in the gap between baseline and the benchmark to qualify for incentive payments. Including improvement targets has turned out to be a key component of the program's success. CCOs are able to meet incremental improvements regardless of where they started in terms of initial performance. If the quality pool program relied on meeting aspirational benchmarks alone, the risk was that providers would be discouraged because for some metrics, the gaps were significant and we wanted to provide incentives for annual improvement. Because the annual performance improvement targets are attainable, the state has seen consistent year over year improvement.

## The Methodology

Quality pool award amounts are determined through a two-stage process. In stage 1, the maximum dollar amount for which a CCO is eligible is allocated based on performance on all 17 CCO incentive measures and benchmarks identified by the Metrics and Scoring Committee. In stage 2, any remaining quality pool funds that are not disbursed in stage 1 are distributed to CCOs meeting criteria on four "challenge" measures. Challenge measures were selected by the Committee as those that are most "transformational," and have an inferred additional weighting by defining the second stage of the quality pool distribution.

In 2015 challenge pool measures were:

- Diabetes: HbA1c poor control;
- Depression screening and follow-up;
- Alcohol and drug misuse (SBIRT);
- Developmental screening.

## Stage 1 Distribution

For 16 of the CCO incentive measures (all but the PCPCH enrollment measure), the portion of available quality pool funds that a CCO receives is based on the number of measures on which it either achieves the absolute benchmark or meets its improvement target over its own previous year performance. If the benchmark is met or the improvement target reached for a specific measure, the CCO receives full credit for the measure, regardless of performance on other measures. For the PCPCH enrollment metric, performance is measured according to a "tiered formula" reflecting the primary care home certification level, there are three (OHA, 2015). The Committee wanted to provide incentives to move CCOs toward enrolling members in the most robust or highest tier of PCPCH to get full credit. The results of the tiered formula are added to the number of measures on which a CCO meets the benchmark or improvement target. As a CCO meets more benchmarks or improvement targets, it receives a higher payment. A CCO can receive 100% of the quality pool funds for which it is eligible if it:

- meets or exceeds the benchmark or the improvement target on at least 75% of the incentive measures (12 of 16 measures);
- meets or exceeds the benchmark or improvement target for the EHR adoption measure as 1 of the 12 measures above; and
- scores at least 0.60 on the PCPCH enrollment measure using the tiered formula.

**TABLE 11.2 Quality Pool Distribution**

| Number of Targets Met (Benchmark or Improvement) | Percent of Quality Pool Payment for Which the CCO Is Eligible |
|---|---|
| At least 12 (including EHR adoption and at least a 0.60 PCPCH enrollment) | 100 |
| At least 12 (not including EHR adoption or less than 0.60 PCPCH enrollment) | 90 |
| At least 11.6 | 80 |
| At least 10.6 | 70 |
| At least 8.6 | 60 |
| At least 6.6 | 50 |
| At least 4.6 | 40 |
| At least 3.6 | 30 |
| At least 2.6 | 20 |
| At least 1.6 | 10 |
| At least 0.6 | 5 |
| Less than 0.6 | No quality pool payment |

The EHR and PCPCH measures act as an absolute hurdle. If they are not met, then the maximum payment that the CCO can receive is 90% of the maximum funds. If a CCO does not meet the improvement target or benchmark on any of the 16 measures and scores below 0.60 on the tiered formula for the PCPCH measure above (a total score less than 0.60), the CCO would not receive any quality pool funds (see Table 11.2).

## Stage 2 Distribution

In the second stage, remaining quality pool funds that have not been allocated to CCOs in stage 1 become the "challenge" pool—these funds are to be distributed to CCOs that qualify based on a subset of four incentive measures, listed above. Through this stage, all quality pool funds are distributed. As per the agreement with CMS, no quality pool funds roll over into a subsequent year. OHA determines the number of instances in which CCOs have met each of the four challenge measures according to specific parameters and those second stage funds are distributed.

## Retiring and Refreshing the Measure Set

It was important from the onset of the Quality Pool Program that metrics were annually evaluated for effectiveness and metrics that met certain criteria would be retired and new metrics would be added. Criteria for retirement include when a metric:

- lacks currency; no longer addresses the area of concern or focus;
- loses its national endorsement, or a measure that is unique to OHA now has a similar nationally vetted measure available;
- tops out for performance (e.g., the benchmark is attained and remains stable over time);
- is not sensitive enough to capture improved performance;
- creates a burden of data collection that outweighs value; or
- is unintentionally duplicative of another measure.

All metrics that are retired continue to be measured and reported even though they are no longer part of the incentive program. At this writing, the program is 4 years old and the following metrics have been retired:

In 2015 the children's ADHD metric was retired because the state had consistently out-performed national benchmarks, and at the CCO level, sample size was an ongoing reliability issue. The metric aimed at reducing elective deliveries before 39 weeks was retired because Oregon hospitals had adopted a hard-stop policy just prior to the adoption of this metric resulting in a fairly rapid and significant improvement. After reaching a rate of about 3%, the Committee believed that there was little room for additional improvement. In 2016 the EHR adoption metric was retired because the impact of including three metrics that were EHR-based contributed to an increased EHR adoption rate meeting the goals of the Metrics and Scoring Committee. Retiring the EHR adoption measure necessitated a change in the quality pool methodology.

In each year that measures are retired, new measures are added. For instance, we had committed to CMS that a dental metric would be added to the measure set as dental care was fully integrated into the CCO model. In 2015 a metric focused on improving dental sealant rates for children was recommended by an oral health subcommittee and adopted by the Metrics and Scoring Committee. In addition, as part of an increased focus on women's and early child health, a metric for effective contraceptive use was added.

Changes have been made to the challenge pool measures as well. PCPCH enrollment was removed as a challenge pool measure because every CCO was well above the established target; the rate for developmental screenings was added in its place. This metric was long seen as transformational by the Committee and had achieved a fair amount of success that was seen as mostly the result of better measurement rather than change in actual practice.

The Committee wanted to promote real improvement beyond better measurement. In 2016 measures of cigarette smoking prevalence and childhood immunization status were added as well.

As part of sustaining a meaningful and robust measurement program, the Committee makes carefully considered modifications to strengthen existing metrics. Among others, changes have been made to reflect that oral health was integrated into the CCO model, the age range for SBIRT rates has been expanded to include children as young as 12 to reflect that issues with substance abuse begin in adolescence, and changes have been made to the measure of follow-up care after hospitalization for mental illness to reflect emerging best practice.

## Implementing the Program

After receiving final approval from CMS, a high-level definition of the incentive metrics program was developed, including the 17 incentive metrics and quality pool payment methodology. However, much work lay ahead to operationalize the program and develop a transparent, valid measurement system. We established a set of principles that guided implementation:

- The program would be fully transparent allowing CCOs and the public to have access to metric specifications, quality pool methodology, and results.
- The state would actively engage CCOs in the process to build the program, for instance, asking CCOs to validate programming.
- The Metrics and Scoring Committee would stay actively involved in monitoring the program.
- CCOs would not have to take on the administrative burden of calculating metrics.

While one of the tenets of selection was that measures would be consistent with state and national quality standards, the Metrics and Scoring Committee chose a number of metrics that required either home-grown development of specifications or development of infrastructure to collect the necessary data. The 17 incentive metrics were not all claims or encounter-based.

It was clear to us that CCOs had staff resources and expertise in measurement, claims, and clinical systems that would extend state staff capacity and benefit the development of the program. We convened a Metrics Technical Advisory Group (Metrics TAG) to help guide the state on the development of metric specifications and to adhere to the principles of transparency and collaboration. The Metrics TAG holds monthly public meetings and is comprised of Medical Directors, IT staff, quality improvement staff, and others from the CCOs as well as state staff. It is an open group in terms of attendance and many times the attendees change based on the areas of discussion.

The Metrics TAG makes recommendations to the Metrics and Scoring Committee regarding topics, such as specifications, benchmarks, measure development, the phase-in of the EHR-based metrics, and other technical issues. The Metrics TAG has been invaluable and key to the success of Oregon's quality incentive program—by expanding the available subject matter expertise and by creating a space for a transparent, collaborative process.

## Reporting and Transparency

At the outset, the state created a metrics web page where specifications, the quality pool methodology, and quality pool dollar amounts are posted (Oregon Health Authority, 2016). In addition, the state committed to providing CCOs monthly progress reports on metrics. Dashboards are created and updated monthly. The dashboards provide detailed information to each CCO about their progress in meeting metrics as well as member-level information so that the CCOs know which members are included in the measurement. Many CCOs utilize this information for quality improvement monitoring and to follow up with members needing services. CCO is also encouraged by the state to ask questions, request technical assistance, and provide feedback on the monthly data provided to them.

Since 2013, Oregon has been publishing regular public reports on statewide and CCO performance on metrics and CCO quality pool annual payments (Fig. 11.1). Oregon engaged a data visualization expert to advise on improving the user-friendliness of reports so that data displays are understandable to the public, Oregon Medicaid clients, policy makers, legislators, and the press (see Fig. 11.2 for an example). The public reports include metrics broken out by subcategories such as race and ethnicity, severe and persistent mental illness, and disability. This allows for an analysis and reporting of potential disparities in health care for these populations.

# STORY OF A METRIC

As described earlier, SBIRT is an evidence-based, effective method to treat alcohol and drug misuse, and the literature indicated that it reduced alcohol consumption and heavy drinking episodes in adults. The Metrics and Scoring Committee chose to measure and pay for SBIRT screening as part of the incentive metrics program to bolster the number of behavioral health and alcohol and substance use measures. Since there was no national standard for measuring SBIRT screening, we had to develop measure specifications and data collection methods to create a new measure. The Metrics TAG wrestled with where to measure and collect data in the process of an initial screen, a secondary screen, and with a positive secondary screen, an intervention. After extensive input and additional guidance from the academic medical

## 2015 Performance overview

### CCO incentive measures

| ■ CCO achieved Benchmark in 2015<br>■ CCO achieved Improvement Target in 2015<br>* Highest performing CCO in each measure | AllCare | Cascade | Columbia Pacific | Eastern Oregon | FamilyCare | Health Share | IHN | Jackson | PacSource Central | PacSource Gorge | PrimaryHealth | Trillium | Umpqua | WOAH | WVCH | Yamhill |
|---|---|---|---|---|---|---|---|---|---|---|---|---|---|---|---|---|
| Access to care (CAHPS) | | | | | | | | * | | | | | | | | |
| Adolescent well care visits | | | | | | | | | | | | | | | | * |
| Alcohol and drug misuse screening (SBIRT) 12+ | | | | | | | | | | | | | | | * | |
| Ambulatory care – Emergency department utilization | | | | | | | | | | | * | | | | | |
| Colorectal cancer screening | | | | | | | | | | | | | * | | | |
| Controlling high blood pressure | | | | | | | | * | | | | | | | | |
| Dental sealants for children | | | | | | | | | | | * | | | | | |
| Depression screening and follow-up | | | | | | | | | | | | | | | | * |
| Developmental screening | | | | | | | | | | | * | | | | | |
| Diabetes HbA1c poor control | | * | | | | | | | | | | | | | | |
| Effective contraceptive use (ages 18–50) | | | | | | | | | | | | | * | | | |
| Electronic Health Record (EHR) adoption | | | | | | | | | | | * | | | | | |
| Follow up after hospitalization for mental illness | | | | | | | | | | * | | | | | | |
| Assessments for children in DHS custody | | | | | | | | | | * | | | | | | |
| Patient-Centered Primary Care Home (PCPCH) enrollment† | | | | | | | | | | | * | | | | | |
| Prenatal and postpartum care: Prenatal care | | | | | | | | | | * | | | | | | |
| Satisfaction with care (CAHPS) | | * | | | | | | | | | | | | | | |

†CCOs earn payment for this measure if at least 60 percent of members are enrolled in a patient-centered primary care home.

**FIGURE 11.1 Performance overview.** *From Oregon's Health System Transformation: CCO Metrics 2015 Final Report. Oregon Health Authority, Office of Health Analytics. Available from: http://www.oregon.gov/oha/Metrics/Pages/HST-Reports.aspx.*

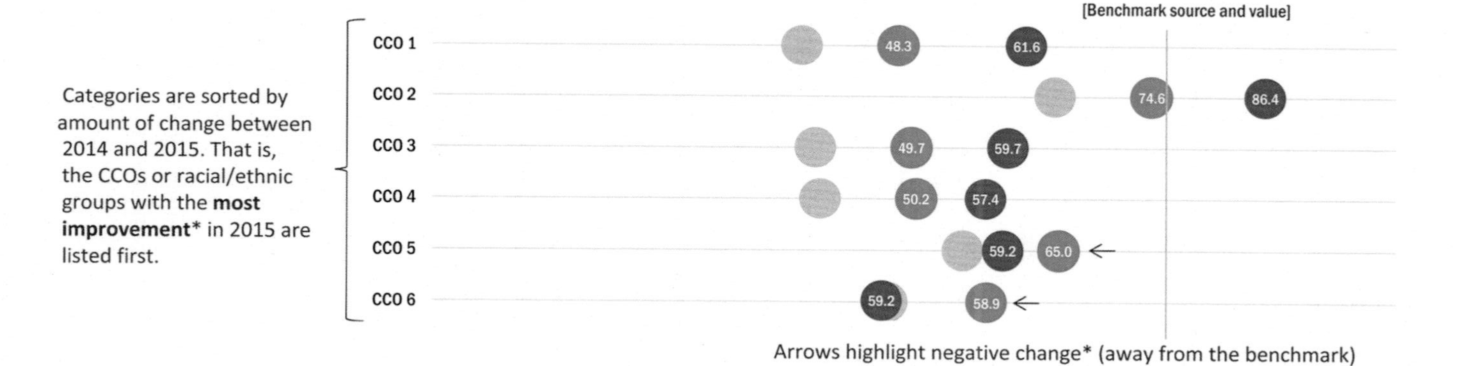


**FIGURE 11.2 How to read the graphs.** *From Oregon's Health System Transformation: CCO Metrics 2015 Final Report. Oregon Health Authority, Office of Health Analytics. Available from: http://www.oregon.gov/oha/Metrics/Pages/HST-Reports.aspx.*

center, the TAG recommended that the measure be claims-based. To start the SBIRT measure would reflect the proportion of adults with a primary care visit who received the secondary or full screen for alcohol use. Since roughly a third of adults do not use alcohol at all, it was expected that at most two-thirds would be given the secondary screen. In addition, selecting an appropriate benchmark was challenging and was established after researching the literature, obtaining experience from the academic medical center, and starting with a very modest benchmark since baseline was essentially at zero.

Once the metric specifications were developed, the work began to engage providers in establishing new workflows and coding and billing practices. The new coding requirements were new and complex. We created an SBIRT work group to provide education, guidance, and technical support to CCOs and their providers. For example, many commercial health plans would not pay for an SBIRT screen. If a commercial patient was screened and their health plan denied the claim, the patient was responsible for the out-of-pocket expense. Providers needed a work flow that could be used for all patients regardless of coverage. Through guidance from the SBIRT work group, the state made modifications in its billing data systems so that providers could submit SBIRT billing codes with a zero for the paid amount. This allowed providers to receive credit for the screen without inadvertently causing the patient to pay an out-of-pocket expense.

The Transformation Center was engaged to develop learning collaboratives with CCOs to spread best practices and shared learning. The improvement in the rate of SBIRT screenings is the result of strong collaboration and sharing with community partners, primary care practices, and CCOs. Oregon has seen a steady increase by all CCOs in screening for alcohol misuse since implementation of the SBIRT measure in 2012. The Metrics TAG is now working on developing an EHR-based SBIRT metric (Fig. 11.3).

## Developing Capacity for EHR-Based Measurement

Three of the initial 17 incentive measures required clinical data: Controlling high Blood Pressure, Diabetes HbA1c Poor Control and Depression Screening, and Follow-up. In 2012 the technology did not exist across the state to collect the data electronically in a cost effective manner. However, we did not want to rely on an outdated, expensive method, chart review, and wanted to begin to build the capacity to collect EHR data from CCOs for purposes of payment.

We found a solution by phasing in requirements across 3 years with the goal to leverage Meaningful Use requirements, promote HIT development within CCOs, and build a clinical quality metrics repository to house data within OHA. In the first year, each CCO submitted a technology plan that described their road map for capturing and submitting clinical EHR-based

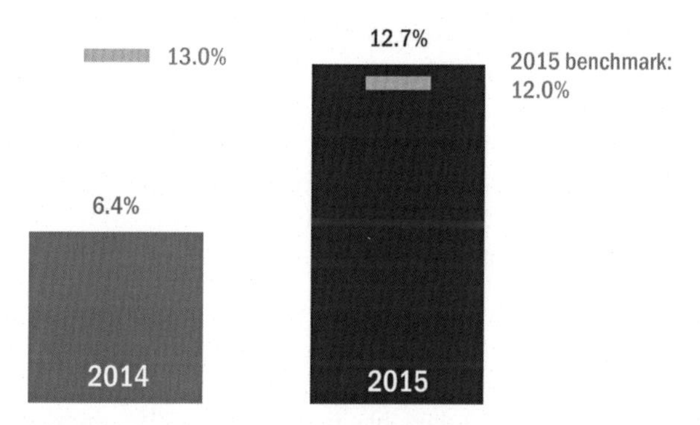


**FIGURE 11.3 SBIRT 2015 graphic.** *From Oregon's Health System Transformation: CCO Metrics 2015 Final Report. Oregon Health Authority, Office of Health Analytics. Available from: http://www.oregon.gov/oha/Metrics/Pages/HST-Reports.aspx.*

data in the future year. Each CCO also submitted proof of concept data that consisted of a convenience sample of 10% of the members for each measure. The CCOs received incentive payments for submitting their technology plans and for submitting the proof of concept data.

In the second year, CCOs completed an expanded technology plan and submitted more complete data. Again, CCOs received incentive dollars for reporting. Since two of the measures (Diabetes HbA1c Poor Control and Depression Screening and Follow-up) were challenge pool measures, the CCOs were paid for performance on these metrics as part of the challenge pool pay out.

In the last year of the phase-in, CCOs submitted data using Meaningful Use EHR technology (QRDA) for two-thirds of their members (and 50% for depression screening and follow-up) and were paid their incentive dollars based on performance of the measures. Since there was no prior performance data to create improvement targets the benchmarks were set to be reasonably attainable.

This phased-in approach required additional approvals from CMS. It was successful in providing funding for CCOs and a clearly laid out bridge strategy to phase in EHR reporting. Challenges remain for CCOs to provide member-level reporting through EHR technology and the development of a clinical metric repository is still underway.

## RESULTS

In the last 5 years, Oregon has made significant progress in transforming its Medicaid delivery system. This new health delivery model has led to better health, better care, and lower per capita costs, saving the federal and state government over $1.7 billion. Even while bending the cost curve, there have been significant improvements in quality, access, and health outcomes according to data from Oregon's robust quality measurement program. Oregon has shown that when CCOs have a meaningful percentage of their payment at risk for performance, they achieve performance improvement and affect transformative change.

OHA just completed the third year of its quality incentive payment program. For measurement year 2015, 4% ($168 million) of total capitation payments were distributed to CCOs based on quality measures performance. Fifteen of the 16 CCOs earned 100% of the possible funds. CCOs have shown impressive improvement:

- Emergency department visits by CCO enrollees have decreased by 39% since 2011.
- The percent of adults who had a hospital stay and were readmitted for any reason within 30 days has improved by 33% since 2011.
- Hospital admissions for short-term complications from diabetes decreased 29% since 2011. Admissions for COPD (decreased by 63%), congestive heart failure (decreased by 32%), and asthma (decreased by 50%) have all shown significant decreases from the state's 2011 baseline.
- The percent of children and adolescents who had a visit with their primary care provider in the past year has increased from 2014. This may be due to the increased focus on improving childhood immunizations and developmental screenings. Adolescent well-care visits have also increased 38% since 2011.
- The percent of children aged 6–14 who received a dental sealant on a permanent molar in the past year increased 65% since 2014.
- The percent of women aged 15–50 who are using an effective contraceptive increased almost 9% since 2014, even with the addition of thousands of new OHP members in 2014.
- CCOs continue to increase the proportion of members enrolled in PCPCHs. PCPCH enrollment has increased 69% since 2012.
- The percent of CCO members who report they received needed information or help and thought they were treated with courtesy and respect by customer service staff has increased almost 10% since 2011 baseline.

These improvements translate directly into better health for Medicaid enrollees and savings for Oregon and the federal government.

## LESSONS LEARNED

Through the development of Oregon's robust measurement program, many lessons have been learned. We have outlined some following key lessons learned:

- Methodologies for measurement and payment need to be as clear, straightforward, and transparent as possible. You cannot expect your partners (plans and providers) to make changes if they do not understand what is being measured. And they cannot trust what they do not understand.
- Metrics cannot stand alone: multiple levers are critical to move health transformation forward. Supports, technical assistance, quality improvement strategies, and learning collaboratives to share best practices are all critical to the success of an overall metrics and reporting program.
- Create glide paths for developing and implementing transformational measures. Be prepared to phase in requirements and expectations. Creating new measures takes more time than expected and requires active participation from stakeholders.
- Do not be afraid of stakeholder involvement. Active participation and problem solving by those being measured are critical to success.
- Set attainable and meaningful performance goals—improvement targets keep everyone in the game and provide a means for incremental improvement over time.
- Regular feedback and transparency is critical. Many measures of health care processes and outcomes are annual measures. By the time the measure is created and reported, it is often a full year or more since the event being measured occurred. Providers and health plans cannot respond effectively with long lags between data collection and reporting. But a metrics program can be set up to regularly report numerators and the denominators for each metric, which allows plans and providers to validate, see trends, and respond quickly.
- Aligning with other purchasers' metrics, e.g., Oregon's exchange, public employee benefits, commercial carriers, strengthens everyone's efforts and moves transformation forward.
- Metrics need to be connected to the state's health policy goals. For example, if a state simply adopts a federal measure set, it may not reflect the unique goals of the state.
- Aligning around a model (Coordinated Care Model in Oregon's case). Incentive metrics alone will not measure success of the model; they are essential, but not adequate. In the case of Oregon, a core component of the Coordinated Care Model is the integration of behavioral and physical health care, but the level of integration is not easily reflected in metrics.
- Perhaps less tangible than the metrics themselves, partnerships formed between providers and the CCOs as they strive to achieve benchmarks and performance improvement goals translate beyond the metrics themselves to drive health system transformation even further.

## THE FUTURE

Because of its success, Oregon plans to continue its incentive metrics program to further support health system transformation. The program intends to push quality improvement in Medicaid forward by developing and adopting more transformational and outcomes-based measures in place of traditional health care quality process measures. Oregon is examining changes to the challenge pool that would better support priority areas such as issues around health equity; it is also exploring the adoption of community-health measures, such as kindergarten readiness, which will require incorporating methods for shared accountability. As the state moves more toward true outcome measures, there will be continuing challenges in determining the right metrics to use, developing methodologies for risk adjustment and attributing accountability for shared outcomes as we move outside of clinic, health center, or hospital walls and toward community health.

None of this is simple, but it is necessary to determine whether we are effectively and adequately improving health, making quality care accessible, eliminating health disparities, and controlling costs for the populations that we serve. The ability to measure health outcomes—to see what is working, to make adjustments as we go, and to share best practices is an essential component of transforming the health care delivery system to drive down costs, improve health care, and improve health.

## REFERENCES

Minnesota Department of Health, Health Economics Program. (2016, June). Minnesota Statewide Quality Reporting and Measurement System: Quality Incentive Payment System. Available from: http://www.health.state.mn.us/healthreform/measurement/QIPSreport2016.pdf.

Oregon Health Authority. (2015, November). Patient-Centered Primary Care Home (PCPCH) Enrollment-2016 (revised Nov 2015). Available from: http://www.oregon.gov/oha/hpa/csi-pcpch/pages/index.aspx.

Oregon Health Authority. (2016, June). Oregon Health System Transformation Coordinated Care Organizations Performance Reports. Available from: http://www.oregon.gov/oha/Metrics/Pages/HST-Reports.aspx.

Chapter 12

# The Oregon Transformation Center: How and Why a State Agency Can Support Change

**Cathy Kaufmann[1], Chris DeMars[2] and Ron Stock[2,3]**
*[1]Health Management Associates, Portland, OR, United States, [2]Oregon Health Authority Transformation Center, Portland, OR, United States, [3]Oregon Health & Science University, Portland, OR, United States*

"Change is hard" is a common refrain whether the topic is New Year's resolutions or organizational restructuring. It is also an accepted truth when it comes to changing the health care system. The difficulty of implementing even small-scale change in health care has been well documented. Despite overwhelming evidence to support needed changes, many attempts to change practice have failed (Berwick, 2003; Cabana et al., 1999; Grol & Wensing, 2013; IOM, 2001). As Dr. Berwick points out, "In healthcare, invention is hard, but dissemination is ever harder" (Berwick, 2003). In other words, in health care as in life, a good idea is not enough to spur people to change.

When Oregon began its efforts to transform its health care delivery system, starting with the implementation of the Coordinated Care Model (CCM) in Medicaid (Chapter 2: State-Level Design: The Coordinated Care Model), state health leaders recognized that changes in payment and policies could only do so much. Real transformation would only occur if real people, not just systems, directly engaged in the change, believed in it and were willing to work for it. Improvements to policies and payments needed to be supported by behavior change on the ground, among providers, communities, the Coordinated Care Organizations (CCOs), and consumers of care. For this reason, the state created a new office housed within the Oregon Heath Authority (OHA) called "The Transformation Center."

## WHY A TRANSFORMATION CENTER?

When it comes to system change, much of our understanding about why some innovations are successfully implemented and spread while others

Health Reform Policy to Practice. DOI: http://dx.doi.org/10.1016/B978-0-12-809827-1.00012-4

fail comes from Everett Rogers. Rogers was a sociologist who developed the "Diffusion of Innovation" theory, his interest in the topic sparked by his father, a farmer, who resisted using a new drought-resistant hybrid seed only to then lose his crop in a drought. Rogers' theory has been used successfully in the Institute for Healthcare Improvement's (IHI's) Breakthrough Collaborative Series work (IHI, 2003). After working with Dr. Don Berwick, President Emeritus and senior fellow at IHI and, at the time, Administrator of the Centers for Medicare and Medicaid Services (CMS), along with staff at the Center for Medicare and Medicaid Innovation (CMMI), leaders in the state were eager to adapt the approach to support the success of Oregon's efforts.

Rogers' theory suggests that there are eight critical components (Table 12.1) in the successful spread of any innovation (Rogers, 2003). Some of these components focus on the innovation itself and how it is perceived.

**TABLE 12.1 Components of the Diffusion of Innovation**

| Component | Description |
|---|---|
| Relative advantage | People must believe that the innovation is an improvement over current practice and its benefits outweigh any risks. |
| Compatibility | People must understand how the innovation fits in with their current beliefs and value system and community needs. Innovations that are not perceived as compatible are less likely to be adopted. |
| Simplicity | The simpler an innovation is to implement, the faster it is likely to spread. More complex innovations are difficult to understand and harder to implement—and therefore harder to spread. |
| Trialability | People need to be able to try out an innovation with minimal investment before moving to full implementation. |
| Observability | Demonstrated evidence that an innovation works helps adoption and spread. People need to see that the innovation works with their own eyes. |
| Reinvention | Successful innovations require constant reinvention by the users. Innovations that allow for appropriate adaptations to serve local community needs are likely to take hold and spread. |
| Change leaders | There must be respected and trusted individuals who can serve as key messengers and champions for the innovation. |
| Active learning networks | Peer-to-peer networks are critical for learning, sharing, and communication needed to support the implementation of the innovation. |

Source: Rogers, E. (2003). *Diffusion of innovations* (5th ed.). New York: Simon and Schuster.

Any change must be viewed as an improvement over the current way of doing things, be compatible with current systems and needs, and not be too complicated. Other components focus on the way the innovation is disseminated. Innovations that people can test before fully adopting are more likely to be attempted. It also helps when people can see the results of the innovation themselves, as well as adapt the innovation to meet their needs or the needs of the community. The last two components focus directly on the relationship side of change. Change does not occur in a vacuum. We are social creatures and the importance of social connections and relationships in the success or failure of any innovation should not be underestimated. People adopt changes when they learn and share with each other about the change, and there is active leadership supporting the change.

The Oregon Transformation Center was created to support the rate and spread of innovation necessary to fully implement the CCM and achieve the triple aim. The Center would work to create the conditions and characteristics Rogers identified as necessary to support the implementation and spread of innovation. There would be a heavy emphasis on developing change leaders, building active learning networks, and encouraging local solutions and adaptations. The Center would support innovation in one region and help it spread quickly to the rest of the state, in other words, the aim would be to "help good ideas travel faster."

The Center would also represent a new way of doing business for the state and the start of a different kind of relationship between the CCOs and the OHA, the state's Medicaid agency. The goals of the Transformation Center were to:

- Champion and promote health system transformation in partnership with CCOs, providers, and communities;
- Build an effective learning network for CCOs; and
- Foster the spread of transformation beyond Medicaid.

## BUILDING THE NEW TRANSFORMATION CENTER

With funding from the CMMI State Innovation Model grant award to the state, the Transformation Center was created as a new office within the OHA, with a direct line of reporting to the agency director. Although the Division of Medical Assistance Programs (DMAP) was the part of the agency accountable for managing the Medicaid program and CCO contracts, a conscious decision was made to house the Transformation Center outside that division. This organizational placement was important because it (1) demonstrated to the rest of the agency, CCOs, and other stakeholders that support for the spread of innovation was an agency priority; and (2) allowed the Center to build a relationship with the CCOs outside of the traditionally rigid and bureaucratic interactions typical in Medicaid managed care plan contract management. Internally, this was often referred to as a "good cop, bad cop" setup. DMAP's role was to hold the CCOs accountable to their contractual requirements, oversee the rate setting process, make payments, measure and manage quality performance, and ensure program integrity. The Transformation Center was accountable for providing support and resources, and for facilitating learning across the CCOs. If DMAP found a CCO to be underperforming in a certain area, the Center was there to help the CCO meet performance goals. If the Transformation Center was viewed as an oversight body, the CCOs would be less willing to share their challenges along with their successes, and engage with the Center in a trusting working relationship. Without this trust established, the Transformation Center could not be successful.

The idea that a state agency could be a partner in innovation and play a role in helping outside, private organizations innovate faster was not universally accepted with open arms. Although CMS and CMMI were very supportive of and invested in the approach, the CCOs themselves were initially skeptical. OHA took the concerns of the CCOs seriously and worked hard to address the skepticism head on. Oregon has a culture of seeking community input and working collaboratively with stakeholders. This collaborative approach is, in large part, why health system transformation and Medicaid expansion were passed nearly unanimously in an evenly divided Oregon State Legislature, despite the starkly partisan rhetoric surrounding President Obama's Affordable Care Act at the national level. OHA maintained an open dialog with the CCOs about its plans for a Transformation Center from the start, seeking input and feedback. For example, OHA asked CCO representatives to participate in the interview panel and hiring decision for the Transformation Center Executive Director position. To succeed, the Transformation Center had to be created *with* the CCOs, rather than be something done to them. The CCOs did not all agree that a Transformation Center was needed nor that it would succeed, but engaging them directly in its formation helped make the Center's success possible.

## STAFFING STRUCTURE

The Transformation Center needed to develop a staffing structure to not only carry out the activities it pledged to do in the waiver and the SIM model grant proposal, but also allow it to successfully serve the dual role of promoting transformation internally while building a network of learning and supporting transformation externally. The external work to support transformation also needed to serve community stakeholders and consumers and their advocates, in addition to clinicians and providers. Consequently, a structure was created that consisted of:

- An executive director with a strong track record of implementing change within the state agency; a generalist perspective of and a strong commitment to the CCM vision; experience in health policy and political advocacy in the state; established working relationships with the state Medicaid health care community; and who worked with the Governor's office and served as part of the leadership team for OHA.

- A clinical champion to connect and communicate the model to providers, including a deep understanding of the delivery system of care in both the outpatient and inpatient settings, as well as expertise in developing leadership and building systems of learning for clinicians.
- A community health champion, with expertise in moving the health system to "upstream," population-based interventions, as well as an advocate of addressing the social determinants of health, community organizing, and engagement.
- Staff known as Innovator Agents (IAs) with a broad array of expertise and backgrounds to serve as direct liaisons to the CCOs and community-based champions of the CCM model.
- Staff with expertise in quality improvement (QI) methodology, learning collaboratives, innovation and dissemination science, and the ability to facilitate groups of providers to change their clinical behavior to match the needs of the CCM.
- Staff to assist in the collection of process and outcome data on the effectiveness of the change interventions and, in an iterative way, translate these data into meaningful feedback to the CCOs and internally, including the Center.
- Other roles to support the work of the Center, including communications, grant writing, contracting and grants management support.
- Outside subject matter experts were brought in on contract to provide focused expertise on an array of topics, including patient engagement, behavioral health integration, health literacy, trauma-informed care, dental health integration experts, alternative payment methods, and leadership training.

The organizational chart in Fig. 12.1 shows the staffing structure of the Center and number of positions put in place to support the work.

## INNOVATOR AGENTS

Once an Executive Director was hired, the first order of business for the new Transformation Center was to fill the "Innovator Agent (IA)" positions. In accordance with the terms of Oregon's waiver agreement with CMS, each CCO was to be assigned an IA. The IAs were intended to serve as a primary point of contact between the CCO and OHA to improve and streamline communication with the agency. IAs were also to provide support for the CCOs' transformation efforts through data-driven feedback and assistance in the adoption of innovations in care, as well as gauging the impact of health

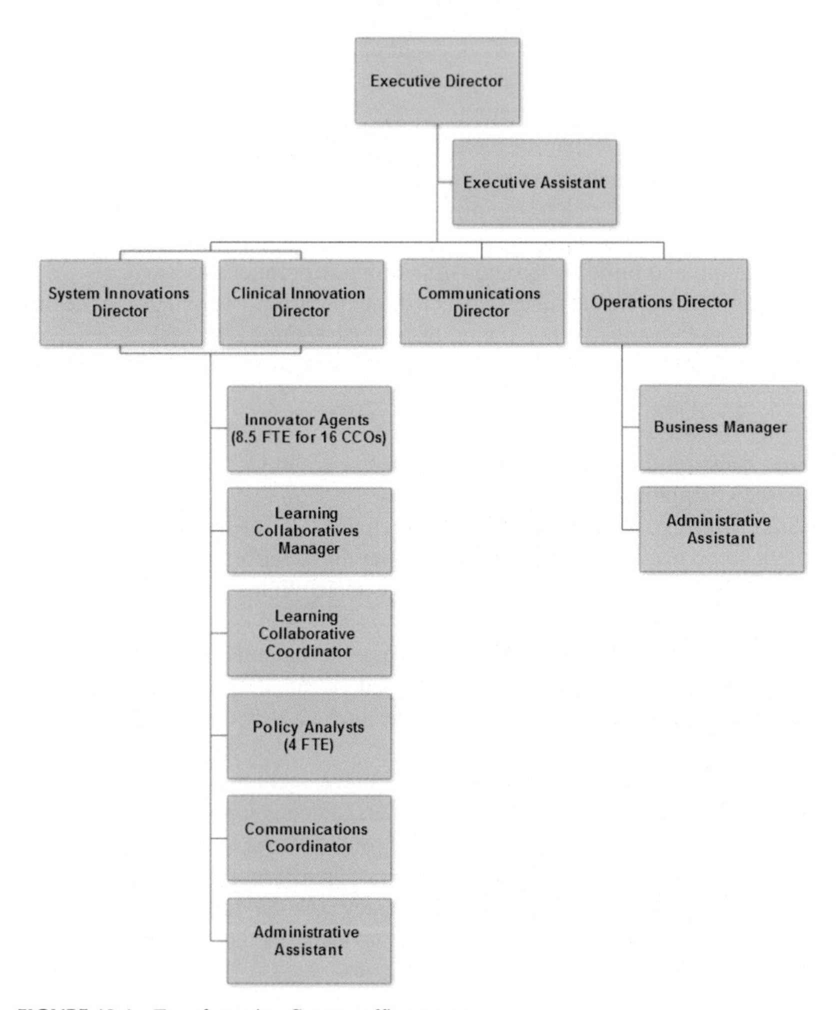


**FIGURE 12.1** Transformation Center staffing structure.

systems transformation on community health needs (see Table 12.2). This role draws on the "change leader" requirement in Rogers' theory.

CCOs were even more skeptical of the role of the IAs than they were the concept of the Transformation Center. Only one CCO—run by a pediatrician who helped develop the concept of the IA—embraced the idea of being assigned an IA. The others felt it was an unnecessary state agency position that would only add another layer of bureaucracy and interfere with the hard work of transformation that lay ahead of them. Once again, intentional collaboration helped navigate these concerns. CCOs were invited to be part of the hiring process for the IA assigned to them. The Center provided a list of prescreened candidates who met the position requirements and encouraged

**TABLE 12.2 Role of the Innovator Agents**

- Serve as a single point of contact between OHA and CCOs
- Identify and facilitate resolution on CCO questions and issues with OHA
- Identify for OHA opportunities and obstacles related to system and process improvements for the agency
- Support innovation within CCOs, doing this work at the direction of the CCO
- Observe meetings of the Community Advisory Councils (CACs) and provide support as needed or requested
- Support the activities of the Transformation Center
- Collaborate with the other IAs, including participating in a peer-learning collaborative with them to discuss ideas, projects, and creative innovation planned or undertaken by their assigned CCO(s)
- Help build and participate in learning collaboratives with other IAs, CCOs, community stakeholders, and/or OHA

the CCOs to interview candidates and provide feedback into the hiring process. CCOs were only assigned IAs whom they felt were a good fit for their needs—a logistical challenge given that most IAs had to work with two CCOs, the only exception being the rural eastern Oregon CCO, which covered a large geographic area (12 counties) of the state. In the end, the Transformation Center had 9 IA positions to serve the state's 16 CCOs. The IAs were "embedded" positions—they were expected to spend more time out in their CCOs' offices and the CCOs' service area than they were in the agency office. A long-standing criticism of the state agency was that staff did not understand what it was like for the Medicaid health plans nor their providers. For example, the Managed Care Organizations (MCOs) in place before CCOs were formed often complained that rules and reporting requirements were developed without sufficiently taking into account potential unintended consequences or the operational implications for plans or their provider networks. IAs were intended, in part, to break down that barrier and build stronger feedback loops. In the new world of health system transformation, the success of the state reform depended upon the success of the CCOs.

IAs played a crucial role for the Transformation Center in the early days. Given the initial resistance to the Center, the IAs served as boots-on-the-ground staff who could forge stronger relationships with the CCOs. In fact, for the first 6 months, the IAs were told their only job was to establish relationships and build trust with the CCOs. Because they spent their time out in the community with CCO leaders, staff, and providers, the IAs had a better understanding of the CCOs' needs and perspectives than agency staff in the state capital. They brought this information and perspective back into the agency, which helped with OHA's internal continuous improvement efforts. The Transformation Center used IAs as a mechanism to share information with and gather input from CCOs and their communities. IAs were also able to help build support for—and provide feedback on—other Transformation Center activities.

## INTERNAL TRANSFORMATION: SUPPORTING CULTURE CHANGE WITHIN THE AGENCY

In the early months after Medicaid transformation was first implemented, most of the needs of the new CCOs centered on operational and policy issues with OHA. For example, Medicaid members with complex health care needs were transitioned into CCOs more slowly than the general population, as were some services, such as non-emergent medical transportation and dental health care. How these transitions occurred and the policies developed to support them took a significant amount of time and attention from both the CCOs and OHA. The IAs helped escalate critical issues during the early days of implementation of the model and, with the help of the Transformation Center Executive Director and other OHA leadership and staff, rapidly resolve problems as they arose.

CCOs pointed to challenges in working with OHA as a barrier to their ability to implement transformation. Time and resources that could be spent on improving the delivery of care needed to be spent on tracking down information from the agency, clarifying policies and procedures and trying to navigate red tape. When policies were issued that resulted in some unintended consequence or reporting requirements seemed duplicative, CCOs could raise the issue with their IAs and learned to rely on these staff for answers and resolutions to problems. This helped build trust in and support for the IA role, as well as for the Transformation Center overall.

OHA embraced the idea that, in order to support health system transformation, the agency needed to transform itself, too. OHA needed to move beyond the traditional Medicaid agency role of just regulating the CCOs—it needed to be a supportive partner. The Transformation Center was always intended to serve as the hub of innovation within OHA as well as the state, working to streamline agency systems and processes, reduce administrative burdens, and assist the agency in better coordinating the work of its different divisions in support of health system transformation. Over time, "bureaucracy busting" became an increasingly greater role for the Transformation Center Executive Director as well as the IAs. Listening to the CCOs' input, addressing their frustrations, and developing solutions in partnership were all a part of this internal state agency transformation, and the TC and IAs played a significant role in facilitating this work.

## BUILDING A NETWORK OF LEARNING AND SUPPORT FOR TRANSFORMATION

### Learning Collaboratives

Beginning to build peer networks and learning communities was the next critical task of the Transformation Center. These communities—or "learning collaboratives"—would become a cornerstone of the Center's activities. Within a few months of its launch, the Transformation Center convened its first learning collaborative of the CCOs. As with the IAs, there was initial resistance from the CCOs about participating. Fortunately, CCO participation in at least one learning collaborative was required in Oregon's waiver agreement with CMS as well as in the CCOs' contracts with OHA, which meant the Transformation Center had a captive audience. The task then was to develop the first learning collaborative in a way that could create an appetite among the CCOs for more opportunities for peer learning beyond the minimal requirements.

Once again, it was important to take a collaborative approach. Because the CCOs initially did not want to commit to any additional meetings, the Center identified an existing meeting in Salem, the state capital, that could serve as the first forum for peer learning. A regular standing monthly meeting with the CCO Medical Directors and OHA staff became the first learning collaborative. Participants decided the focus should be on the 17 CCO quality incentive measures put in place as part of the Waiver agreement, which provided a concrete starting point for learning and sharing across the CCOs. Bringing the CCOs together to discuss how their incentive metrics approaches was an important accomplishment particularly given that some CCOs have overlapping regions and are in competition with each other. What started as a chore eventually became an activity in which the CCOs actively participated. In fact, over time, participation in this learning collaborative grew as CCOs found value in openly sharing successes and challenges with each other. Many of the discussions about the incentive measures led to broader conversations about the delivery of primary care and the needs of Medicaid beneficiaries.

This initial learning collaborative focused on the incentive measures spurred additional learning communities with an expanded group of participants. For example, the Center convened a Complex Care Collaborative of CCOs, made up of clinical practice leaders and staff working to improve health outcomes for Medicaid members with complex health care needs. It consisted of two day-long meetings in which participants shared their approaches to identifying and caring for beneficiaries with complex health needs, as well as a three hour training on trauma-informed care. Over 120 participants representing all 16 CCOs attended each session, and more than double that number participated in the training.

The Center also developed a collaborative for the Community Advisory Councils (CACs) early on. CACs are statutorily required bodies for each CCO, which must be made up of consumers and community representatives and are required to ensure that the health care needs of the consumers and the community are being addressed. The focus of this collaborative was to develop the leadership skills of the members, given that a majority were required to be consumers, many of whom had little experience in doing committee work. The collaborative also gave CAC members an opportunity to learn from one another and develop strategies for strengthening the voice of consumers in their regions. Topics of these monthly learning collaborative meetings ranged from learning about areas relevant to the CAC's work, such as public health and trauma-informed care, to how to hold productive meetings and communicate effectively with CCO leadership. The Transformation Center also hosted annual in-person convenings for CAC representatives from all CCOs to further support sharing and knowledge-building.

In addition to the in-person or video conference meetings, each collaborative had a space to connect online on the Transformation Center's online collaboration tool, Groupsite, where members could network, post resources, and share best practices. Recordings of the collaboratives, along with materials and meeting minutes, are housed on the Transformation Center website (see www.transformationcenter.org).

## Coordinated Care Model Summit

In addition to the learning collaboratives, the Transformation Center convened annual summits to bring together CCOs, CACs, providers, community stakeholders, health leaders, consumers, and lawmakers for learning concrete, innovative strategies, and sharing what health system transformation looks like on the ground. Hundreds of people from across the state attended these summits each year, which included multiple breakout sessions highlighting emerging best practices in Oregon, as well as presentations from national experts. The Governor and OHA Director also made a point to speak at these summits, reiterating the importance of the CCM to the state. Attendees regularly reported that they planned to implement at least one innovative practice that they learned about at the summit, and that they made new connections with colleagues and other organizations on which they planned to follow up. These summits provided an opportunity not only for learning and sharing, but for reflecting on the successes to date and maintaining momentum.

## Council of Clinical Innovator Fellows Program

In July 2014, the Center launched the Clinical Innovator Fellows program, a statewide, multidisciplinary cadre of 14 emerging innovation clinician leaders who are actively working with local project teams to implement health care transformation projects in their community. During the year-long program, fellows develop skills in health care improvement, implementation, and dissemination science that results in a network of expertise supporting the Oregon CCM.

The meetings have focused on the following topics: leadership development, systems science and QI, CCOs and the CCM, project management and measurement, health equity, public health, behavioral health integration, organization and individual resiliency, and community health (see Chapter 14: Emerging Models to Prepare the Workforce for Health System Change). Now completing its third cohort of fellows, this program has exceeded expectations in developing a local community cadre of leaders that are familiar with the state CCM approach, trained to lead innovation projects and mentor other leaders within their community or organization.

## Transformation Fund Projects

To jump start innovation and in recognition of the fact that CCOs would need time to achieve cost savings, the Oregon legislature created a $30 million Transformation Fund in 2013 to be distributed as grants to the CCOs. The Transformation Center oversaw these awards. The 125 Transformation Fund projects represented a broad cross-section of topics, from electronic health records to primary care infrastructure to public- and community-health initiatives. (A publicly available report on these projects is available from: http://www.oregon.gov/OHA/HPA/CSI-TC/Pages/Transformation-Funds.aspx.) In the grant award process, it became clear that most of the CCOs—or the community projects they funded through the grant dollars—needed technical assistance and training on measurement and the science of innovation. The Transformation Center saw the opportunity to use these grants to build internal capacity for QI and project management, and required each CCO assign a portfolio manager to oversee the projects funded with the Transformation Fund grant award. Teams from every CCO, led by the portfolio manager and project staff, were provided with targeted technical assistance and required to participate in a 3-day IHI in-person training on QI methodology. Portfolio managers and other CCO QI staff, including some internal OHA staff, participated in an 8-month QI Manager training and development of a statewide virtual "community of practice."

## Transformation Plan Oversight and Support

As part of their contract with the state, CCOs are required to create and report on "Transformation Plans." These plans provide a window into the CCOs' efforts at implementing health system transformation in eight critical areas, including integration of care, alternative payment methods, health information technology and health information exchange, culturally and linguistically competent care, and community health improvement. The Transformation Center is responsible for reviewing the CCOs' Transformation Plans, progress reports, and the CCOs' updates to these plans every 2 years. Center staff uses the results of this analysis to inform Transformation Center supports for CCOs, and to make connections between the CCOs and other areas of support available within OHA.

## Community Health Improvement Plan Review and Implementation Support

The Transformation Center is accountable for reviewing the CCOs' community health improvement plans. Center staff not only ensured CCOs' contractual requirements were met, but also identified priorities across plans. Approximately 60% of community health improvement plan priorities fall in the areas of public health, social determinants of health, and health equity, and the remaining priorities fall into clinical areas such as mental health/substance abuse, oral health, and access to health care. The Center has begun to make connections between the CCOs and OHA entities that could provide support for community health improvement plan implementation, such as the Office of Equity and Inclusion, the Public Health Division, and the Early Learning Division. Community health improvement plan implementation support is also available through the Center's Technical Assistance Bank.

## Technical Assistance Bank

Based on requests from CCOs and their CACs, the Transformation Center began offering CCOs and their CACs the opportunity to receive technical assistance in key areas to help foster health system transformation. Each CCO was designated 35–50 hours (it varied over the years) per year of free consultation from outside consultants on contract with the Transformation Center. The designated hours included 10 hours of consultation to support CACs and other community-based work. The Technical Assistance Bank includes consultants who can provide CCOs expert support in areas such as alternative payment methods, behavioral health integration, oral health integration, health equity, health information technology, QI and measurement, primary care transformation, and patient and family engagement.

## Online Resources

The Transformation Center created an easy-to-navigate and attractive website (e.g., not the typical state government agency website) to house resources and catalog its efforts and of CCOs across the state. Recordings of learning collaboratives and summit breakout sessions are available on the site, along with reports and other resources. Creating a site outside the state structure also allowed the Center to quickly and easily add and change content, rather than submitting requests for additions and changes through a different department. However, the site recently migrated back to the standard state agency website structure, though it continues to provide important resources.

## TRANSFORMATION CENTER 2.0

During its first two years, the Center focused on building and fostering a culture of innovation across CCOs, working to promote health system transformation and build an effective peer-learning network for CCOs and their providers. In mid-2015, the Center engaged in a robust strategic planning process that incorporated the Center's experience working with CCOs as well as input from across the OHA. As a result, the Center's goals evolved to reflect the provision of more targeted technical assistance based on CCOs' performance metrics and evaluation outcomes. Not only did this shift complement OHA's movement toward supporting the long-term success of the CCM, but also it reflected the Transformation Center's natural evolution. Specifically, the Center had spent the first 2–3 years establishing relationships with the CCOs, fostering a culture of sharing between them, and clinical "best practice" knowledge transfer; it was now time to more intentionally help the CCOs achieve their intended outcomes. In addition, during the first few years, CCOs repeatedly shared that they wanted deeper dives on topics.

At the beginning of 2016, the Center launched "Transformation Center 2.0," with a variety of new strategies aligned with its targeted technical assistance goal. These strategies illustrate the Center's evolution from a focus on trust, relationship building, and knowledge transfer to being perceived as a resource for, and responding to the requests of, CCOs. For example:

- The development of an online Behavioral Health Integration resource library originated at a breakout session brainstorm at the Center's June 2015 "Innovation Café" convening.
- A focus on helping CACs recruit and engage members came from evaluations from participants of the CAC learning collaborative.
- The targeted QI technical assistance the Center provides around incentive metrics resulted from a request for more robust assistance per the broad metrics overviews delivered at the incentive metrics learning collaborative identified above.

The topic areas for the Center's targeted technical assistance work include behavioral health integration, CACs, alternative payment methods, CCO Incentive measures, health equity, clinical delivery supports, and oral health integration (see Fig. 12.2).

In 2016 the OHA Patient-Centered Primary Care Home program—which oversees the certification of Oregon's primary care medical homes—was folded into the Transformation Center. As a result, the Center updated its goals to reflect a focus not just on the health system, but on supporting innovation at the practice and community levels. Going forward, the Center will continue work in the following key areas: primary care (including the PCPCH program), behavioral health integration, oral health integration,

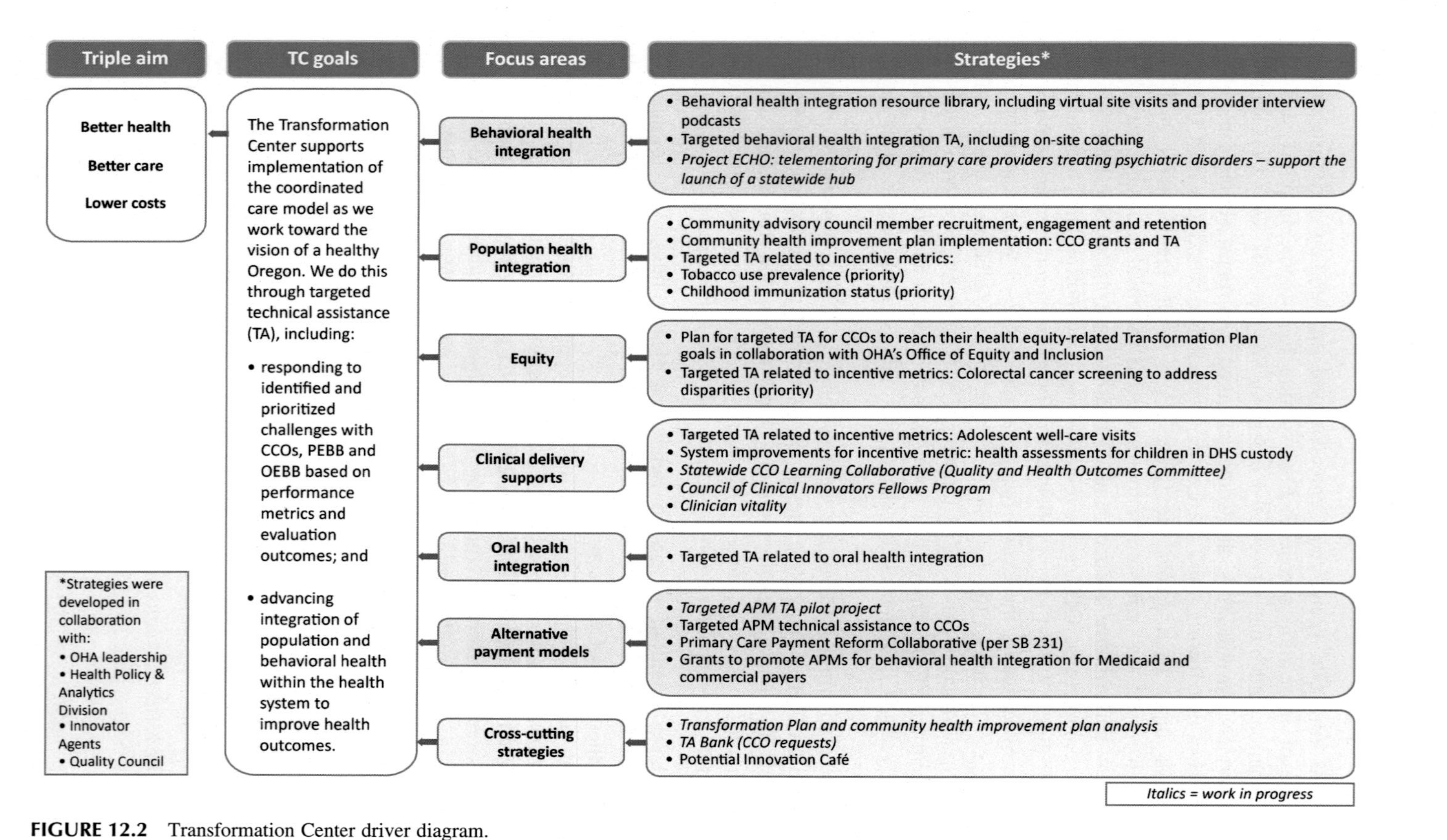


**FIGURE 12.2** Transformation Center driver diagram.

population health, and alternative payment models, and will continue to offer targeted technical assistance, hold key collaboratives, and organize convenings. In addition, while the CMMI State Innovation Model grant that initially supported the Center has come to an end, the Center has received sustainable funding from the Oregon legislature for positions and programs.

## KEY LESSONS LEARNED

We learned many important lessons developing the Transformation Center, including lessons about how to run effective learning collaboratives, engage clinicians in leadership, develop leadership skills in consumer representatives, and manage internal politics while pushing for change. Following are what we believe to be the key lessons:

*Trust matters most*: The most important lesson learned is how critical the need to build trust and develop relationships is in the success of any effort like this. This kind of relationship building takes time and the ability to listen and be responsive to feedback. It requires a way of operating that state agencies typically do not have the luxury of time to achieve; however, its importance cannot be underestimated. States considering the development of a transformation center of their own should plan for a heavy upfront investment of time and resources for establishing trusting relationships; this investment must include funding for state agency staff to travel out to communities to share plans and gather feedback, rather than requiring community stakeholders always come to them.

*Mandate a little, inspire a desire for more*: No matter how firmly we believe in transforming the health care delivery system, most of us—whether we are state policymakers, plan administrators, or providers—find it hard to make time to learn and share about the changes needed in the face of all the other work piling up. All too often, important changes fail to take hold because the day-to-day responsibilities outweigh the aspirational vision of the future. Mandating a minimal level of participation helped us force the CCOs and their providers to the table. It was an effective strategy for getting their foot in the door. But to get their true engagement, we could not mandate more than minimal participation, as requiring more would only build resentment and increased resistance. Instead, we had to make sure the learning collaboratives were worth the time CCOs and providers spent on them, and run them well enough that the participants would ask for more. It was clear that the financial incentives affiliated with the CCO quality measures provided a platform for this type of engagement to flourish.

*Do not reinvent the wheel*: The Transformation Center staff spent a significant amount of time and resources making sure they learned what was already happening across Oregon before undertaking new activities. Part of building trust and demonstrating value is showing that you are not just providing more of the same. We conducted (or contracted for) environmental scans for topics such as health equity, complex care initiatives, integration of behavioral health with primary care, alternative payment methods, and CCOs' use of funds to pay for health-related expenses to better understand existing state and community resources and projects in order to build on existing work rather than duplicating it.

*Money and free help works*: Distributing the Transformation Fund grants was another effective strategy at not only getting the CCOs to try new approaches they might hesitate to try without a demonstrated return on investment, but also encouraging them to engage the Transformation Center. In order to receive the grants, CCOs needed to establish a "transformation portfolio manager" who would work with the Center to develop measurement plans and share information on the impact of the projects that were piloted. The free hours of technical assistance provided through the Technical Assistance Bank also proved successful. They helped the CCOs see that they were getting something of value from the Transformation Center, instead of just giving the Center their time without anything in return.

*Run it like a start-up*: In the first year alone, the Transformation Center hired staff, traveled around the state to gather feedback, facilitated a total of 37 learning collaborative sessions and 5 day-long learning events, and launched four online networking communities on a web-based collaboration platform. Although a heavy workload is not unusual in state agencies, the speed with which that work was accomplished was unique. The Transformation Center, like any start-up, needed to prove its value quickly. A start-up mentality meant that not only did a lot of work get accomplished very quickly, but it was done with an eye to keeping the "customer," Oregon's CCOs, engaged. It meant that the Center was responsive and flexible, willing to change plans or strategies in response to feedback and to provide the CCOs the kind of help they needed and wanted. This rapid growth and flexibility sent a strong signal that the Center was not state agency business as usual and meant the Center got to live to see another day. The continuation of this culture of flexibility and responsiveness has helped ensure the Center's relevance to Oregon's health system transformation needs.

*Do not let internal transformation take over*: The Transformation Center successfully used internal agency transformation efforts—the "bureaucracy busting"—as a means of establishing trust with the CCOs and supporting needed culture change within the agency. However, those same efforts were sometimes at cross-purposes with the true goal of the Center, which was to support change externally. Internal transformation was a tool, not an end goal, but the amount of staff resources it required limited our capacity for the important external work. It was a tricky balance to find. Without any efforts at internal transformation, it would have been hard for the Center to establish positive working relationships with the CCOs. However, the internal work resulted in some positions within the Center being focused almost entirely on the agency instead of the CCOs' innovation work. This had the greatest impact on the IAs. These positions were meant to serve as champions of change and to facilitate learning within their CCOs. Instead, their focus was almost entirely on facilitating the CCO–OHA relationship.

*People, not systems, make change*: The biggest take-away from our experience developing and implementing a Transformation Center in Oregon is that change is hard because it requires working with people, not just systems—and the operative word there is *with*. David Sedaris writes, "I haven't the slightest idea how to change people, but I still keep a long list of prospective candidates just in case I should ever figure it out." We did not figure out how to make anyone or anything change, not the CCOs, the providers nor the state agency. But we did figure out that they were all willing to change of their own accord if they were given the right incentives along with the support and resources they needed to learn from each other and become champions of that change, rather than bystanders in it. Peer networks and the development of change leaders are critical components of system change. The success of change ultimately depends on the champions, networks, and relationships that support it. No attempt to transform the way care is delivered will be successful without acknowledging this and building the necessary infrastructure to support it.

## CONCLUSION

The Transformation Center demonstrates state agencies can and should be a partner in innovation. Despite the initial external reluctance to believe an office within a state agency could support change, we believe that the work has and will continue to make a difference. Not everything the Transformation Center tried worked, but its efforts showed that states can support change at the state and regional scale while also supporting the critical work that takes place in local communities. The state's approach, through the Center, was to provide training and skills development to the community so that specific interventions could be addressed locally at the practice level and that "best practice" learnings could spread both within and between communities and organizations. We encourage other states or local governments to consider building their own version of a transformation center to support their efforts at health system transformation.

## REFERENCES

Berwick, D. (2003). Disseminating innovations in health care. *JAMA*, *289*(15), 1969–1975.

Cabana, M., et al. (1999). Why don't physicians follow clinical practice guidelines: A framework for improvement. *JAMA*, *282*, 1458–1465.

Grol, R., & Wensing, M. (2013). Implementation of change in healthcare: A complex problem. In: *Improving patient care: The implementation of change in health care* (2nd ed., pp. 1–17). Oxford, UK: John Wiley & Sons, Ltd

Institute for Healthcare Improvement. (2003). *The breakthrough series: IHI's Collaborative Model for achieving breakthrough improvement.* IHI Innovation Series white paper. Available from: http://www.ihi.org/resources/Pages/IHIWhitePapers/TheBreakthroughSeriesIHIsCollaborativeModelforAchievingBreakthroughImprovement.aspx.

Institute of Medicine. (2001). *Crossing the quality chasm: A new health system for the 21st century.* Washington, DC: National Academy Press.

Rogers, E. (2003). *Diffusion of innovations* (5th ed.). New York: Simon and Schuster.

Chapter 13

# Supporting the Coordinated Care Model Through Technology

**Susan Otter**
*Oregon Health Authority, Portland, OR, United States*

Oregon is on an extraordinary path to transform the delivery of health care to improve health outcomes, quality of care, and reduce costs. This "health system transformation" effort is premised on a model of coordinated care that includes new methods for care coordination, accountability for performance, and new models of payment based on outcomes and health. To succeed, the coordinated care model relies on new systems for capturing, analyzing, and sharing information about patient care and outcomes, quality of care, and new models of sharing care information among all members of care teams.

In 2012 the Oregon Health Authority (OHA) focused first on its Medicaid population, implementing the coordinated care model through new Coordinated Care Organizations (CCOs). These regional care networks bring all types of health care providers (physical health, behavioral health, and dental) together to deliver coordinated care, while being held accountable for outcomes. CCOs now operate in every county in Oregon, and cover more than 90% of Oregonians on Medicaid. Moving forward, Oregon is working to accelerate and spread the coordinated care model beyond the Medicaid population to public employees, Medicare, and private payers. To be effective, Oregon's transformed health care system increasingly relies on access to patient information and the Health IT (health information technology) infrastructure to share and analyze data. *Health IT (HIT)* refers to a wide range of products and services—including software, hardware, and infrastructure—designed to collect, store, and exchange patient data to support patient-centered care. *Health information exchange (HIE)* is the electronic movement of health information among organizations following national standards. HIE facilitates sharing of health information across technological and organizational boundaries to enable better care. HIE can be used

**Health Reform Policy to Practice. DOI: http://dx.doi.org/10.1016/B978-0-12-809827-1.00013-6**

as a verb as well as a noun—organizations that provide HIE services are referred to as "HIEs." HIT impacts nearly every aspect of coordinated care, including:

- sharing patient information between providers and care team members for care coordination;
- analyzing health information in useful ways to manage populations and identify target populations for interventions;
- supporting accountability by assessing the quality and impact of care on patients;
- tracking outcomes and metrics to enable alternative payment methods and quality incentives; and
- supporting patient/consumer engagement via transparent access to health information and access to providers electronically.

This chapter describes the history of Oregon's HIT efforts as Oregon prepared to launch health system transformation; the development of Oregon's HIT strategic plan and key levers to change; implementation successes, challenges, and lessons learned; and the future of HIT as Oregon embarks on the next phase of transformation.

## HISTORY: A 10-YEAR EVOLUTION OF OREGON'S HIT EFFORTS

### Legislative Action and Federal Investment

HIT has been included as a key component of Oregon's evolving legislative and policy landscape for more than 10 years. Oregon's legislature has established a series of health policy committees to develop and oversee Oregon's health reform efforts, and, in turn, a series of parallel HIT committees informing the policy recommendations. Since as early as 2005, Oregon's HIT goals have been clear and strategies to achieve these goals have been largely consistent. The Oregon Health Policy Commission's Electronic Health Records & Data Connectivity Subcommittee included the following goals in its 2005 report to the legislature (Oregon Health Policy Commission, 2005):

- Oregonians' health record information is available to them and their health care provider anytime and anywhere that it is needed.
- Oregonians' health records are confidential and secure at all times.
- These goals are best achieved through widespread adoption of robust, secure, and interoperable electronic health records (EHRs) that support the delivery of high-quality efficient health care.

Similar goals and clear strategies were presented in the 2008 report to the Oregon Health Fund Board by the Health Information Infrastructure Advisory Committee (Governor's Health Information Infrastructure Advisory

Committee, 2008). Both the Health Policy Commission and Oregon Health Fund Board incorporated the work of these HIT committees into their formal reports to the Governor and Legislature (Roadmap for Health Care Reform, 2007; Aim High: Building a Healthy Oregon, 2008).

In 2009 the legislature established the HIT Oversight Council (HITOC) as recommended by the Health Information Infrastructure Advisory Committee in 2008, as part of a package of major health reform legislation. For the first time, there was a dedicated position and small budget for HIT efforts. The HITOC began meeting in the fall of 2009 as new federal HIT investments were announced, and it quickly adjusted to focus on leveraging these opportunities to meet its legislative objectives.

The Federal Health Information Technology for Economic and Clinical Health (HITECH) Act, enacted as part of the American Recovery and Reinvestment Act of 2009, provided billions of dollars in federal investment and programs to promote the adoption and meaningful use of HIT. In particular, Oregon hospitals, physicians, nurse practitioners, dentists, and others stood to gain significant federal incentive payments for adoption and "meaningful use" of EHRs under the Medicare and Medicaid EHR Incentive Programs. Each state was eligible for significant HIE investments (Oregon received nearly $8.6 million) under the Office of the National Coordinator for HIT (ONC) State HIE Cooperative Agreement program. These investments significantly drove forward Oregon's HIT objectives and set Oregon hospitals and physical health providers on the path to wide-scale EHR adoption.

In 2010 the Oregon Health Policy Board's 2010 "Action Plan for Health" (OHA, 2010) included recommendations from HITOC Strategic and Operational Plans. Oregon's HIE Strategic and Operational Plans and various supporting documents can be found here: http://www.oregon.gov/oha/HPA/OHIT-HITOC/Pages/Reports.aspx; see http://www.oregon.gov/oha/HPA/OHIT-HITOC/Documents/HIEStrategicPlanOR.pdf, for 2010 Strategic Plan; and http://www.oregon.gov/oha/HPA/OHIT-HITOC/Documents/HIEOperationalPlanOR.pdf, for Operational Plan including a call for the development of a sustainable funding model and legislation to create a public/private "state designated entity" for HIE. This recommendation was not ultimately supported for the 2011 legislative session, largely due to the higher priority of passing legislation establishing a health insurance exchange public corporation. There were significant concerns that the two proposals would become confused due to the similarities between the insurance exchange and the HIE concepts. The 2011 Oregon legislature did pass groundbreaking legislation to implement the Action Plan for Health, including the framework for implementing Medicaid CCOs and the insurance exchange public corporation entity, which later became known as Cover Oregon. It would not be until 2015 that Oregon saw legislation implementing recommendations allowing the state participation in a public/private collaborative and the express creation of an Oregon HIT Program that could sustain itself through fees if needed.

Also in 2011, Oregon created the OHA and established the Office of HIT as recommended in the Action Plan for Health. The Office of HIT was to be a consolidated office for both planning and implementation and serves as a resource for state and other public and private users of health information.

## Challenges to HIT Efforts

When OHA launched health system transformation in 2012, with the advent of the Medicaid CCOs, Oregon's HIT environment had already improved greatly in a few short years. Health care organizations in Oregon have been early adopters of EHRs, accelerated by Medicare and Medicaid EHR Incentive programs providing significant financial support for providers' adoption and meaningful use of certified EHRs. By 2012, nearly all Oregon hospitals and the majority of Oregon's physical health providers had adopted or upgraded to an EHR that met federal standards. OHA was implementing its HIE strategic and operational plans, which were founded on an evolving landscape of regional HIE—several of which had already failed, but a few were struggling.

Despite these investments, many gaps remained. Providers experienced challenges and frustrations with their EHRs functionality and interoperability with the myriad other systems in use. By mid-2013, more than 100 different EHRs had been adopted by Oregon providers. HIE was occurring, but providers' experiences varied greatly across Oregon. Whether a provider was effectively able to exchange health information was more likely if they practiced in a region with a community HIE, were affiliated with a large health system's EHR, or had resources including federal incentive payments and capable IT staff. Furthermore, some types of health information were easier to exchange—depending on whether information is contained in highly aligned standardized data formats or not, and whether it is limited by complex policies meant to protect particularly sensitive health information.

While launching our broad scale health system transformation, progress on state-level HIT efforts in Oregon had slowed. The ONC had concerns that the OHA had not made significant progress forward in implementing its strategic HIE plans. OHA's Office of HIT had been reorganized under the Office of Information Services within OHA and became further removed from the core work of OHA in launching transformation. Efforts to secure legislation to launch a HIT-focused public/private governance entity and establish authority for either a provider tax or user fees to ensure sustainability had been unsuccessful. OHA's state HIT services launched under the program name "CareAccord" in 2012—offering a first service of web-based Direct secure messaging (www.careaccord.org). Although CareAccord was the first state

HIE organization to receive national accreditation, it consistently had slower uptake than expected. OHA's ONC Cooperative Agreement was waning—and would end in early 2014, which meant CareAccord was likely to end as well. With no ability to charge fees or secure funding, CareAccord was officially put on hold and staff were unable to promote CareAccord and secure new users.

The impetus for a renewed focus on HIT came as Medicaid CCOs formed and CCO leaders had a much better sense of their needs for HIT. CCOs had come together quickly in 2012 and began to realize that to coordinate care across physical, behavioral, and oral health, providers must have efficient tools to share health information. CCOs needed clinical data to ensure that they could manage to their fixed global budgets and earn significant financial incentives related to quality outcomes metrics. Oregon's largest CCO sent an urgent letter to OHA encouraging the state to consider the investments needed in HIT and act where appropriate to make these investments at the state level ensure efficiencies and economies of scale.

## DEVELOPING OREGON'S PATH TOWARD A HIT STRATEGY

Given the changes brought by the advent of CCOs and the new intense need for the right HIT to support the coordinated care model, OHA needed to reevaluate its approach to HIT to ensure its priorities and activities were well aligned. In particular, the OHA needed to reset its HIT strategic plan, stabilize funding, determine the future of its one HIT service, CareAccord, and reestablish credibility with federal and state leadership and reinvigorate work with external stakeholders. In early 2013, OHA leveraged Robert Wood Johnson Foundation funding for a HIT consultant to bring an objective assessment and recommendation to OHA for moving forward with HIT.

The first step was a listening tour. The purpose of the listening tour was to identify stakeholder and especially CCO HIT needs and to understand what the right role would be for the State, and for statewide services. Office of HIT consultants and leadership met with each CCO as well as other internal leadership and external health care stakeholders. It was challenging to synthesize the disparate information from these sources. For example, in one community, the CCO was partnering with the local hospital to reinvigorate a regional community HIE organization. The CCO leadership informed us that under no circumstances should the state develop a statewide HIE platform and require all providers to participate. In another community, the CCO, independent physician association (IPA), hospital, and providers had repeatedly come together to attempt to develop a regional community HIE organization, without any success—running up against long-standing political issues stemming largely from mistrust between organizations. The CCO leadership indicated that the only way HIE would ever come to their community is if the state set up a statewide HIE and required all providers to participate. It seemed that there were clear differences in technology capabilities across communities and a lack of consensus on the scope or role that state-level HIT services should support. However, there were a few themes that came across consistently among stakeholders: the need for HIT to support care coordination across all members of a care team, the support of data aggregation, and analytics that incorporate clinical data.

There was further need to clarify the appropriate role that the State government should play, and where it was important for CCOs, health plans, hospitals, and providers to play their part. In discussions with internal executives, we found a "highway" metaphor to be particularly useful in deciphering the proper role for the State. While it was clear that the State should provide infrastructure or the "highway," it was also clear that it was not the State's role to provide every health care stakeholder with the tools or "car" to operate on that highway. Unfortunately, with HIT, it was difficult to tell the difference between the "highway" and which aspects of HIT is a "car." Leadership agreed that there may be exceptions for supporting those health care organizations who face significant barriers—it may be the State's role to provide an option where there is not one, like public transportation. This metaphor provided some clarity as the myriad technology needs for CCOs and providers became better known through the listening tour and as the Office of HIT staff and consultants worked to synthesize and create a "straw model" for our strategic plan.

## Oregon's Current HIT Strategic Plan

Despite the wide variety of perspectives, there were a few common threads we found in the course of the listening tour—and these formed the basis of our HIT straw model, which ended up forming the basis of our HIT strategic plan. Oregon's HIT/HIE Business Plan Framework (2014–17) was the resulting strategic plan, and includes themes from the listening sessions. Technology consultants helped prepare straw models for the scope of the state's role and a specific roadmap for technology, governance, and funding to provide a starting point for stakeholder input. We worked with OHA leadership and the Governor's office to refine the model, and the Governor's office convened a brief stakeholder advisory panel to advise it on moving forward with HIT.

With the straw model in hand, OHA recruited stakeholders to participate in an "HIT Task Force" to synthesize stakeholder input and develop a HIT/HIE Business Plan Framework to chart a path for statewide efforts over the next several years. OHA sought a diverse membership for the Task Force, representing health plans/payers, health systems, hospitals, providers, local HIE efforts, advocates/consumers, and HITOC members. Members were geographically and technologically diverse as well (e.g., at least one representative was included from a hospital system that did not use Epic, which is the predominant EHR system for Oregon hospital systems). Several influential HIT stakeholders were very interested in participating but unwilling to serve longer term on HITOC, so the HIT Task Force was set up as a separate, time-limited advisory group. With an ambitious timetable, the HIT Task Force created a strategic plan containing vision and goals, roles for the state, principles/considerations, and a HIT Roadmap, which are highlighted below (HIT Task Force, 2014).

In the strategic plan, Oregon set a vision of a transformed health system in which HIT efforts ensure that the care Oregonians receive is optimized by HIT, where:

- *Providers* have access to meaningful, timely, relevant, and actionable patient information to coordinate and deliver "whole person" care.
- *Systems* (health systems, CCOs, health plans, providers) effectively and efficiently collect and use aggregated clinical data for quality improvement, population management, and incentivizing health and prevention. In turn, *policymakers* use aggregated data and metrics to provide transparency into the health and quality of care in the state, and to inform policy development.
- *Individuals* and their families access their clinical information and use it as a tool to improve their health and engage with their providers.

Consistent with national HIT efforts, Oregon's plan references the ONC strategy statement that "All patients, their families, and providers should expect consistent and timely access to standardized health information that can be securely shared between primary care providers, specialists, hospitals, behavioral health, Long-Term Post-Acute Care, home and community-based services, other support and enabling services providers, care and case managers and coordinators, and other authorized individuals and institutions" (Office of the National Coordinator for HIT, 2013).

In order to achieve the goals outlined earlier, the strategic plan identified the following roles for State government:

- *The State will coordinate and support community and organizational HIT/HIE efforts.* Recognizing that HIT/HIE efforts must be in place locally to achieve a vision of HIT-optimized health care, the State can support, facilitate, inform, convene, and offer guidance to providers, communities, and organizations engaged in HIT/HIE.
- *The State will align requirements and establish standards for participation in statewide HIT/HIE services.* To ensure that health information can be seamlessly shared, aggregated, and used, the State is in a unique position to establish standards and align requirements around interoperability, privacy, and security. State efforts should rely on already established national standards where they exist.

- *The State will provide a set of HIT/HIE technology and services.* New and existing state-level services connect and support community and organizational HIT/HIE efforts where they exist, fill gaps where these efforts do not exist, and ensure that all providers on a care team have a means to participate in basic sharing of information needed to coordinate care.

In addition, the stakeholder group recommended the following considerations or principles in pursuit of state-level services:

*Leverage existing resources and national standards, while anticipating changes*:

- Consider investments and resources already in place.
- Leverage federal Meaningful Use requirements and national standards; anticipate standards as they evolve.
- Monitor and adapt to changing federal, state, and local environments.

*Demonstrate incremental progress, cultivate support, and establish credibility*:

- Advance through relentless incrementalism: define a manageable scope, deliver, and then expand.
- Communicate frequently with measurable progress. Demonstrate optimal value for patients and providers toward the triple aim of better health, better care, and lower costs.
- Provide public transparency into development and operations of statewide resources.
- Be a good steward of limited public resources.
- Establish long-term financial, leadership, and political sustainability. These are interdependent.
- Seek broad stakeholder involvement and support. Statewide resources cannot be developed alone.

*Create services with value*:

- Maximize benefits to Oregonians while considering costs. Do not disenfranchise ("do no harm"), and be inclusive of providers that face barriers to participation.
- Support provider participation in HIT-optimized health care; meet providers where they are. Recognize the challenges especially for smaller, independent providers and providers who are not eligible for federally funded EHR incentives.
- Prioritize efforts to achieve a common good and that local entities could not do on their own.
- Cultivate and communicate value at the individual, provider, system, and state levels.
- Champions and personal stories can be very effective.
- Support new models of "HIT-optimized" health care that result in better quality, whole person care and improved health outcomes and lower costs for all.

| | 2010–13<br>Phase 1 | 2014–16<br>Phase 1.5 | 2016 Forward<br>Phase 2.0 |
|---|---|---|---|
| Governance, operations, and policy | **Oregon Health Authority (OHA) With HIT Oversight Council (HITOC) and HIT Task Force**<br>• Strategic planning, oversight, transparency, policy, accountability<br>**OHA**<br>• Implementation, operations | **OHA with HITOC**<br>• Strategic planning, transparency, policy<br>- State HIT Legislation in 2015<br>**Steering Committee/CCO HITAG**<br>• Phase 1.5 oversight, accountability<br>• Planning for HIT Designated Entity<br>• Develop standards/compatibility program<br>**OHA :** Implementation, operations | **OHA With HITOC and Steering Committee**<br>• Strategic planning, oversight, transparency, policy, accountability<br>•Standards/compatibility program<br>**HIT Designated Entity**<br>• Implementation, operations |
| Technology and services | **CareAccord**<br>• CareAccord direct secure messaging (launched May 2012)<br>• Trust/Interstate efforts (National Association for Trusted Exchange, Direct Trust) | **CareAccord**<br>• Direct secure messaging; access to enabling infrastructure. Trust/interstate efforts.<br>**Enabling infrastructure**<br>• Provider directory/information services<br>• Statewide hospital notifications, EDIE<br>• Common credentialing<br>**Services for Medicaid**<br>• Clinical Quality Metrics Registry<br>• Technical assistance to eligible providers | **CareAccord**<br>• Direct secure messaging; access to enabling infrastructure. Trust/ interstate efforts.<br>**Enabling infrastructure and Medicaid services**<br>• Enhanced statewide enabling services and record location<br>• Supporting query and data analytics<br>• Patient/provider attribution |
| Finance | **Office of the National Coordinator for HIT (ONC)**<br>• ONC Cooperative Agreement (2010—February 2014) | **CMS/State Match/Investors**<br>• Planning broad–based financing model<br>• CMS funding for Medicaid implementation; with legislation—expand to private (non-Medicaid) users<br>• State/CMS contribute ongoing funding for services that support state Medicaid operations | **Public/private partnership**<br>• Broad-based financing model provides financial stability<br>• State/CMS contribute ongoing funding for services that support state Medicaid operations |

**FIGURE 13.1** Oregon's Roadmap for HIT. *From Oregon Health Authority (OHA), 2014. Oregon's Business Plan Framework for Health Information Technology and Health Information Exchange (2014-2017). Health Information Technology Task Force Recommendations. Available from: http://www.oregon.gov/oha/HPA/OHIT-HITOC/Documents/HIT_Final_BusinessPlanFramework_ForRelease_2014-05-30.pdf.*

- Protect the health information of Oregonians.
- Ensure information sharing is private and secure and complies with federal requirements (such as HIPAA) and other protections.

In some cases, the strategic plan identified significant recommendations or specific technology strategies, and in some areas, such as governance and finance, the HIT Task Force developed a high-level approach and principles to follow without specifics. The resulting plan was presented to, and endorsed by, the HITOC, and work began to implement the recommendations. See Fig. 13.1 for a depiction of Oregon's HIT Roadmap, which summarizes the strategies identified in the plan.

## IMPLEMENTING THE HIT STRATEGY: SUCCESSES, OPPORTUNITIES, AND CHALLENGES

### Leveraging State Transformation Funds

In the summer of 2013, Oregon's legislature awarded $30 million of general funds to the OHA to support Medicaid transformation efforts, intended to be innovation project seed funding to CCOs. These "Transformation Funds" were to be distributed according to a formula to each CCO through an application process. As the CCOs considered their needs, several CCOs requested further information about the state's HIT plans and learned of the opportunity for 90% federal (HITECH) matching funds for Medicaid's share of statewide HIT services. OHA proposed it retain some portion of the $30 million to invest in a set of statewide HIT services that were critical to support the coordinated care model. After much deliberation, all 16 CCOs agreed that the OHA should use $3 million of state Transformation Funds to secure federal matching dollars to invest in a set of statewide HIT services.

CCOs requested that OHA establish an advisory group to represent CCOs' HIT interests and advise OHA on the use of Transformation Funds to support the implementation of key HIT services and initiatives. The HIT Advisory Group is charged with identifying major requirements for technology, such as scope, priorities, timelines, and milestones; monitoring and advising on implementation and communications; and reporting back to CCOs on progress made.

In addition to state-level investments, each CCO chose to invest a portion of their Transformation Funds in HIT—typically investing in both care coordination/HIE tools and population management/analytics tools. In 2014, Office of HIT staff and consultants conducted in-depth meetings on HIT with each CCO to identify CCO investments and to ensure that OHA's efforts continued to be in support and alignment of CCO efforts. Table 13.1 illustrates how CCO HIT investments were utilized.

In 2013 and 2014 the OHA received Centers for Medicare & Medicaid Services (CMS) approval for matching funds for its proposed efforts. These funds support the following HIT efforts. A sixth effort, supporting administrative efficiencies related to credentialing, was legislatively mandated, and will be supported by fees (for further information about each effort, please see the OHA Office of HIT website: http://healthit.oregon.gov):

- *Statewide hospital event notifications*, to support care coordination for CCOs, health plans, and providers when their patients or members have a hospital event. This service was launched as the Oregon Emergency Department Information Exchange (EDIE) Utility and affiliated product, PreManage.
- *Statewide Direct secure messaging*, to ensure that all health care-related entities can share information about the patients they have in common, supported by two efforts: (1) CareAccord (described earlier, launched 2012, ongoing) and (2) the "Flat File Directory," a statewide Direct secure messaging address book, collected and shared monthly, to be phased out in 2018 with the launch of a more robust Provider Directory.
- *Technical assistance for Medicaid providers to support the adoption and meaningful use of their certified EHRs* (January 2016–May 2018). The Oregon Medicaid Meaningful Use Technical Assistance Program

**TABLE 13.1 CCO Investment in HIT Efforts (as of Spring 2015)**

| | No. of CCOs | Overview |
|---|---|---|
| HIE | 14 | One regional HIE organization offering closed loop referrals and a consolidated community health record (five CCOs) |
| | | Two regional HIE organizations in development (two CCOs) |
| | | One community-wide EHR (one CCO) |
| | | PreManage (real-time hospital event notifications) (nine CCOs) |
| Case Management and Care Coordination | 10 | One social services focused tool (two CCOs) |
| | | Case management tools (nine CCOs) |
| Population Management, Metrics Tracking, Data Analytics | 15 | Population management tools (nine CCOs) |
| | | Business intelligence (BI) tools (six CCOs) |
| | | Other health analytics tools (11 CCOs) |
| Hosted EHR via Affiliated IPA | 3 | |

Source: Oregon Health Authority (OHA), 2015. Oregon Coordinated Care Organizations' Health Information Technology Efforts. Avaialble from: http://www.oregon.gov/oha/HPA/OHIT/Resources/CCO%20HIT%20Summary%20Report%20July%202015.pdf.

is supported with Medicaid funding (90% federal matching funds). Technical assistance will help providers effectively use their EHR technology and realize the benefits of their investments. It will also help support CCO efforts related to care coordination, quality improvement and metrics, and clinical quality data reporting required for the CCO quality incentive program.

- *A statewide Provider Directory*, critical to supporting care coordination and HIE, analytics and population management, as well as operational efficiencies (to be launched 2018). The Provider Directory will leverage authoritative provider information from a *new Common Credentialing Program*. This program includes a database and processes to centralize obtaining and verifying of Oregon health care practitioner credentialing information (legislatively mandated and fee funded—to be launched 2018);
- *A Clinical Quality Metrics Registry*, to capture clinical quality metrics from EHRs, with an initial focus on required CCO EHR-based quality metric reporting and Medicaid EHR Incentive Program reporting. OHA envisions this registry could be extended to other pay for performance

programs, supporting a "report once" strategy for providers currently burdened by myriad reporting requirements.

## Taking a Chance on EDIE

In 2013 the Oregon Health Leadership Council (OHLC; www.orhealthleadershipcouncil.org/edie), a collaborative representing Oregon's commercial health plans, hospital systems, and a few CCOs, identified reducing unnecessary emergency department utilization as a strategic objective. It recommended an approach taken by Washington's hospital association, in particular the use of a tool for bringing high-value information to emergency department providers on their patients who were high utilizers of hospital services. The Emergency Department Information Exchange (EDIE) provides hospitals with real-time notifications (via standardized Admit, Discharge, Transfer data feeds) about their patients who are frequent users of emergency department services. EDIE notifications provide limited but critical information such as date and location of recent patient hospital visits, key care recommendations, and known care providers.

The OHLC contacted OHA to discuss a novel approach—the two entities would sponsor the first year's costs to connect all hospitals in Oregon to EDIE, and take a chance that Oregon could replicate the kinds of improvements and savings that Washington had seen. OHA leveraged federal State Innovation Model (SIM) funding from the Centers for Medicare & Medicaid Innovation, to cover half the implementation costs, with the caveat that the vast majority of Oregon's hospitals had to agree to participate. This requirement recognized that the true value of this investment would be greatest only if all or nearly all participated. OHLC agreed, and worked closely with the Oregon Association of Hospitals and Health Systems to recruit hospital CEOs, and succeeded in getting all hospitals to agree to participate.

Implementing their connection to EDIE was a different story, and in some cases hospital chief information officers (CIOs) were concerned that they were left out of the process, and expected to adjust their IT schedules to fit in something that they were not sure had value. OHLC engaged a retired CIO with credibility and relationships who was able to engage reluctant CIOs and help ensure their concerns were mitigated where possible. A subsidy was offered to critical access hospitals to cover their implementation costs, which also helped considerably. OHLC circulated a monthly implementation status list of each Oregon hospital (which was later dubbed the "list of shame") to identify the stage of implementation each hospital had achieved, and which some hospitals still had work to do. This peer pressure helped garner action from even the most challenging hospital environments, and eventually all of Oregon's eligible hospitals had made their emergency department and inpatient data available in EDIE, adding Oregon's data to the hospital event data from Washington state hospitals.

Whereas EDIE provides alerts to professionals within the hospital system, PreManage (http://collectivemedicaltech.com/what-we-do-2/premanage/) is software that accesses the information in EDIE and makes it available to health care organizations outside the hospital system in real-time when their patient or member has a hospital event. Those organizations include CCOs, providers, clinics, and health plans. At the end of 2016, nearly all of Oregon's 16 CCOs and about a half dozen commercial health plans have subscribed to PreManage and most are extending their PreManage subscription to their key primary care and behavioral health practices. Subscribers can add key care coordination information into the system, such as notes about prescribing practices for a patient on pain medications, cooccurring issues, contact information for critical care coordinators, and other concise information to inform emergency department providers and other PreManage users. Together, EDIE and PreManage help improve care coordination across Oregon, which can result in fewer hospital visits and duplicative services.

EDIE is supported by a unique public/private partnership, called the "EDIE Utility" (for further information and the EDIE Utility Business Plan see OHLC's website: www.orhealthleadershipcouncil.org/edie). Formed as a 3-year demonstration, the costs for EDIE are shared between hospitals and payers (health plans and OHA on behalf of CCOs). The OHLC serves as fiscal agent and partners with the hospital association who supports data aggregation for performance reporting to the utility members. The EDIE Utility Governance Board makes decisions and includes representation from hospitals, health plans, CCOs, providers, OHA, the hospital association, and ad hoc members.

The reception of EDIE and PreManage by users has been extremely positive. The success of EDIE led to stakeholder support for a legislation change in 2016 to allow hospitals to access Oregon's Prescription Drug Monitoring Program (PDMP) information from EDIE or other HIT systems, to automatically view controlled substances that had been prescribed and filled for high utilizing patients. One emergency department provider said, "I don't remember how I used to do my job without EDIE and PDMP. It's taken the profiling out of what I do—takes the guesswork out."

CCOs serving hard-to-reach homeless populations have created cohorts within PreManage and prepped case managers to reach out in person when notified real-time by PreManage that these members arrive at the emergency department. Care guidelines are shared with emergency department providers and other PreManage users for challenging, complex patients, which allow providers to align treatment. A primary care clinic representative said, "when care guidelines are shared, patients feel that 'wherever I go, everyone is going to help me in the same way'. And they will say 'you all are talking to each other'" (OHA, 2016a).

## Sustaining the Oregon HIT Program

Bolstered by the success of EDIE, Oregon stakeholders supported significant new HIT legislation. With the support of many stakeholder associations and vocal proponents, and none opposed, Oregon's legislature passed House Bill 2294 (2015: https://olis.leg.state.or.us/liz/2015R1/Downloads/MeasureDocument/HB2294/Enrolled), to update Oregon's original HIT statute (establishing HITOC in 2009) to account for changes since 2009. This legislation included three major components to formalize and support OHA's HIT efforts:

- Establish the Oregon HIT Program within OHA, allowing the agency to offer services statewide (beyond Medicaid to the private sector). Participation in the Oregon HIT Program's services is not mandated and OHA may charge user fees for such services to cover costs and ensure sustainability.
- Provide OHA greater flexibility in working with stakeholders and partners. The statute allows OHA to enter into partnerships or collaboratives when other entities in Oregon are establishing statewide HIT infrastructure tools.
- Move HITOC under the Oregon Health Policy Board to ensure statewide HIT efforts align and support health system transformation.

The OHA established the Oregon HIT Program, formalizing the efforts underway, and reporting to the legislature on progress. As of September 2016, significant progress has been made on its HIT promises to the CCOs related to Transformation Funds and leveraging significant federal match to support major improvements in Oregon's HIT infrastructure. OHA also leveraged two major federal grants to support new initiatives focused on telehealth, patient access to full clinician notes (Open Notes), and a model consent process and technical implementation for sharing protected addictions treatment information subject to federal privacy and consent restrictions (under federal rule 42 CFR Part 2) (for 2016 status on Oregon's HIT environment, as well as the Oregon HIT Program efforts and initiatives, see OHA report "Oregon Health Information Technology Program: Annual Report to the Legislature", 2016b).

## State Office of HIT Structure

Today, the Office of HIT has a staff of about 25, with a Director overseeing two teams: the HIT Policy and Meaningful Use team, and the HIT Program/Implementation team. Functionally, the Office of HIT plays multiple roles. It relies on technology subject matter expert consultants, who advise on the changing technical and federal/policy environment, as well as a systems integrator/prime vendor who manages technology vendor contracts and ensures

*The Office of Health Information Technology develops and supports effective health IT policies, programs, and partnerships to enable improved health for all Oregonians.*

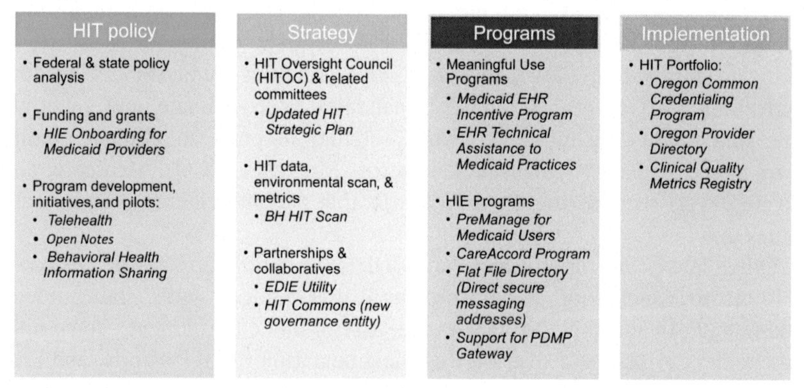


**FIGURE 13.2** State Office of HIT Organizational structure, December 2016.

successful implementation and operations of many of our HIT services. Office of HIT staff and contractors are funded largely through 90% federal HITECH match, but have significant base state funding. The one-time investment of $3 million in state Transformation Funding continues to support the implementation and operations of our core services, and will likely be spent by the end of the next state budget cycle in mid-2019 (Fig. 13.2).

## FUTURE OF HIT IN OREGON

Oregon has seen many successes in supporting the vision of "HIT-optimized" health care, albeit with many bumps along the way. EHRs are being used more throughout Oregon provider practices—improving providers' ability to access medical records across systems. Oregon providers and hospitals are in the top tier of states accessing millions of federal EHR "meaningful use" incentive dollars each year. CCOs are playing a leading role in HIT investments across the state—supporting efforts to share health information electronically to enable all members of the care team to coordinate effectively. In several regions, CCOs and their community partners have supported robust community HIE infrastructure to ensure Oregon providers are achieving meaningful health outcomes for their patients—looking at outcomes as opposed to processes—and rewarding improved outcomes with financial incentives. New statewide innovations such as the EDIE and PreManage are bringing real-time hospital event information to providers, CCOs, health plans, and emergency departments across the state, making a real difference in getting people the right care, in the right place, at the right time.

Where Oregon has been able to move the needle on HIT, it has benefitted from a number of critical levers. Perhaps, the most critical lever has been the federal and state health reform environment that has created financial incentives and penalties driving adoption, coordination, and the need to seek efficient HIT supports. Key among these, federal HITECH investments and meaningful use incentives to providers and hospitals; Center for Medicare & Medicaid Innovation SIM and state Transformation Fund dollars seeding innovation; CCOs' global budgets, mandates to coordinate and integrate care, and quality incentive payments relying, in part, on clinical quality metrics; and aligned payment models across Medicaid, CCOs, Medicare, and private payers recognizing the critical role that patient-centered primary care homes play.

Other levers and approaches have led to valuable support for Oregon's HIT efforts, including partnership and involvement with stakeholders. Taking significant time to listen and assess the environment, convening stakeholders to plan and oversee the implementation of HIT efforts, and providing transparency and checking in along the way have led to support for legislation and funding and stakeholder support. Selecting a few, focused efforts, demonstrating success, and leveraging existing investments, while anticipating that changes will occur, help build trust and credibility, which provide the foundation for current and future successes. Partnering with hospitals and health plans under a shared governance and financing model has proven to be incredibly effective for the EDIE Utility—recognizing that state HIT efforts are stronger with the resources and aligned interest of external stakeholders, and recognizing the important role that local investments and community collaborations play in successful use and implementation.

Although we are nearing the end of the current HIT strategic plan and the EDIE Utility demonstration, Oregon is well poised to adjust and set strategic and governance direction for the next 3–5 years. Under the direction of the Oregon Health Policy Board and Oregon's HITOC, priorities include a focus on "real world interoperability," behavioral health information sharing, and an update to the HIT/HIE Business Plan Framework. As part of its strategic planning work, HITOC has reconfirmed the state role identified in 2013, and renewed interest in developing a public/private governance partnership, to deliver on Oregon's statewide HIT efforts and its HIE networks. This new governance model will seek to build on the success of the EDIE Utility.

Oregon's coordinated care model is evolving, and the HIT needed to support that model must evolve as well. Many widely recognize the importance of social factors on health outcomes, and the need to expand the definition of what information is relevant to share, what metrics are relevant to hold entities accountable to, and what is needed to effectively manage populations and improve health outcomes. Furthermore, payment models are evolving such that providers are increasingly at risk for those outcomes and will have a critical need to coordinate care with new partners. The OHA is considering these opportunities and needs as it seeks to leverage new federal funding available for connecting Medicaid providers of all types to HIE, and will create a roadmap for the next 5 years for HIE priorities through the end of the federal HITECH funding in 2021. OHA and HITOC look forward to wrestling with these new expectations and demands, and creating new partnerships with payers, providers, hospitals, social services, and others to meet these needs and support improved health and health care for Oregonians.

## REFERENCES


Oregon Health Policy Commission. (2005, March). *Report to the 73rd legislative assembly: Electronic health records & data connectivity*. Available from http://library.state.or.us/repository/2009/200910201102202/.

Governor's Health Information Infrastructure Advisory Committee; Office of the Governor, State of Oregon (2008). *Executive Order No. 08-09*. Available from https://www.oregon.gov/gov/Documents/executive_orders/eo0809.pdf; Report to the Oregon Health Fund Board — Health Information Infrastructure Advisory Committee Recommendations. Available from http://library.state.or.us/repository/2008/200812311523333/.

Oregon Health Policy Commission. (2007). *Road map for health care reform*. Available from http://library.state.or.us/repository/2007/200710221009581/.

Oregon Health Fund Board. (2008). Aim high: Building a healthy Oregon. Available from http://library.state.or.us/repository/2008/200812311555003/.

OHA (Oregon Health Authority). (2010, December). Oregon's action plan for health. Available from http://library.state.or.us/repository/2011/201101100846561/.

OHA (Oregon Health Authority). (2014). *Oregon's business plan framework for health information technology and health information exchange (2014–2017). Health Information Technology Task Force Recommendations. Available from* http://www.oregon.gov/oha/HPA/OHIT-HITOC/Documents/HIT_Final_BusinessPlanFramework_ForRelease_2014-05-30.pdf.

OHA (Oregon Health Authority). (2015). Oregon coordinated care organizations' health information technology efforts. Available from http://www.oregon.gov/oha/HPA/OHIT/Resources/CCO%20HIT%20Summary%20Report%20July%202015.pdf.

OHA (Oregon Health Authority). (2016a, January). *Emergency department information exchange: Collaboration helps coordinate care across Oregon*. Available from http://www.oregon.gov/oha/OhIT/Documents/exchange-newsletter-Jan-2016.pdf.

OHA (Oregon Health Authority). (2016b). Oregon health information technology program: Annual report to the legislature. Available from http://www.oregon.gov/oha/HPA/OHIT/Resources/Annual%20Report%20to%20the%20Legislature.pdf.

HIT Task Force. (2014, May). *Oregon's business plan framework for health information technology and health information exchange (2014–17): Health information technology task force recommendations*. Oregon Health Authority. Available from http://library.state.or.us/repository/2014/201406241336072/.

Office of the National Coordinator for HIT. (2013). *Principles and strategy for accelerating health information exchange (HIE)*. Available from https://www.healthit.gov/sites/default/files/acceleratinghieprinciples_strategy.pdf.

Chapter 14

# Emerging Models to Prepare the Workforce for Health System Change

**Ron Stock[1,2], Emilee Coulter-Thompson[2,3], Leann R. Johnson[4] and Evan Saulino[2,5]**

*[1]Oregon Health & Science University, Portland, OR, United States, [2]Oregon Health Authority Transformation Center, Portland, OR, United States, [3]University of Michigan, Ann Arbor, MI, United States, [4]Oregon Health Authority, Portland, OR, United States, [5]Providence Health and Services, Portland, OR, United States*

*Will I do better if my care is delivered by a team? Will my chance of having a better outcome improve? This is what I value. This is the bottom line.*

Jessie Gruman (1953–2014), founder and president of Center for Advancing Health (Okun et al., 2014).

Oregon's reform effort is superimposed upon a multifaceted health care system and a growing number of patients with complex health needs which stresses the capacity of the health care system and the skills of those delivering it. This complex environment includes individuals with multiple chronic conditions, behavioral health issues, preventable health conditions requiring "upstream" interventions, and the challenges of an aging population. The solutions are many, and the breadth of the changes occurring simultaneously in Oregon is an example of a disruptive system of care that will require a multiprofessional caregiver workforce to work together collaboratively with individuals, families, and communities using an interprofessional team approach albeit with new skills and knowledge. For this collaborative approach to work effectively, health professionals in all disciplines (including traditional health workers, social workers, nurses, pharmacists, oral health, physical therapists, dieticians, physicians, public health professionals, and so forth) and settings (within clinics, hospitals, and community-based organizations) must be trained to work together. This workforce must be trained to deliver team-based, patient-centered, community-oriented care that is effective, efficient, and based on value, not volume. But this type of care does not just happen, nor can we rely on the current workforce to figure out

**Health Reform Policy to Practice. DOI: http://dx.doi.org/10.1016/B978-0-12-809827-1.00014-8**

on their own how to work together without providing them the skills and incentives to be successful.

Critical to the success of health reform will be matching needed individual and team behaviors, clinical skills, communication skills, ethical considerations, and community engagement skills required with the educational approaches utilized in our academic health professional institutions and clinical practice. Health professions education must move beyond the professionally segregated classroom and workshop-oriented continuing education approaches into hands-on, innovative training venues and models, such as community clinical "laboratories" that facilitate ongoing peer networking and training paradigms using an interprofessional community approach within the context of real-life clinical care experiences. The workforce training approach should not just attend to new students from different disciplines but must also address the training needs of existing practices and clinical teams that aspire to change and develop these new approaches. New continuing education training paradigms will be needed to transform these clinical delivery systems and develop the health care leaders needed to mentor and sustain the model of care for generations.

Like most states, Oregon's legislative and health care leaders have made an effort to address a prepared health care workforce strategy that not only supports recruitment and retention of health care professionals needed to practice in the state, but also ensures a workforce that will work in new and different ways, requiring new types of health care workers and competencies. This chapter explores the health care clinician workforce needs in terms of recruitment and retention of trained professionals to support the team as well as the efforts to develop the competencies needed to support a reformed system of care. Three examples of training interventions that show early promise for success in Oregon, and could be replicated elsewhere, are described.

## CLINICIAN SUPPLY AND DEMAND

The need for more clinicians, particularly physicians, to meet an ever-increasing demand for their services has been debated for years and hindered by various methodologies to determine the need and a changing health care environment that impacts clinical delivery. The importance of this debate is clear for state and federal health reform efforts since reform success will be highly dependent on the capacity of the health care workforce to meet the demands for clinical services and preventive interventions with health insurance expansion and care delivery redesign. Understanding this dynamic in Oregon to recruit and retain a diverse clinician workforce prepared to support this model well into the future is critical to sustainability.

Nationally, there is general consensus that the future demand for physicians will be greater than the supply. The American Association of Medical Colleges (AAMC) estimates that a shortage of 45,000 primary care physicians and 46,000 specialists by 2020, mostly a result of overall population growth, an aging demographic, and retiring "baby boomer" physicians (Association of American Medical Colleges, 2010). The American Medical Association (AMA) predicted that the primary care physician workforce would require a 24% growth to meet demand in 2015 (Petterson et al., 2012). Unfortunately, what these alarming statistics do not reflect is an inadequate distribution of the physician workforce with rural and underserved populations experiencing a much greater need. Most medical schools and Graduate Medical Education (GME) programs reside in urban areas. A number of new medical schools, both allopathic and osteopathic, have emerged over the past decade, many in communities or regions of the state where need is greatest. Many existing medical school programs have expanded their classes to meet this future demand; however, GME residency positions available to students have not kept pace due to capping of new positions since the Balanced Budget Act of 1997 and funding restrictions through Medicare GME legislation that limit expansion of GME programs.

This is of particular importance in Oregon where demand for primary care clinicians is high in rural and underserved areas. Historically, Oregon has been an importer of physicians with Oregon ranked 38th in the nation in number of all resident physician positions per population and 40th in the country in primary care resident positions. To address this need, Oregon Health & Science University (OHSU) has increased the number of medical students admitted each year and a new osteopathic medical school, College of Osteopathic Medicine of the Pacific-Northwest, was established in 2012. In Oregon, 46% of students trained in the state stay, or return, to the state to practice, 56% of primary care residents stay in the state after training and if both student and resident training occurs in the state then 70% remain (Oregon Healthcare Workforce Committee, 2011; The Projected Demand for Physicians, 2014). To address this need for more resident physician training positions, in particular in Family Medicine, a public–private grant was initiated to form a GME Primary Care Consortium to develop a plan to expand GME positions available in the state and promote resident and student rotations in rural/underserved regions (Health Care Workforce Committee, Oregon Health Authority, 2014).

In Oregon, a 2014 report to the Oregon Health Authority (OHA) examined the question of the projected demand for physicians, nurse practitioners, and physician assistants between the years of 2013 and 2020 at both the state and county level (The Projected Demand for Physicians, 2014). Multiple levels of data were analyzed including the Oregon All Payer, All Claims database, and the Oregon Health Care Workforce Licensing Database. Unlike a lot of analyses, this analysis took into account projected rates of utilization, payer types, the influence of team-based care, and information technologies in the proposed new model of care. Results showed that clinician demand varies widely depending on a number of scenarios. At baseline, it is projected that between the years 2013 and 2020, there will be a 16% increase of need over current demand. It varies considerably by county and geographically from as low as 9% in some counties to 28% in others with a generally higher demand in rural areas where uninsured populations appear to influence the need. However, this demand is reduced to 14% if the expected Medicaid utilization of services declines in the new model. If team-based care scenarios are examined, the demand for physicians reduces to 12%, but nurse practitioner (NP) and physicians assistant (PA) demand increases to 31%. Finally, implementation of information technologies to its fullest extent could reduce demand to 11% above what would be expected from historical data, and the combination of that with team-based care reduces the demand even more. This report highlights the importance of taking delivery system redesign, provider access, and workforce capacity into account when assessing future workforce needs and reliance on old assumptions may lead to an inaccurate assessment of need. However, for other disciplines needed in a team-based approach to care delivery, it is unclear how many health care professionals will be needed to support such a model.

Increasing the number of trainees in health care fields takes a long time, anywhere from 3 to 7 years for a trainee to be available to the workforce and so longitudinal recruitment strategies to attract professionals from other regions of the country, or internationally, are necessary. Many practices, health systems, and health care facilities have used incentives to draw applicants to their employment opportunity and then retain them with varying success. From 2002 to 2015, more than 300 physicians have been placed in rural or underserved communities in Oregon through the J-1 Visa Waiver or Physician Visa Program (Oregon Primary Care Office). Administered through the state Primary Care Office, this program allows state's up to 35 international graduates per year who complete residencies in the United States to remain here to practice in federally designated shortage areas. Oregon law requires that 80% of the waivers be given to primary care physicians. Physician trainee J-1 visa holders are typically required to return to their home after training for 2 years; however, federal legislation enacted by Congress in 1994 by Senator Conrad of North Dakota and known as the "Conrad 30 Waiver Program" (https://www.uscis.gov/working-united-states/students-and-exchange-visitors/conrad-30-waiver-program), allowed these visa recipients to stay in the United States to practice in an underserved area fulfilling a 3-year employment with a health care facility. In Oregon, more than 88% of physician trainee J-1 visa holders that started their commitment completed their contractual agreements and by all accounts have been a successful program. However, the program faces a number of challenges including the increased national vigilance by immigration authorities, the "fit" of an international physician from different cultures into a rural community, and ability of the physician to match their clinical skills to the needs of the health care community.

Critical to a successful recruitment and retention plan is the availability of both loan forgiveness and loan repayment programs. These programs increase the likelihood that a provider will practice in needy areas, with uninsured populations, and more likely to continue practicing in an underserved area after their obligation is completed. Several programs are available in Oregon both to the physician community and multiple other clinician disciplines, including nursing, behavioral health consultants, physician assistants, and dentists. Oregon spends approximately $30 million a year for incentive programs. Table 14.1 outlines the state and federal funding sources for Oregon students, resident trainees, and practicing clinicians (Health Care Workforce Committee, Oregon Health Authority, 2014).

Each state in the country has its own unique approach to providing financial support to its clinicians and Oregon has a diverse array of programs available, both state and federal. However, because these programs were developed independently over time there are continuing concerns that the programs lack a complementary approach to the statewide need. There is no common governance or administration and so responsibility of the programs is spread across a number of state agencies. Federal programs are managed separately and so a recipient may be held accountable to a number of different program rules, eligibility criteria, reporting mechanisms, and all applied independently. Due to lack of data to measure the effectiveness, such as postprogram retention and impact on local workforce capacity, it is difficult to know how effective these programs meet the needs of the clinicians and the state health care workforce. Although important to have a diverse menu

**TABLE 14.1 Oregon Financial Incentive Programs to Support Clinician Recruitment and Retention[a]**

Programs in Oregon using *state* funding include:

- Rural Medical Practitioners Insurance Subsidy Program
- Medicaid Primary Care Loan Repayment Program
- Scholars for a Healthy Oregon Program (Loan Forgiveness)
- Oregon State Loan Forgiveness Program
- Rural Practitioner Tax Credit
- Emergency Medical Technician (EMT) Tax Credit

Programs in Oregon using *federal* funding include:

- Oregon State Partnership Loan Repayment Program (SLRP[b]; HRSA[c])
- National Health Service Corps (NHSC) Loan Repayment
- National Health Service Corps (NHSC) Scholarship Program
- Nurse Corps Education Loan Repayment Program (NELRP)
- Federal Faculty Loan Repayment Program

[a]*Health Care Workforce Committee, Oregon Health Authority (2014).*
[b]*State Loan Repayment Program administered under the state Office of Rural Health.*
[c]*Health Resources & Services Administration grant funding to state.*

of programs available, it is even more important that the greatest number of providers has access to them and actually take advantage of their availability. By most accounts, these programs have been successful. However, to address the previous concerns, state legislative House Bill 3396 (https://olis.leg.state.or.us/liz/2015R1/Downloads/MeasureDocument/HB3396/Introduced) was enacted in 2015 and calls for a report to the Oregon Health Policy Board (OHPB) to examine whether current spending is the best use of state tax dollars and to make recommendations to strengthen and coordinate the programs. Specifically, the report will focus on (1) financial assistance for students in both public and private institutions; (2) loans, grants, other financial incentives to hospitals and health systems to expand GME training sites; (3) subsidies to providers practicing in rural/underserved areas; (4) creation of retirement plans for rural clinicians; (5) funding for Type A, B, C hospitals at risk for closure; and (6) development of tax credit criteria for clinicians in rural/underserved communities.

Finally, there are many employer-funded and professional-specific financial grants, many of which are private, that support recruitment and retention of providers within their organization and capturing these programs could help align efforts with the state strategy. Although financial and grant opportunities are a key component to retaining clinicians, there are a number of nonfinancial incentives that can be equally as important. These include the practice and community environment, work autonomy, schedule flexibility, career development opportunities, and educational and career opportunities for children and spouses. Most comprehensive strategies take all of these components into consideration and are critical to addressing the clinician availability and vitality of the workforce.

## COMPETENCY TRAINING FOR A NEW SYSTEM OF CARE

In 2011 the OHPB asked the Oregon Healthcare Workforce Committee to assess the statewide core competency needs for health reform and identify workforce models and strategies to best support the development of those competencies. The Healthcare Workforce Committee, established in 2010 by House Bill 2009, Section (7)(a), reports directly to the OHPB and is directed to advise the Board on workforce needs to recruit, retain, and train for the new system of care. Reviewing existing literature and recommendations from national convenings (Institute of Medicine, 2001, 2003; Interprofessional Education Collaborative Expert Panel, 2011), the committee then interviewed key state and national health care professionals, administrators, and policy experts. It became clear to the committee that it strongly endorse a workforce training strategy that embraced interprofessional, team-based care particularly within the context of the emerging development of the Patient-Centered Primary Care Home (PCPCH) and locally, accountable Medicaid-funded Coordinated Care Organizations (CCOs). The report identified the key competencies to include individual skills in a collaborative practice, health information technology, and communication, as well as an organization and system-level responsibility to support this model of care, flexible reimbursement methods, operational and managerial supports, and community engagement. Although identification of the competency needs is necessary, it is insufficient to affect change. The committee supported action-oriented recommendations for team-based, interprofessional care in three distinct areas: policy, education, and practice (Table 14.2) (Oregon Healthcare Workforce Committee, 2011). Using this framework as a means to foster interprofessional, team-based care, Oregon has developed and is moving forward in developing new approaches and programs that support the model of care. Although this has been described and segmented into these three areas of focus, there is considerable overlap among them, thus generating a need for a comprehensive holistic approach in order to attain a successful competency-based workforce.

**TABLE 14.2 Oregon Healthcare Workforce Committee Recommendations for Building a Workforce for New Systems of Care, 2011[a]**

Policy recommendations
- Establish and expand pilot programs to test alternative payment models
- Develop job descriptions for new positions such as care coordinators, navigators, and community health workers
- Provide opportunities for multipayor alignment around promising alternative models of reimbursement
- Revise job descriptions for existing categories of health care workers to reflect the nature of interprofessional, team-based care

Education recommendations
- Set expectations for collaboration between education communities and health care employers
- Collaborate across disciplinary boundaries to develop and implement the same set of interprofessional competencies
- Develop shared methods for training and assessment of interprofessional competencies
- Provide opportunities for faculty to gain experience with interprofessional practice and new models of care
- Increase opportunities for interprofessional training, especially in clinical settings

Practice recommendations
- Foster a collaborative, egalitarian workplace culture to assure the successful implementation of team-based care in existing practices
- Identify successful early adopters of team-based care models to assist practices with transition
- Prioritize investment in information technology infrastructure
- Revise hiring and human resources practices to enable recruitment, retention, and evaluation of professionals engaged in interprofessional and team-based care

[a]*Oregon Healthcare Workforce Committee (2011).*

## Policy

Advocating a strong belief that how we pay for health care services influences practice behavior, a number of alternative payment methodologies (APM) were initiated and supported by the state to move away from fee-for-service and toward value-based payment. For example, these projects included redesign of practices to address performance quality measures—some of which were incentivized, risk-adjusted capitation for certain populations, and "per member per month" payments to practices that integrate physical and behavioral health care thus creating new roles and services. For many practices and institutions, this has required new workforce personnel and team-oriented practice workflows. Although many of these projects were supported through the CMMI Statewide Innovation Model (SIM) grant funding and Coordinated Care Organization (CCO) investment, legislative infusion of general funds was also used to encourage innovation to support several projects within the CCO to prepare for, and initiate, alternative payment models (Transformation Funds: http://www.oregon.gov/OHA/HPA/CSI-TC/Pages/Transformation-Funds.aspx). Additional opportunities to support multipayer alignment around promising APMs were supported through state Medicaid participation in the CMMI Comprehensive Primary Care initiative (CPCi), requiring some of the elements of the care model in the state employees' health insurance program. More recently, Oregon became 1 of 14 regions chosen to participate in the national CPC Plus primary care payment and care delivery redesign program thus further aligning with state objectives. In 2016, state legislation (Senate Bill 231) established a multistakeholder learning collaborative to engage in a conversation addressing payment methodology for primary care (http://www.oregon.gov/OHA/HPA/CSI-TC/Pages/SB231-Primary-Care-Payment-Reform-Collaborative.aspx) with the intent of creating a practice milieu in which payment methods drive behavior change, thus creating new relationships and roles within and outside of the practice setting.

Development of teams requires clarity of team roles, thus new job descriptions were developed for traditional and nontraditional health care workers, and behavioral health providers in primary care in order to maintain some degree of quality standards to support documentation and reimbursement for roles not typically supported in the practice team. Although work is needed to further refine these roles, there is much to be learned from national consensus-oriented manuscripts to provide a framework for these policy and training discussions. An example in Oregon that has assisted policymakers is the framework that describes behavioral health integration through the AHRQ Behavioral Health and Primary Care Integration Lexicon (https://integrationacademy.ahrq.gov/sites/default/files/Lexicon.pdf). This national expert consensus-driven document allowed scope of practice policy discussions to move forward once there was agreement on standard, agreed-upon definitions.

## Education

Engagement of our states' higher education community has been an important component to prepare students for a profession in the new health care delivery environment. For example, Oregon Health & Science University (OHSU) School of Medicine has implemented a major revision that moves toward a competency-based curriculum and evaluation for undergraduate and GME. One component of the revision is the OHSU Interprofessional Education (IPE) curriculum initiative (http://www.ohsu.edu/xd/education/student-services/about-us/provost/interprofessional-educatio-ipe.cfm) that creates a collaborative training environment between nursing, medical, pharmacy, and allied health professions (OSHU, 2016). Core to accomplishing this vision is a newly built OHSU Collaborative Life Sciences Building that provides learning spaces to support team-based learning across multiple disciplines and academic institutions including an interprofessional clinical simulation laboratory and surgical training center. Students from multiple disciplines are also required to participate together in a year-long case-based active learning course entitled "Foundations of Patient Safety and Interprofessional Practice." More than 2000 students and 250 faculty representing all of OHSU health care professional schools have participated thus far. Outside of the states' academic medical center, other state higher education institutions have received support to advance training for their health profession students. At the Portland State University, Clinical Social Work Department, a HRSA/SAMHSA grant is being used to support the "Integrated Care Project" to equip a new generation of social workers to gain skills in care management and provide behavioral and mental health support in a team-based practice environment for children, adolescents, and transition age young adults. These 30 social work students apply for admission to the program, receive a stipend and engage in both seminars and advanced placement into the clinical practice setting (https://www.pdx.edu/ssw/integrated-care-project). Multiple other health profession institutions have developed programs to train students to be Behavioral Health Consultants in primary care. These include Doctor of Clinical Psychology programs with an emphasis on health psychology and practicum placements in primary care practices at George Fox University and Pacific University-Oregon (http://www.pacificu.edu/future-graduate-professional/colleges/college-health-professions/areas-study/psychology-clinical-psychology-psyd/tracks/health-psychology-track).

## Practice

Fostering a healthy, collaborative workforce culture requires clinical leadership prepared to lead a multidisciplinary staff through the journey of clinical delivery transformation. Supported through State Innovation Model (SIM) grant funding and Transformation Funds stemming from state general funds, a number of programs have been initiated. Two of these programs are illustrated as examples in this chapter: (1) the Clinical Innovation Fellows Program that provides leadership and project management skills training in the context of a multidisciplinary team and (2) the PCPCH Program, a state and SIM-funded "medical home" program that serves the dual role of certification and technical assistance for practice redesign using a practice facilitation and a peer-to-peer learning paradigm, which has also served the purpose of identifying exemplar practices to promote across the state. Over the past 4 years, Oregon has been a participant with six other states/regions in the CMMI CPCi for redesign of practices to address the needs of Medicare beneficiaries. Many of the 67 primary care practices in the CPCi have been actively involved in the Medicaid CCO development and all are part of the PCPCH Program, creating considerable synergy between programs yet also highlighting the challenges of implementing multiple funded programs that are not synchronized. Many of the projects supported by state general fund "Transformation Fund" grants to all CCOs led to innovative clinical practice redesign and training, such as complex patient care, care management supports, IT system support and redesign, quality measure data management, community outreach from practices, and use of traditional health workers as team members (http://www.oregon.gov/OHA/HPA/CSI-TC/Pages/Transformation-Funds.asp). Through facilitation by the OHA Transformation Center, learning collaboratives and the intentional organization of forums were initiated to support sharing and peer-to-peer learning of leaders and staff (Transformation Center, Oregon Health Authority: http://www.oregon.gov/oha/HPA/CSI-TC/Pages/index.aspx).

## PROMISING MODELS OF TRAINING

Three workforce training interventions that have emerged as catalysts to implementation of health reform in Oregon are described in the following section of this chapter: (1) the Clinical Innovation Fellows Program, (2) the Traditional Health Workers (THWs) Program, and (3) the PCPCH Program. In the first example, early success of Oregon's Clinical Innovation Fellows program has demonstrated the value of investing in local clinician leadership development. Clinicians working to reform delivery systems at the practice level face increasing demands, including voluminous quality measures, integration of behavioral and physical health care, payment reform, and change fatigue. Leadership development programs that provide local clinician leaders with vital resources and support to lead transformation are critical to the success and sustainability of health system transformation.

The second example will demonstrate that although community health workers have existed for decades, mostly in an informal capacity within the health care system and more formally internationally, there are few examples of a structured approach to standardize the role, document competencies, and certify the discipline. The THWs Program example will describe Oregon's approach and process to legitimize a community health role that is integrated into, and supportive of, a reformed health care system.

Finally, establishing a primary care infrastructure requires a clear understanding of providing primary health care within the community that is responsive to patient and family needs. Core to that system is the primary care home providing a prepared, proactive team approach to the delivery of care both within the practice and across the medical neighborhood. In this last example, a primary care practice team intervention that combines state certification of the medical home and supports for practice redesign and quality improvement is described that focuses on workforce development at the clinical practice delivery level.

## The Clinical Innovation Fellows Program

**Emilee Coulter-Thompson**
*University of Michigan, Ann Arbor, MI, United States*

The Oregon Clinical Innovation Fellows program is a statewide, multidisciplinary leadership development program in which clinical innovators actively work with teams to implement health system transformation projects in their local communities. Through participation in a year-long learning experience and mentorship, a cohort of 12–15 Clinical Innovation Fellows develop and refine skills in leadership, quality improvement, implementation, and dissemination science. The program's goal is to create a network of clinical leaders supporting the Oregon coordinated care model and State health system transformation efforts.

The program was created by the OHA Transformation Center, a hub of innovation and learning within state government. When the Transformation Center launched in 2013, it convened a series of learning collaboratives to develop peer learning networks and promote the rapid sharing of promising practices among key CCO stakeholders driving Medicaid transformation. To complement these larger, payer-level convenings, the Center invested in innovation projects and leadership development for individual clinicians, who could be potential change champions testing and implementing innovations at the delivery system level. The idea to form a Clinical Innovation Fellows program was inspired by the OHA leadership's early discussions with Dr. Don Berwick, founder of the Centers for Medicare and Medicaid Services (CMS) Innovation Advisors Program (https://innovation.cms.gov/initiatives/Innovation-Advisors-Program/), which Dr. Berwick created in 2011. The Transformation Center conducted an environmental scan of leadership programs, including the Practice Change Leaders for Aging and Health program at the University of

Colorado, which is jointly supported by the Atlantic Philanthropies and the John A. Hartford Foundation, and interviews with executives from the Institute for Healthcare Improvement. With funding from the Centers for Medicare and Medicaid Innovation (CMMI) State Innovation Model (SIM) grant, the program recruited four, paid faculty and established $15,000 annual grants to support the selected fellows' innovation projects and program participation.

### Fellows and Innovation Projects

In the first two years, 58 people applied to the program by written application, including letters of support from CCOs and organizational sponsors. A committee of CCO, government, and faculty representatives selected 29 midcareer professionals, and 28 graduated the program (13 in cohort 1, 15 in cohort 2). The selected fellows represented Oregon's geography and a broad array of clinical disciplines: seven physicians, six social workers, three nurses, three social scientists, two expanded practice dental hygienists, two public health professionals, and one dietician, pharmacist, physical therapist, physician's assistant, and psychologist. Innovation projects varied widely, from the development of a therapeutic garden with HIV positive residents, the integration of behavioral health services into primary care, screening for adverse child events in practice, development of tele-dermatology using miniipads, to improving care transitions from the hospital to outpatient settings and expanding access to oral health services with tele-dentistry.

### Faculty

The faculty each had years of professional experience, had completed prior leadership training, and demonstrated a strong commitment to leadership development and mentorship. The faculty included three physicians, two social workers, and their expertise spanned family medicine, geriatric medicine, behavioral health, public health, and social science. Faculty were selected based on a generalist orientation, previous training in quality improvement and dissemination methodology, and strong understanding of prevention models. All faculty shared a common vision of the need to better prepare clinicians to lead and transform health systems of the future.

### Curriculum

The faculty developed the curriculum based on results of preliminary skills assessments of the fellows in each cohort, from faculty knowledge and expertise in health system transformation and leadership, and periodic surveys of the fellows to assess their interests and feedback. Popular topics included strategic leadership, project management, measurement, health equity, and spokesperson training (see Table 14.3). The program held 12 sessions each year, including four day-long, in-person meetings; monthly online

**TABLE 14.3 Clinical Innovation Fellows Program Overview of Curriculum Topics and Activities**

| Format | Topics and Meeting Activities |
|---|---|
| Face-to-face meetings | • Opening meeting: program introductions, networking activities<br>• Leadership development<br>• Project presentations and sharing activities<br>• Panel of Clinical Innovation Fellow graduates<br>• Complex adaptive systems<br>• Envisioning the ideal health care system<br>• The Alaska Southcentral Foundation Nuka System of Care<br>• Developing a project charter<br>• Health equity<br>• Public narrative: effective engagement through storytelling<br>• Poster presentation practice session activity<br>• Working in teams<br>• Policy advocacy<br>• Spokesperson/media training<br>• Reflecting on accomplishments activity |
| Online meetings | • Introduction to CCOs<br>• Quality improvement measurement<br>• Community health: CCO Community Advisory Councils<br>• Organizational trauma and resilience<br>• Project management<br>• Leadership development<br>• Alternative payment methods<br>• Health literacy<br>• Business planning |
| Assignments | • Project charters<br>• Interview a CCO leader<br>• Midyear and final progress reports<br>• Blog posts on the group's online networking site<br>• Project posters to present at a statewide conference<br>• Group presentations at the final meeting |
| Mentorship | • Mentor/fellow 1:1 meetings<br>• Mentor group meetings<br>• Faculty visits to fellows' work sites |

meetings; and a gathering and poster session at Oregon's annual Coordinated Care Model Summits. Sessions were presented by the faculty and local and national experts. Fellows met monthly with their faculty mentors, both one-on-one and in small groups. In year two, the program offered Continuing Medical Education credits.

## *Lessons Learned*

Lessons learned are grouped into three areas: (1) the fellows' experience, (2) reflections on program effectiveness, and (3) impact on Oregon's health system transformation.

### The Fellows' Experience

At the end of their fellowship year, all 28 fellows rated the CCI program as valuable in their growth as a leader and in supporting their work. Multiple fellows described their experience as life-changing. A qualitative analysis of midyear and final progress reports showed several common themes. Fellows identified gaining several valuable skills from the program, including messaging, delivering presentations, networking, leadership, project management, and program development skills. Faculty mentorship was the highest rated program component. The mentors served as sounding boards for discussing new ideas; they provided encouragement, especially when support was lacking within a fellow's organization; they offered feedback and problem-solving assistance for innovation projects; and they connected fellows with other fellows and faculty.

While the fellows were very diverse with respect to discipline and project topic, the fellows reported similar challenges. Fellows struggled to balance the project with existing work responsibilities and competing priorities. Some attributed this to a lack of organizational support; others underestimated the amount of time required to implement the project. According to one fellow, "I have been organizing the whole program in addition to my 30 + hours/week of clinical care and it can be a bit overwhelming."

Leadership development was especially important for some fellows to begin to identify and develop their own sense of power as leaders. One fellow, a dental hygienist, reported that she did not see herself as a leader before participating in the program. Then throughout the program year, she began to realize her own expertise and potential to spearhead innovative solutions within her community. Another fellow described how in a large health system, it was very rare for a social worker in a case manager role to have the opportunity to manage a large, complex project. In essence, the program helped some fellows to challenge and transcend discipline-related barriers in hierarchal systems in which dentists, doctors, and administrators historically have held the most power.

### Reflections on Program Effectiveness

The multidisciplinary nature of the program was important for leveling the clinical playing field. No single discipline had a majority in either cohort. The program served as a microcosm of an integrated, team-based health system. The program ran most smoothly when the cohort was in a similar stage of professional development (mid-career professionals with at least five years of professional experience and who were recently emerging as leaders).

Given the diversity of the fellows' professional backgrounds, it was especially important to tailor the curriculum to meet their individual needs. Through an iterative process facilitated by weekly faculty meetings, the faculty continuously refined the curriculum by gathering and responding to data at multiple points in the program—the skills assessment at the start and end of the program, session evaluations and discussions, monthly mentorship conversations, and midyear and final progress reports.

The preliminary skills assessment identified health equity as a priority topic to include in the curriculum for both cohorts. Each cohort indicated that they had little prior knowledge or experience with health equity. However, fellows did not prioritize health equity when ranking curriculum topics. The faculty reviewed and discussed the assessment results with the fellows, then intentionally added a health equity session near the beginning of the program year. The impact was significant. One fellow went on to participate in a year-long health equity leadership training program. Other fellows reported that this session expanded their world view and influenced specific changes in their project plans.

Overall, the innovation project component of the program was important for fellows to test the skills they learned throughout the year. The program provided fellows resources and increased visibility and credibility within their organizations and communities, which enabled their projects to progress more quickly. One fellow indicated, "I feel certain that my project would not have progressed to its current stage without the backing of my [fellowship] experience."

## Impact on Oregon's Health System Transformation

The program had several positive impacts on Oregon's health system transformation efforts. First, the grant funding jump started local innovation projects across the state, and the majority of these projects are being sustained.

Second, the program built a peer network of clinical innovation leaders who supported one another during the ups and downs of advancing change. Fellows discussed how transforming health systems can feel like a risky and lonely endeavor, especially if one is a solo voice within an organization advocating for innovation. The support of like-minded individuals allowed fellows to stretch and take more risks.

Third, the program helped to strengthen the leadership and quality improvement capacity and increase the national visibility and civic engagement of clinical leaders in Oregon. The program provided clinical leaders with a vocabulary and contextual orientation to be effective champions for change and policy advocates in their local communities. It also prepared fellows to advance their careers and assume higher leadership positions. Several fellows were promoted, accepted into higher education programs, or featured in national news media during the fellowship year.

While most health care leadership development programs are national in scope, state-based programs, such as Oregon's Clinical Innovation Fellows program, offer the advantage of leadership training and professional support within a local context. Amidst multiple competing priorities facing Oregon's health care organizations, the Clinical Innovation Fellows program demonstrated the value of investing in workforce and leadership development. Within a short-time period, the state of Oregon made rapid changes to advance health system transformation. Mobilizing clinical leadership was, and will continue to be, essential to effectively sustaining this change.

## Training the Traditional Health Worker in a New Model of Care

**Leann R. Johnson**
*Oregon Health Authority, Portland, OR, United States*

As Oregon embarked upon health care reform, the model for delivery and care was paramount in how Oregon transformed its health delivery system. As a proven intervention in providing more culturally appropriate, specific, and responsive services and care (American Hospital Association, 2013; Betancourt & Green, 2010; Bliss, Meenoo, Ayers, & Lupi, 2016; Fisher, Burnet, Huang, Chin, & Cagney, 2007), the THW model was identified by community-based organizations, health equity advocates, and the state as imperative to the success of health system transformation. THWs integrated into one's care team, working on behalf of the patient, client or member side by side with care clinicians was a vision for the intervention.

In Oregon, communities of color, immigrant, and refugee populations have carried the torch for the THW model. The OHA's Office of Equity and Inclusion (OEI) provided research and guidance navigating public sector culture and requirements. A series of legislative actions and rule making codified the work both supporting and requiring the development and adoption of the THW model in the state of Oregon. In 2011 House Bill 3650 legislation framed the THW model explicitly identifying and establishing criteria, education, and training requirements for four primary types of THWs:

1. *Community Health Workers (CHW)* assist the member, patient, client, or community member to get the health care necessary to support optimal health.
2. *Peer Wellness Specialists (PWS)* provide assistance, support, and encouragement to help individuals address both physical and behavioral health issues.
3. *Peer Support Specialists (PSS)* provide assistance, support, and encouragement to help individuals with addictions and behavioral health concerns.
4. *Personal Health Navigators (NAV)* provide care coordination for members in the health system.

The responsibility to develop and manage the THW program statewide is housed in the OEI, a division of the OHA. The division provides leadership, policy development and implementation, and technical assistance and consultation within the Authority and throughout Oregon's health system to address health inequities and disparities to achieve greater health equity. The division also manages workforce inclusion programs and civil rights related compliance within OHA and for the public interacting with OHA programs and contractors, including CCOs. Once legislation was passed, OEI developed additional standards through the state's required rules-making process.

### *The Traditional Health Worker Training Program*

The OHA OEI certifies the credentials of THWs who are then placed on a THW Registry and approves THW training programs throughout the state. The registry is available on the OHA OEI website where CCOs and health care providers can search for THWs available in their service areas or regions (http://www.oregon.gov/OHA/OEI/Pages/Traditional-Health-Worker-Program.aspx). The individuals on the registry are certified for 3 years and subject to additional training and 20 hours of continuing education in each 3-year period. As of October 2016, more than 1000 traditional health workers have been trained and certified by the THW program, exceeding the goal of 300 set by the CMS during the CMS Waiver demonstration approval period from July 2012 to June 2017.

The THW program includes one full-time coordinator and one full-time administrative specialist. The role of the administrative specialist is to review THW applications for compliance with the required standards, submit materials for the mandatory criminal background check and enter the approved workers into the registry. The coordinator manages the overall program including the THW Commission, established through House Bill 3407 and a cornerstone of the THW program. The primary role of the THW Commission is to make recommendations to the OHA and advise the Authority to ensure that the program is responsive to the health needs of consumers and diverse communities. The Commission, by design, is comprised of representatives/stakeholders from all categories of THWs, health care providers, local and state governmental agencies, community advocates/community-based organizations, CCOs, and members from organizations that manage THW training programs including higher education.

The Commission has three subcommittees, which involve additional members from the aforementioned areas. The subcommittees are Scope of Practice, THW Systems Integration, and Training Evaluation Metrics and Program Scoring:

1. The Scope of Practice subcommittee develops and monitors the roles, expectations, and supervisory relationship of THWs related to the care team.

2. The THW Systems Integration subcommittee works to integrate THWs into the larger health system, which includes assessing and addressing barriers, opportunities, and improving access for traditionally underserved populations.
3. The Training Evaluation Metrics and Program Scoring (TEMPS) subcommittee reviews THW training programs and developed the metrics and standards to guide the review process and the approval of training programs and continuing education requirements for THWs.

### *Implementing Traditional Health Worker Training Programs*

Successful implementation of the THW program has been contingent upon the appropriate review and certification of programs that train THWs. The Training Evaluation Metrics and Program Scoring (TEMPS) subcommittee is central to this process. As of October 2016, the OHA via the TEMPS subcommittee of the THW Commission has approved 45 training programs in the state of Oregon. There is a cost to individuals who wish to become certified by one of these programs; however, there is no additional fee charged by OHA to register individuals wishing to be certified by the state once they pass training program requirements.

An approved certification program may charge anywhere from approximately $560–2000 for the training required by the THW program. To be certified the CHWs, Peer Wellness Specialists and Peer Health Navigators must complete at least 80 hours of training. The Peer Support Specialist must complete at least 40 hours of training. Doulas require fewer training hours, approximately 40. The programs must adhere to Oregon Administrative Rules, OAR 410-180-0300 through 410-180-0380 (http://arcweb.sos.state.or.us/pages/rules/oars_400/oar_410/410_180.html).

Programs must also demonstrate that they actively collaborate with at least one community-based organization and involve experienced THWs in developing and teaching multiple aspects of the program. Training formats are to be accessible, inclusive, and use diverse teaching methodologies to consider learning style, culture, and language differences. Curriculum, as well, must consider said differences yet maintain standardized curricular content. Flexible enrollment requirements must accommodate disproportionate degree attainment for culturally diverse populations. In addition, the program must assess for THW competencies, provide a feedback mechanism for trainees for continuous improvement purposes, and provide records of attendance and completion.

Because the use of THWs preceded legislation and the state's THW program, provisions were made to "grandparent in" individuals with "on the job" experience. In other words, individuals who could demonstrate 2000–3000 hours of work or volunteer experience as a THW can be certified by one of the OHA-certified training programs. The process requires precourse assessment, evaluation of competencies, and incumbent worker training. Another option is provisional certification where the THW can complete any 40-hour or more THW training program even if it is not certified by the OHA process. The provisional THW then has 1 year to complete the remaining required training via an OHA-certified training program.

Trainings are offered by organizations and institutions ranging from community-based organizations, not-for-profits and county health departments to state community colleges. One example of a training program is housed at the Central Oregon Community College (COCC) in Bend, Oregon. The college's CHW program prepares CHWs to "facilitate access to a wide range of services through advocacy, outreach, referral, community education, informal mentoring, and social support." The program requires that trainees achieve the following objectives (https://www.cocc.edu/continuinged/community-health-worker/):

1. Students will be able to demonstrate effective listening skills and interview techniques with clients and community practitioners to assist individual and families to engage in healthy behaviors.
2. Students will be able to identify and effectively bridge cultural, linguistic, geographic, and structural differences, which are barriers to individual's capacity to access health care and/or promote and adapt healthy behaviors.
3. Students will be able to apply a holistic approach to facilitate system-level changes that promote and support health needs, which reduce duplication, unnecessary, or harmful interventions that result in costly ineffective services.
4. Students will identify and assist individuals and families in making desired behavioral changes to acquire behaviors and practices that promote positive health outcomes, which are collaboratively and mutually agreed upon.
5. Students will learn the elements of community organizing; community needs assessment, as well as group facilitation skills and popular education methods to promote self-efficacy and empowerment for individuals and families addressing health care issues.

The COCC program requires a high-school diploma or Graduate Educational Development (GED) test, good communication skills, and the ability to be self-directed with an interest in the promotion of healthy behaviors. This program is comprised of five online learning modules, which amount to 50 hours of the training online and 33 hours of classroom instruction.

### *Successes and Challenges: Integrating the Traditional Health Worker Into Health Systems*

While the certification process for THW has experienced a backlog primarily due to the lack of adequate staffing for the program, the greatest challenge regarding THWs is the actual utilization of the workers in the health system. THWs bring great benefit to the system and the people involved. The workers are often trusted community members and leaders, are often trained in Chronic Disease Management, and can provide ongoing support and navigation for health care consumers, especially individuals from culturally diverse and traditionally disadvantaged populations. From the perspective of the CCO, the THW model provides an avenue, particularly in partnership with community-based organizations, to develop specific interventions and address specific health disparities among populations that reside within their service area.

While the THW structure works and the mutual benefits are clear, one of the greatest challenges seems to be attitudinal. Whether it be a lack of understanding of the role of the THW or resistance to incorporating a nonclinician into a care team, in retrospect key players in the health system such as providers, could have been better educated in terms of how to integrate and optimize THWs into the system and care teams. That said, Oregon is recognized nationally for its development and implementation of its THW Program. Most recently, CMS tapped Oregon, Arizona, Minnesota, Rhode Island, and New Mexico as model programs and is working with these five states to better understand how the THW model can be adopted by and developed in other states.

## Patient-Centered Primary Care Home Program—A Strategy to Promote On-the-Ground Primary Care Transformation

**Evan Saulino**
*Oregon Health Authority, and Providence Health and Services, Portland, OR, United States*

As a key part of Oregon's broader health transformation efforts, the Oregon Legislature established the PCPCH Program in 2009 through passage of House Bill 2009. In 2010, the OHPB charged the OHA with providing access to patient-centered primary care, and set an aggressive goal that 75% of all Oregonians would get care in a PCPCH by 2015. Building on the best of traditional primary care as evidenced through the research of Barbara Starfield (Starfield, Leiyu Shi, & Macinko, 2005) and others (Nutting et al., 2011), Oregon's advanced primary care "medical home" model was defined through work with diverse groups of Oregonians from across the state, the PCPCH Standards Advisory Committee, and utilizing best available evidence. The Program recognized the first Oregon PCPCH in October 2011. By 2016 over 600 clinics were recognized as PCPCHs, providing care for three out of four Oregonians (nearly 3 million lives)—reaching the aspirational goal set out 5 years prior. The third iteration of the model launched in 2017.

The rapid and widespread engagement of nearly three-fourths of all Oregon primary care clinics in the voluntary PCPCH Program has provided valuable opportunities to learn from the diverse rapidly changing health care environment, to identify and spread best practices, and to address common barriers to transformation. In 2014, guided by a SIM funded evaluation of PCPCH policy implementation, clinic experience (Rissi et al., 2015), and on-the-ground model implementation (Saulino, 2013), the PCPCH verification site visit was restructured to better support transformation efforts. A primary goal was to utilize the verification process important for Program integrity, to provide clinics an opportunity for reflection and assessment, to support clinic redesign, and to foster connections with other practices. These engaged clinics could then use the process as a springboard in their "medical home" transformation. The site-visit structure and process was also designed to address the fact that although clinics were often able to implement specific components of the PCPCH model, the restructuring of team roles, workflows, technology, and clinic culture was much more difficult. At the same time, no matter how advanced the clinics were in their transformation work, they expressed the need for additional financial and infrastructure support. They routinely requested direct technical assistance and connections with "someone like us" to "avoid reinventing the wheel."

New PCPCH Program site-visit staff were hired by the OHA, chosen for their diverse clinical care, managerial, Quality Improvement (QI), and health policy experience. They were provided Practice Coach training, and received "on-boarding" through direct PCPCH clinician and staff interaction. To promote the spread of best practices, primary care clinicians referred to as Clinical Transformation Consultants (CTCs) were recruited from "high performing" PCPCHs around the state to act as consultants and clinician peer–mentors for the clinics being visited. The Practice Coach and CTC peer–mentor clinicians were also made available free of charge to clinics for up to 6 months following the visit to work on anything related to practice support or the PCPCH model of care. This assistance included, but was not limited to, helping the clinic successfully complete a 90-day Improvement Plan to implement PCPCH care standards that were not verified at a site visit. Due to the demonstrated importance of clinician/operations dyads in successful PCPCH clinic transformation, CTCs were paired with clinic operations leaders (often the CTC's own clinic manager) to provide a second potential peer–mentor for postsite-visit technical assistance. Follow-up assistance provided by the Practice Coach and/or CTC/operations leader was based upon the clinic's request. At the same time, the scope of anticipated need for assistance was estimated through the PCPCH Team's postvisit use of a simple scale built on core components of more complex "medical home indices" such as the PCMH-Assessment tool.

While the PCPCH Practice Coach and Compliance-focused team members had the job of verification primarily, and technical assistance secondarily, the CTCs were specifically not involved with verification of PCPCH measures at site visits. The CTCs were trained to focus on making the connection between PCPCH Program intent and on-the-ground implementation given the clinic's population and organizational realities. They were advised to avoid taking an "expert" approach; instead they connected with clinics as a fellow learner, asking open-ended questions, and providing perspective to nudge the clinic further along their improvement journey.

In order to maximize peer–mentor relationships, the initial cohort of 15 CTCs included actively practicing primary care physicians representing a variety of geographies, populations, and practice types across Oregon. CTC physicians were contracted to work with 3–6 clinics annually. They were supported with federal SIM grant dollars, through a deliverables-driven contract at a rate approximately equivalent to market hourly wages as a primary care physician. The CTCs received webinar training, and a "training site visit" where they worked 1:1 with an experienced CTC during an actual clinic site visit, and completing a postsite visit "CTC Report" for the clinic. The "CTC Report" routinely included links to resources to help the clinic meet identified gaps and goals, as well as structured assessments of the clinic's strengths, challenges, and opportunities in PCPCH transformation.

Nearly 95% of clinic respondents (73/77) identified in postvisit surveys indicated that they wanted follow-up assistance. The PCPCH site-visit process has been effective in promoting frontline implementation of PCPCH care practices, as 92% of clinics requiring an Improvement Plan were able to implement the structures and activities needed to meet the intent of the unverified PCPCH standards. Clinics also received assistance to learn about and implement changes that supported primary care workforce and team function (e.g., "scribes") but that were beyond the official scope of the PCPCH model. These experiences demonstrate the ability of clinics to transform their model of care when offered assistance to do so. They also show the potential effectiveness of an improvement strategy to push rapid implementation of PCPCH care innovations compared to typical regulatory, compliance-only verification strategies that provide a "report card," but not an opportunity or pathway for improvement.

## *Lessons Learned*

### Reflections on Program Effectiveness

The innovations in PCPCH site visits have provided a "proof of concept" that pairing compliance with technical assistance roles and incorporating peer–mentor clinicians into a verification process that clinics find valuable and meets Program integrity needs can be effective. Although they played distinct roles for verification purposes, the PCPCH Practice Coach, and Compliance-focused team members interacted significantly with all levels of staff, clinicians, and patients throughout the site-visit process. They found this presented a multitude of potential learning opportunities before, during, and after the visits. Despite the fact that it remained common for clinics to have unverified PCPCH standards during the site visit, triggering the Improvement Plan compliance process, clinics consistently provided positive ratings and comments in anonymous postsite visit surveys.

Uptake of postsite visit assistance was variable, and depended in part on the existing assistance or support infrastructure that clinics could access. Uptake also differed based upon clinic culture and engagement with the PCPCH model intent and site-visit process overall. Small- and medium-sized independent clinics and small organizations (2–5 clinics) were more likely to engage with postsite-visit assistance, but there were instances of large health system clinics/leadership taking advantage of postsite-visit assistance as well.

Some examples of postsite-visit follow-up activities facilitated by Practice Coaches and CTCs included:

- *Direct assistance to address unmet patient population needs* identified at the site visit, such as work to integrate Behavioral Health in the PCPCH.
- *Follow-up visits* from the PCPCH Practice Coach to help the clinic work through issues specific to implementing PCPCH components and structural/cultural transformation.
- *"Mentor clinic" visits* by the clinic leadership/staff to the CTC's practice to learn from peers.

### Experience of Clinics Receiving Site Visits

Although they were experiencing stress in a rapidly changing environment, clinics reported that PCPCH Program verification site visits supported their vision for improvement focused on making care better for patients. Clinics who at site visits universally expressed change fatigue, workforce challenges, and lack of resources to support innovation, routinely also provided anonymous postvisit feedback like, "The best site visit by the state I have ever experienced in my professional career. The support from the OHA site visitors was game changing."

### Experience of Peer–Mentor Clinical Transformation Consultant Clinicians

The peer–mentor CTC clinicians also reported positive experiences, and frequently unexpected benefits from participation, as they learned important things from the clinics they visited that helped them with their own clinic's transformation journey. Although CTCs were often apprehensive about their ability to assess and mentor other clinics at first, they ended up valuing the work. However, for clinics needing a high level of support, the CTCs sometimes felt that they did not have access to the necessary resources, and conversely could become frustrated when clinics did not engage. Some struggled to balance the additional work with busy clinical practice demands. The CTCs consistently reflected that they would like more structure to support their work—for social networking, and to better share learning. Only one CTC stopped the work because of dissatisfaction with the site-visit process while one other declined to continue visits in protest over lagging primary care payment reform efforts.

### Impact on Oregon's Health System Transformation

Overall, Oregon's PCPCH Program implementation has provided a foundation for the state's transformation efforts. Robust evaluation data (Gelmon, Wallace, Sandberg, Petchel, & Bouranis, 2016), demonstrating positive impacts on care quality, utilization, and cost across Oregon's population, provide convincing evidence that the PCPCH Program and clinical model of care are effective. The quantitative evaluation data show:

- An estimated total savings to the Oregon health care system of $240 million over 3 years (2012–14) due to the PCPCH program implementation.
- Clinics mature in this model of care. The longer a clinic is certified as a PCPCH, the greater the cost savings per patient—savings double over 3 years.
- As spending in PCPCHs increases, spending for specialty care, emergency department, and inpatient care decrease.

Designed to promote effective and sustainable transformation within the stressed primary care system, innovations in PCPCH site-visit function and structure have demonstrated ways to assist clinics in robust implementation of the PCPCH model. The site-visit innovations have promoted the spread of best practices, helped clinics identify and overcome barriers, and supported the development of peer learning networks.

While it is considered unusual to pair verification and assistance roles, and to include outside "experts" in a compliance process, the openness of the PCPCH Program to learn from those it serves has unlocked potential value in clinic and clinician engagement to effectively promote on-the-ground transformation across the diverse geographic, clinical, and population environments in Oregon. In cooperation with community partners such as the Patient-Centered Primary Care Institute, the PCPCH program is exploring strategies to expand this model into an extension network (Mold, 2016) supported through a shared funding mechanism to sustain and further Oregon's health care transformation efforts.

## REFERENCES

American Hospital Association. (2013, June). *Becoming a culturally competent health care organization*. Equity of care. Institute for Diversity in Health Management; Health Research & Educational Trust.

Association of American Medical Colleges. (2010, June). *Physician shortages to worsen without increases in residency training*. Available from: https://www.aamc.org/download/150584/data/physician_shortages_factsheet.pdf.

Betancourt, J. R., & Green, A. R. (2010). Commentary: Linking cultural competence training to improved health outcomes: Perspectives from the field. *Academic Medicine*, *85*(4), 583–585.

Bliss, D., Meenoo, M., Ayers, J., & Lupi, M. V. (2016). Cross-sectoral collaboration: The State Health Official's role in elevating and promoting health equity in all policies in Minnesota. *Journal of Public Health Management & Practice*, *22*, S87–S93.

Fisher, T. L., Burnet, D. L., Huang, E. S., Chin, M. H., & Cagney, K. A. (2007). Cultural leverage: Interventions in using culture to narrow racial disparities in health care. *Medical Care Research and Review*, *64*(5 Suppl.), 243S–282S.

Gelmon, S., Wallace, N., Sandberg, B., Petchel, S., & Bouranis, N. (2016). *Implementation of Oregon's PCPCH Program: Exemplary practice and program findings*. Available from: http://www.oregon.gov/OHA/HPA/CSI-PCPCH/Documents/PCPCH-Program-Implementation-Report-Final-Sept-2016.pdf.

Health Care Workforce Committee, Oregon Health Authority. (2014, July 1). *Expanding graduate medical education: Policy options memo for the Oregon Health Policy Board*. Available from: http://www.oregon.gov/OHA/HPA/HP-HCW/Documents/GME%20Full%20Final%20Report%207-1-14.pdf.

Health Care Workforce Committee, Oregon Health Authority. (2014, July 1). *Financial incentive programs in Oregon: A report to the Oregon Health Policy Board*. Available from: http://www.oregon.gov/OHA/HPA/HP-HCW/Documents/Provider%20Incentives%20Full%20Final%20Report%20July%202014.pdf.

Institute of Medicine. (2001). *Crossing the quality chasm: A New Health System for the 21st century*. Institute of Medicine Committee on Quality Health Care in America: National Academies Press,Washington, DC.

Institute of Medicine. (2003). *Health professions education: A bridge to quality*. Institute of Medicine, Committee on the Health Professions Education Summit. National Academies Press, Washington, DC.

Interprofessional Education Collaborative Expert Panel. (2011). *Core competencies for interprofessional collaborative practice: Report of an expert panel*. Washington, DC: Interprofessional Education Collaborative.

Mold, J. (2016). *EvidenceNOW is a test of primary care extension*. Available from: https://escalates.org/story/primarycareextension/.

Nutting, P. A., Crabtree, P. F., Miller, W. L., Stange, K. C., Stewart, E., & Jaen, C. (2011). Transforming physician practices to patient-centered medical homes: Lessons from the national demonstration project. *Health Affairs*, *30*(3), 439–445 . Available from: http://content.healthaffairs.org/content/30/3/439.

OHSU (Oregon Health & Sciences University). (2016). *OHSU interprofessional initiative: 2012 – March 2016 overview report*. Available from http://www.ohsu.edu/xd/education/student-services/about-us/provost/upload/IPI-Annual-Rpt-5-18-16-2.pdf.

Okun, S., Schoenbaum, S., Andrews, D., Chidambaran, P., Chollette, V., Gruman, J., ... Henderson, D. (2014). *Patients and health care teams forging effective partnerships. Discussion Paper.* Washington, DC: Institute of Medicine. Available from: https://nam.edu/perspectives-2014-patients-and-health-care-teams-forging-effective-partnerships/.

Oregon Healthcare Workforce Committee. (2011). *Improving Oregon's health: Recommendations for building a healthcare workforce for new systems of care.* Portland, OR: Oregon Health Policy Board, Oregon Health Authority, Office of Health Policy and Research. Available from: http://www.oregon.gov/oha/OHPR/HCW/Documents/Workforce%20Compentencies%20for%20New%20Systems%20of%20Care.pdf.

Oregon Primary Care Office. *Physician Visa Waiver Program (2002–2015).* Available from: http://www.oregon.gov/OHA/HPA/HP-PCO/Pages/J1.aspx.

Petterson, S. M., Liaw, W. R., Phillips, R. L., Rabin, D. L., Meyers, D. S., & Basemore, A. W. (2012). Projecting US primary care physician workforce needs: 2010–2025. *Annals of Family Medicine*, *10*, 503–509.

Rissi, J. J., Gelmon, S., Saulino, E., Merrithew, N., Baker, R., & Hatcher, P. (2015). Building the foundation for health system transformation: Oregon's patient-centered primary care home program. *Journal of Public Health Management and Practice*, *21*(1), 34–41 . Available from: http://journals.lww.com/jphmp/Abstract/2015/01000/Building_the_Foundation_for_Health_System.8.aspx.

Saulino, E. (2013). *Patient-centered primary care home site visit evaluation.* Available from: http://www.oregon.gov/oha/HPA/CSI-PCPCH/Documents/2013%20PCPCH%20Site%20Visit%20Evaluation.pdf.

Starfield, B., Leiyu Shi, L., & Macinko, J. (2005). Contribution of primary care to health systems and health. *Milbank Quarterly*, *83*(3), 457–502 . Available from: http://onlinelibrary.wiley.com/doi/10.1111/j.1468-0009.2005.00409.x/abstract.

The Projected Demand for Physicians. (2014, February). *Nurse practitioners, and physician assistants in Oregon: 2013–2020.* Prepared by Office for Oregon Health Policy & Research, OHSU Center for Health System Effectiveness and Oregon Healthcare Workforce Institute. Available from: http://www.oregonhwi.org/resources/documents/ProjectionsReportCORRECTEDFINALfor2-4-14.pdf.

Section III

# Future Implications for State and National Health Reform

Chapter 15

# Evaluating Medicaid: Moving Beyond Coverage to Understanding Program Design

**K. John McConnell and Stephanie Renfro**
*Oregon Health & Science University, Portland, OR, United States*

The roots of many of today's prominent payment reform efforts can be traced to academic research. Some of the most famous health policy work, initiated in Dartmouth, repeatedly demonstrated the persistence of substantial "small area variations" in utilization and spending in the Medicare populations, with little apparent connection to quality or patient outcomes (Fisher et al., 2003a,b; Wennberg & Gittelsohn, 1973). The strength of this research was a concrete example of the potential waste in the health care system and was instrumental in developing the "Accountable Care Organization" (ACO) model, one of the most prominent payment reforms established under the Affordable Care Act.

Research on the commercial insurance market, in contrast, has shown that spending differences are less connected to small area variations in utilization, but highly connected to variations in prices or negotiated reimbursement rates. This emphasis has resulted in different approaches to controlling spending. Commercial insurance products have developed narrow networks as a way of negotiating lower per-unit costs of care. More recent innovations suggest that ACO-type arrangements and global budgets are feasible within a commercial population and can lead to slower health care spending, through combinations of changes in utilization and the use of lower cost providers.

In contrast, the evidence on how to address spending in Medicaid is underdeveloped. The lack of attention reflects the distinctive nature of the program. Medicaid is a federal–state partnership, blurring the lines about who is ultimately accountable for program design and spending. Medicaid is substantially more complex than its Medicare or commercial counterparts, consisting of products for four distinct groups: medical care for young adults and children; medical care for disabled individuals; partial, complementary insurance for low-income seniors; and long-term care for all of these

**Health Reform Policy to Practice. DOI: http://dx.doi.org/10.1016/B978-0-12-809827-1.00015-X**

low-income populations. Medicaid is also characterized by heterogeneity across states in terms of the populations covered, the services covered, the use of managed care, and their approaches to mental health, addiction, and dental services.

With the expansions of the Affordable Care Act, the Medicaid insurance program has become the nation's largest health insurer, providing coverage for approximately 70 million people, accounting for about 22% of the total population. These changes have created a greater demand for delivery system and payment models that offer sustainable health care spending growth rates. The need for models that can deliver on this promise is particularly acute when viewed through the lens of the vulnerable populations served by Medicaid. If states respond to higher Medicaid spending by reducing the coverage or benefit generosity of their programs, the consequences are likely to fall on low-income individuals who may have few alternatives for obtaining coverage or medical care. Thus, the call for policymakers and researchers is to develop and evaluate Medicaid payment and delivery system models that can prove (and improve) the value of the program.

The growing interest in new models for Medicaid helps explain the attention garnered by Oregon's 2012 transition to Coordinated Care Organizations (CCOs). A variety of states, including Oregon, had made modest attempts at Medicaid reforms throughout the years. However, Oregon's 2012 effort may represent the largest attempt of any state to transform the Medicaid payment and delivery system since the program's inception in 1965.

## UNDERLYING FINANCING, ECONOMICS, AND FEASIBILITY OF THE OREGON TRANSITION

Oregon's move to the CCO model was made possible, in part, by years of work at the state, provider, and community-levels. The framework for CCOs as a mechanism for Medicaid transformation was shaped by an activist governor (John Kitzhaber), a newly consolidated state health care agency, and more than 75 public meetings seeking community input. When the final plans were put into place, Oregon moved quickly, setting an ambitious agenda in the scope and the scale of its reform.

From the beginning, the CCO model incorporated many elements on the "wish lists" of health policy researchers. While grounded in payment and delivery system models that had theoretical (and some empirical) support, the CCO model had not been tested in the ways that Oregon proposed. CCOs had similarities to ACOs which, in 2012, were seen as a new model that could control health care spending, improve quality, and minimize the negative experiences associated with managed care models of the 1990s. However, Oregon's model was substantially more ambitious than the typical Medicare ACO model, leading some to call CCOs "ACOs on steroids."

Some of the noteworthy components of the CCO model included governing boards composed of health care providers, community members, and stakeholders in the local health systems; efforts to integrate physical, behavioral, and dental care at the financial and delivery system levels; and a dedicated effort to reduce health disparities. From a payment perspective, one of the most significant components of the CCO model was its global budget, a payment model viewed as among the most promising for improving the value of care.

The global budget, which carried "two-sided" financial risk, created a prime opportunity for research. By 2012, it became clear that the health care system would move—or attempt to move—away from fee-for-service payments and toward "alternative payment models," including global budgets and bundled or episode-based payments. Within a few years, Centers for Medicare and Medicaid Services (CMS) would announce that it planned to move 50% of Medicare payments to value-based payments by 2018. Payment reform was also surfacing in the private market. Blue Cross Blue Shield of Massachusetts created the Alternative Quality Contract (AQC) model, which brought provider groups into global budget contracts with quality-based performance bonuses, with analyses indicating that the program achieved savings in its first year and increased savings in subsequent years (Song et al., 2011, 2012, 2014).

As the commercial and Medicare markets moved toward payment reforms, some states were just beginning to contemplate changes in their Medicaid programs. At the time, Oregon and Colorado were the only states that had implemented ACO-type models. [As of this writing, nine additional states have launched Medicaid ACOs, and six more are actively pursuing them (CHCS, 2016).]

Given the desire to understand the implications of alternatives to the fee-for-service payment model, the initial design of the CCO model alone would have warranted attention. However, interest in the CCO model was considerably enhanced with Oregon's agreement with CMS, which provided $1.9 billion to the state to assist in the transition to CCOs. In exchange, Oregon agreed to reduce its rate of spending growth by 2 percentage points and to achieve those savings without diminishing the quality of care. Furthermore, if Oregon did not meet its spending and quality benchmarks, it would be required to pay hundreds of millions of dollars back to the federal government.

This exchange set the stage for important policy research: a comprehensive reform model, incorporating untested elements of promising payment and delivery system models, with CMS stepping in to set a predefined spending growth rate and penalize the state if it could not achieve results. The research questions were clear: Would Oregon's CCOs be able to "bend the cost curve"—a goal that had become the holy grail for all payment and delivery system innovations? If so, how would this be achieved? Within the experimentation afforded by 16 different CCOs, what models and approaches were most successful? Did any fail? What lessons could be extracted from Oregon to help improve the quality, value, sustainability, and perhaps even the expansion of the Medicaid program beyond Oregon? And, were there elements of the coordinated care model that could be feasibly adopted by commercial insurance groups?

## CAPITALIZING ON THE RESEARCH OPPORTUNITY

Oregon researchers and state administrators had 10 years of experience in working together and developing trust and the mechanisms needed to share data and work through contracting. State administrators and policymakers had prioritized the availability of data, including Medicaid claims and enrollment information. In exchange, the research community worked to provide analyses and results to the state in a timely fashion, typically prioritizing results that could help the state first, with publication concerns secondary. The state had also helped to foster a highly collaborative environment, creating partnerships between Oregon Health & Science University and researchers at Portland State University, the Providence Center for Outcomes Research and Education, and other research centers. These mutually beneficial relationships allowed the state to lean on the expertise available from the surrounding academic community, and provided academics with data (and questions) that were well suited for research pursuits.

The evaluation of the Oregon Medicaid transformation to CCOs necessitated a mixed methods approach, using both qualitative and quantitative methods. While claims and survey data could begin to tell a story of how the state performed, qualitative data would be needed to fill critical gaps. First, qualitative data were necessary in order to clarify what the CCOs were and what they were not doing. While the frameworks for ACOs and medical homes had been established and were relatively well understood, outside observers were not sure what to make of the CCOs. Were they more like managed care organizations (MCOs) or more like ACOs? Qualitative data would help explain the CCO structure as well as tease out the heterogeneity among CCOs. It would also enhance our understanding of the organizational and political context that allowed for their implementation and would create mechanisms for change or, inadvertently, barriers to transformation.

## CONFRONTING THE CHALLENGES OF MEDICAID RESEARCH AND EVALUATION

Medicaid evaluations face challenges that are unique to this population. This phenomenon provides one explanation for the relative prominence of research on individuals covered by Medicare, commercial insurance, and the Veteran's Administration (VA) program. In addition to the usual difficulties associated with obtaining timely and complete data, Medicaid research must also confront challenges associated with a limited pool of potential comparison populations or states, complexities of tracking expenditures in a managed care environment, and multiple populations that are covered by the program.

Data-related challenges are numerous and of several varieties. Administrative claims data are widely used for measurement of health care quality, utilization, and expenditures, and are representative of services for the entire enrolled population. Yet, these data vary in layout, detail, and ease of extraction by state Medicaid programs. Claims data are useful for identifying whether a particular service occurred (e.g., "Did the diabetic member receive their annual glucose screen?") but are less conducive to measurement of health outcomes (e.g., "Was the member's blood glucose well-controlled?"). Electronic health records are better suited for health outcomes measurement since they include laboratory results, but these data structures vary widely among the many systems available, making it difficult to coalesce information across clinics and health systems. Surveys can be crafted to meet very specific research needs but can be expensive to design and field and, unless they have been in place for longitudinal studies, may be difficult to implement for natural experiment-type evaluations.

While administrative claims data represent the best opportunity for reporting on health care expenditures, Medicaid claims have limitations that are not typically found in Medicare or commercial data. Medicaid managed care has grown exponentially over the last 25 years, covering approximately 75% of beneficiaries in 2016, compared to just 10% in 1991. Under managed care capitated arrangements (wherein the health plan pays the provider a set amount per member over a period of time, rather than for individual services provided), encounter claims capture the delivery of services but record paid amounts as $0. These arrangements—which make sense for states, MCOs, and providers—create challenges for research, especially when there is a strong demand to understand the implications of reform choices on overall health care spending. One approach is to "reprice" claims using median fee-for-service prices or the Oregon fee-for-service reimbursement schedule. While this approach has the benefit of attaching an approximate price to each claim, it is imperfect. In particular, repriced claims capture changes in utilization but miss changes in spending that could be attributed to different reimbursement rates. Claims-based evaluations of spending also fail to capture administrative costs, "flexible" (nonmedical) spending, and bonus payments to providers outside of regular health care services (e.g., incentivized quality measures).

Another challenge that can arise in the use of Medicaid claims data is that there may be organizational separation of claims and encounter data for physical health care and mental health care. Mental health and addiction disorders are common among the Medicaid populations; there is a strong need to understand how these services are delivered and the ways in which they can improve value and outcomes. Nonetheless, obtaining data for these services can be more difficult than obtaining data for physical health services. In some cases, states may "carve out" mental health, engaging behavioral health organizations to manage the mental health portion of Medicaid care. This separation may lead to mental health and physical health claims that are housed in different systems, creating a significant obstacle to obtaining a complete claims or expenditure profile. In addition, federal regulations governing how and when information on treatment for alcohol and drug use disorders can be shared have created another barrier to obtaining those data for research.

Quality measurement is also often more difficult in the Medicaid populations than in their commercial and Medicare counterparts. A plethora of nationally endorsed measures exist for quality measurement, yet our experience has shown that required data elements and small sample sizes can introduce barriers to use. Continuous enrollment criteria support efforts to ensure that measures accurately capture services delivered (or not), but the resulting eligible population may not reflect the population at large. Furthermore, quality measures that rely on long enrollment periods work in stably insured populations, but are not well suited for Medicaid, where populations frequently "churn" on and off coverage. In addition, it can be impossible to distinguish between true improvements in care (e.g., increase in screening rates) versus improvements in coding (e.g., adoption of a CPT code because it earns "credit" for a newly incentivized measure). The October 2015 transition from ICD-9-CM to ICD-10-CM diagnosis codes in the United States may further exacerbate efforts to study the performance of some measures over time.

Individuals who are dually eligible for Medicaid and Medicare coverage represent a unique set of barriers when it comes to acquiring data for research, as a full analysis requires linking Medicaid and Medicare claims. Even when researchers are able to access data from both programs, combining the two sources can be prohibitively difficult. Formats may vary and portions of care may be filtered from one data set but not another. Most problematic is that member identifiers (e.g., IDs) may not be consistent between the two data sources.

Oregon's reform demonstrated some of the tradeoffs between the scale and scope of an ambitious policy and the ability to comprehensively and objectively evaluate the outcomes. By moving virtually everyone into the CCO program within 6 months, Oregon simultaneously accelerated delivery system reform while removing the opportunity for a within-state comparison group. Comparison data for Medicaid evaluations are notoriously difficult to obtain. Studies of reforms in Medicare, commercial, and VA populations often benefit from claims or related data that are consistent across sites, allowing for comparison groups when policies are isolated to specific regions or patient groups. In contrast, most Medicaid policies are implemented at the state level, and comparison states that can serve as a true "business as usual" control are rare. Most states are experimenting with their Medicaid program in some way, through combinations of expansions, waivers, ACO-type models, the introduction of copayments, or a shift to managed care models. Furthermore, even if a set of ideal comparison states could be identified, obtaining Medicaid data is not straightforward. In contrast to Medicare and commercial insurance, there are no nationally representative data that can be drawn on to construct comparison data. The one partial exception is the data that can be obtained through the ResDAC–CMS partnership (www.resdac.org). However, these data are limited in terms of the states available, the absence of managed care claims (researchers are generally advised to restrict analyses to fee-for-service claims), and, significantly, a 2–3-year delay. Obtaining data for timely research requires working with state Medicaid agencies on a one-by-one basis.

In Oregon, we took comparison groups where we could find them. This led to partnerships with Colorado and Washington. We also worked with data from individuals in Oregon with commercial insurance, through the state's All Payer All Claims database (APAC, 2016). In our work with the commercial data, we used propensity score methods to reweight this population so that it resembled the Medicaid population on observable characteristics, creating a type of "synthetic" Medicaid comparison population. Although we recognized that the commercial population would differ from the Medicaid population both in the health status (observed and unobserved) and benefit packages, one advantage of these data (representing over 80% of the commercial market, with more than 1 million individuals) was that it provided an assessment of secular, systematic changes across the state. For example, Oregon implemented a Patient-Centered Primary Care Home (PCPCH) Program concurrently with the advent of CCOs. Comparison to Oregon's commercial population helped to account for differences as a result of CCOs rather than the PCPCH Program. We also purchased hospital discharge data through the Agency for Healthcare Research and Quality's "Healthcare Cost and Utilization Project (HCUP)." While these data were limited in that they allowed for assessments limited to individuals who were admitted to the hospital, the HCUP portal provided a simple way to acquire data for multiple states that could serve as potential comparisons without having to work through the political and relationship gymnastics required to obtain Medicaid databases directly from states. Furthermore, since inpatient admissions represent an important component of health care spending, these data provided a potentially useful insight to the changes occurring within CCOs.

## LESSONS LEARNED

Research on Medicaid is difficult. Data are difficult to obtain, pricing and expenditure measures are complex, churn on and off the program can introduce noise in quality measurement, and with widespread experimentation across the nation, finding comparison groups to allow for the study of interventions or natural experiments is challenging. These challenges will continue to make research on Medicare and commercial populations more straightforward, just as the fable of the man who lost his keys in the park but searched for them under the streetlight because "that's where the light is." However, we would encourage researchers to forge into the lesser known world of Medicaid, particularly at this critical juncture in the history of the program.

We need research on Medicaid not only because of the size of the program, but also because it serves a highly vulnerable population of socioeconomically disadvantaged individuals with high rates of disability and behavioral health disorders. Furthermore, the budgetary implications of the Medicaid program should not be overlooked; Medicaid has grown to account for about 16% of total health care spending. Studies and evaluations that shed light on ways to improve the value of the program can benefit the vulnerable populations it serves while helping to make the program sustainable from a long-term financial and budgetary viewpoint.

Research on Medicaid programs draws on questions that are common in evaluations of Medicare and commercially insured populations, including assessments of the ways program and benefit designs control costs or improve quality. However, by necessity, deeper appraisals of the Medicaid program call for questions that are not necessarily relevant for the Medicare and commercial populations.

One of the areas that demands attention from studies of the Medicaid population is access to and treatment of mental health and addiction disorders. While these conditions exist across all populations, their prevalence and importance is of a magnitude larger in the Medicaid program. Medicaid is the largest source of financing for these services, and estimates of prevalence of behavioral disorders in the Medicaid population range from one-third to almost one-half, substantially higher than the uninsured, Medicare, and commercial populations. Given the prevalence and expense of these conditions in the Medicaid population, it is imperative that Medicaid programs are structured to support high-value and high-quality care for these conditions. For example, a large number of studies—including more than 30 randomized trials—have demonstrated that integrating behavioral health with primary care can improve patient outcomes, and suggest the potential for lower health care spending (Woltmann et al., 2012). Yet relatively little is known about how to scale integration or structure reimbursement for Medicaid programs. Typically, Medicaid programs pay separately for addiction care, mental health case management, psychiatric day treatment, and related services. Oregon identified more than 26 different funding streams (most of which related to behavioral health care)—a financial arrangement that ensured funds were spent on behavioral health, but added to the fragmentation and lack of integration in these services. Financial integration through the global budget was designed to promote integration.

Another area that has received increased attention is the role of social determinants of health. While these factors may affect health across all populations, the roles of education, employment, social support networks, and housing may be particularly salient for low-income, vulnerable Medicaid populations. A number of states are exploring approaches to address homelessness in their Medicaid populations, including using managed care plans to cover housing-related services or funding housing-based supports through waivers. The ability for Oregon CCOs to spend Medicaid dollars on "flexible services" that were not traditional "medically necessary" services can be seen as an extension of this approach—using Medicaid dollars to improve health and reduce spending, even if these dollars did not pay directly for medical services.

Under the waiver, flexible services could include, for example, supportive housing, although locally, researchers often referred to this category of spending as "air conditioner" dollars, referencing an example proffered by the Governor: rather than paying for multiple emergency department visits and hospital admissions for a woman whose chronic conditions were exacerbated by a heat wave, CCOs would be free to purchase a $200 air conditioner to solve the problem, thereby reducing costs and keeping the Medicaid patient healthy and out of the hospital. (We watched with some amusement as the much acclaimed "air conditioner" grew from expository example, to apocryphal anecdote, to accepted fact.)

As innovative as Oregon's "flexible" spending was, the change created challenges from a research perspective. Such expenditures were not represented in claims data, and are tracked in a rudimentary way in financial forms. Given the interest in making investments that affect social determinants of health, future research on Medicaid will need more sophisticated tools to track these types of spending and understand more clearly what types of nonclaims-related services were paid for and how they impacted health. Since investments to address social determinants of health represent a departure from the more familiar claims-based expenditures, states and researchers should expect a higher degree of scrutiny from CMS and others looking for justification of these approaches.

Dental care is yet another area that is arguably more important for research on Medicaid than on other populations. Medicaid is the predominant means for providing dental insurance coverage for families and individuals with limited financial resources, a group that has traditionally received substantially less dental care than the rest of the population. Medicaid has generally mandated states to cover dental benefits for children, with adult coverage treated as optional, although a number of states have used the ACA as a vehicle for establishing or enhancing dental benefits for adults. However, Medicaid beneficiaries have routinely faced difficulties obtaining recommended dental care, with two-thirds of children covered by Medicaid receiving no annual dental services, even though the prevalence of dental diseases in low-income children is extremely high and largely preventable. In commercial and Medicare populations, dental care is most frequently treated as isolated from general medical care, but access to dental services and the risk of poor outcomes are less of a concern. However, the separation of oral health care from general health care may be suboptimal in Medicaid populations. Research that assesses the overall health of the Medicaid population may benefit from a greater emphasis on the role of oral health than has been seen in research on other populations.

A long literature describes the vast and pervasive disparities that exist for most racial and ethnic groups across health care services. These disparities are particularly acute among individuals eligible for the Medicaid program, which provides a disproportionate amount of insurance coverage for racial and ethnic minorities. Researchers can help elucidate the potential for Medicaid reforms to address disparities. This line of research is important because there is the potential for some reform efforts to increase, rather than decrease, racial and ethnic disparities. For example, policies targeting high hospital readmission rates may penalize providers who care for lower income patients, and the use of value-based purchasing may adversely impact providers who lack the capital to embark on large-scale quality improvement efforts.

Our experience and assessment of recent developments and reform efforts suggest that the best Medicaid research will evolve out of multistate partnerships. While a few states may be large enough to independently evaluate the impacts of isolated reform efforts, this approach will not be feasible for most programs. There are substantial benefits to developing partnerships across states. In particular, aggregating claims data across multiple states can leverage economies of scale. Once data are loaded and standardized, the measures, cohorts, and analytic approaches developed and coded for one state can frequently be applied to other states. This creates the opportunity for cross sectional comparisons of utilization, spending, and quality across different states and cohorts, as well as the ability to assess trends over time. The economies of scale can also create bandwidth for evaluations that states might not be able to assess on their own. Methodological advances from working with one state's data—i.e., better ways to measure expenditures or quality, or novel methods for tracking the impact of social determinants of health—can be applied to multiple states. In addition, results from an independent entity may be seen as more impartial, with benefits to state administrators who are trying to establish credibility or justify program changes to federal administrators or local politicians. Researchers should benefit from approaches that have greater generalizability and may produce more useful estimates and inferences than could be gained from a single-state study.

Researchers working with state Medicaid programs should also be aware of the tradeoffs and sensitivities involved with these partnerships. A state–academic partnership means that researchers will need to focus on questions that states care about, which may be different than the questions that researchers would pursue independently. Researchers and state administrators need to adjust to different timelines. States will want answers as soon as possible, while researchers will frequently place a higher premium on a more thorough study approach that may be vetted through the time-intensive peer-reviewed journal pathway. Finally, researchers will need to navigate the politics of conducting research that may have program or political implications. The goal of research is to develop objective evidence. But objective evidence that suggests that a favorite state reform is not succeeding may create tensions between researchers and politicians or Medicaid administrators. These may be particularly acute if academic research incorporates new methods or comparisons that suggest findings that contrast a state's own evaluations or public reports. Even if the results are not in question, tensions can also arise if researchers are not careful in the language they use or messages they project, or if state administrators are caught by surprise when new findings are published. These partnerships hold substantial benefits to academics, Medicaid agencies, and to the public health, but sustaining the partnership requires investments in relationships, trust building, and communication.

## CONCLUSION

Medicaid is perhaps the most under-researched insurance program in the nation, even while it provides coverage for the 60 million low-income individuals who have relatively little other recourse for insurance or medical care. As Medicaid has grown, its financial burden has increased. The legacy nature of the program creates challenges for researchers and evaluators unique to the program and substantially thornier than typically found in research on commercial or Medicare populations.

We hope to see an expansion in Medicaid research and evaluation that is on par with the growth of the program over the last few years. Researchers should not expect that this endeavor will come without pain, but the potential implications and ability to influence policy are substantial. Reform efforts should be rigorously evaluated in order to determine their impact on quality, utilization, expenditures, the patient experience, and the impact on health disparities. The identification of essential components of Medicaid programs can help administrators, clinicians, and policymakers understand the types of organizational changes and policies that are feasible and can lead the way to a sustainable, high-value Medicaid program that benefits tens of millions of individuals.

## REFERENCES

APAC. (2016). *Oregon all payer all claims database (APAC): An overview*. Available from: http://www.oregon.gov/oha/HPA/ANALYTICS/APAC%20Page%20Docs/APAC-Overview.pdf.

CHCS. (2016). *Medicaid accountable care organizations: State update*. Trenton, NJ: Center for Health Care Strategies, Inc, December 2016 Fact Sheet.

Fisher, E. S., Wennberg, D. E., Stukel, T. A., Gottlieb, D. J., Lucas, F. L., & Pinder, E. L. (2003a). The implications of regional variations in Medicare spending. Part 1: The content, quality, and accessibility of care. *Annals of Internal Medicine*, *138*(4), 273–287.

Fisher, E. S., Wennberg, D. E., Stukel, T. A., Gottlieb, D. J., Lucas, F. L., & Pinder, E. L. (2003b). The implications of regional variations in Medicare spending. Part 2: Health outcomes and satisfaction with care. *Annals of Internal Medicine*, *138*(4), 288–298.

Song, Z., Rose, S., Safran, D. G., Landon, B. E., Day, M. P., & Chernew, M. E. (2014). Changes in health care spending and quality 4 years into global payment. *The New England Journal of Medicine*, *371*(18), 1704–1714.

Song, Z., Safran, D. G., Landon, B. E., He, Y., Ellis, R. P., Mechanic, R. E., & Chernew, M. E. (2011). Health care spending and quality in year 1 of the alternative quality contract. *The New England Journal of Medicine*, *365*(10), 909–918.

Song, Z., Safran, D. G., Landon, B. E., et al. (2012). The "alternative quality contract," based on a global budget, lowered medical spending and improved quality. *Health Affairs*, *31*(8), 1885–1894.

Wennberg, J., & Gittelsohn, A. (1973). Small area variations in health care delivery. *Science*, *182*(4117), 1102–1108.

Woltmann, E., Grogan-Kaylor, A., Perron, B., Georges, H., Kilbourne, A. M., & Bauer, M. S. (2012). Comparative effectiveness of collaborative chronic care models for mental health conditions across primary, specialty, and behavioral health care settings: Systematic review and meta-analysis. *American Journal of Psychiatry*, *169*(8), 790–804.

Chapter 16

# Expanding Beyond Oregon – Lessons for Other States

**Christopher F. Koller**
*Milbank Memorial Fund, New York, NY, United States*

How applicable is the Oregon Coordinated Care Organization (CCO) model for other states? While often the particularities of a state can be used as an excuse for the status quo, differences in economic resources and social values among states are real. Texas is really different from Vermont and the citizens of each like it that way. Does Oregon's context—a recent history of state health policy innovation, a relatively even balance of political power between parties and a participatory civic culture—make it a uniquely fertile ground for state-led experimentation in delivery system reform?

Differences not withstanding, there are common health policy challenges facing states for which the Oregon model can offer some valuable lessons. This chapter summarizes the most important challenges and then discusses how Oregon's CCO experience can help address them.

## CHALLENGES

### Challenge 1: Medicaid Sustainability

Medicaid is subject to the same per-person inflation rates as other health care payers—rates that have routinely exceeded general inflation. As a result, regardless of whether states have chosen to expand Medicaid eligibility under the Affordable Care Act (ACA), Medicaid expenses consume an increasing share of state spending. In 1995, according to the National Association of State Budget Officers, Medicaid consumed 14.4% of state general fund spending and 6.8% of special fund spending—before any accounting for the matching federal funds, which flow through state budgets. By 2015, those numbers had grown to 19.3 and 10, respectively.

This growth pushes state governments to look for additional sources of revenue and is at the expense of other categories of state services. In his

**Health Reform Policy to Practice. DOI: http://dx.doi.org/10.1016/B978-0-12-809827-1.00016-1**

campaign for Governor of Massachusetts, former Medicare Administrator and longtime health care quality improvement advocate Don Berwick labeled the health care system as "confiscatory," pointing to state expenditure data produced by the state's health policy commission, which showed all categories of state spending in inflation-adjusted dollars to be flat or decreasing between fiscal years 2006 and 2016, *except* for health care. In taking money from other categories of state spending to pay for current medical services, Berwick and others maintain, states have less to invest in the housing, education, and social services, which could improve the social factors that influence health status and reduce the demand for medical services in the future.

The statutorily mandated decrease in the federal participation rate for Medicaid populations added with the ACA, and the prospect of federal laws capping federal funds to each state for Medicaid, as is being discussed in early 2017, only increases the challenge of Medicaid sustainability. Historically, the surest way to deliver cost savings in Medicaid has been to reduce the number of eligible recipients, reduce the scope of benefits, or reduce rates paid to providers. Critics object that these measures are socially unjust and politically perilous and merely shift costs, which will still be incurred to Medicare and private insurance, and look for innovations such as CCOs.

## Challenge 2: Acting on the Social Determinants of Health

Research clearly documents that medical care is not the main contributor to the lifespan of a population. Some 80% of premature deaths in the United States are attributable to factors outside the medical care system—including personal behaviors, social circumstances, genetics, and environmental conditions (Fig. 16.1). Yet the United States is notably out of step with any other country in the relative portion of its GDP that is spent on the services that would improve these factors, relative to its spending on health care (Bradley, 2016) and as a result our lifespans are shorter than countries of similar wealth with different spending patterns.

Why are state governments unable to shift their investments from health care to higher return fields? The returns on such spending could be particularly beneficial for state Medicaid programs, which enroll the high-cost high-need populations, which have experienced the greatest effects of poor lifestyle habits (diet, exercise, tobacco use), social circumstances, genes, and environment. Indeed, part of the vision of CCOs is that locally governed institutions responsible for the total costs of care for populations would do just that—paying for the housing, employment, and coordination services that would reduce the demand for health care services.

State officials, in one structured discussion (Rogan & Bradley, 2016), identified three barriers to acting on the evidence regarding the value of investing in services to act on the social determinants of health.

First, the health of the state's population is not always prioritized relative to other societal goals. Competing societal goals includes maximizing

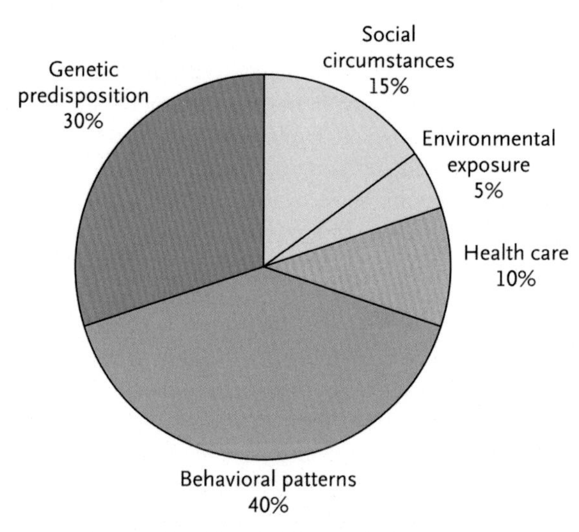


**FIGURE 16.1** Determinants of health and their contribution to premature death. *Adapted by Schroeder from McGinnis JM, Williams-Russo P, Knickman JR. The case for more active policy attention to health promotion. Health Aff (Millwood) 2002;21:78-93.*

personal liberty and minimizing the size of government and its taxes. Advocates for improving population health are forced to acknowledge the legitimacy of these competing values and the reality of the conflict among them.

Second, incentives, including financial and political incentives to improve health, are misaligned. Even if values conflicts are reconciled, economic and political ones remain. In addition to fee-for-service provider payment mechanisms that incent volume over quality, policymakers pointed to the "wrong pockets problem" in government, where expenses incurred by one agency result in savings in another. Population health also does not have individual beneficiaries who testify for laws and fund political campaigns. In addition, the delayed returns from investments in social services and population health—years for early childhood interventions—require a longer time frame than many revenue-strapped governments believe that they can afford.

Finally, there is a lack of consensus regarding who is responsible for health. Is one's health a function of luck (genes and where and to whom one is born) or one's personal actions (behaviors such as diet and exercise)? The answer is not simple, and personal responsibility, a character trait that most feel is the mark of emotionally stable and healthy individuals, is socially desirable. The fear that public policies can discourage personal responsibility marks much of the discussions regarding investing in social services.

## Challenge 3: Weak State Administrative Capacity

States with the political will to take on rising Medicaid expense trends in more systematic ways than cutting benefits, rates, and eligibility requirements, and the desire to focus on long-term population health still face a fundamental problem of administrative capacity in state government. Implementing large and complex policy initiatives across the largest component of many state budgets requires resources, which are hard to obtain and deploy.

Leadership is hard to retain. With salaries that are a fraction of what positions with similar responsibilities in the private sector offer, the average tenure for a state Medicaid Director is less than 2 years. Furthermore, Medicaid Directors need a senior management team with the skills and capacity to implement complex policy initiatives. Managing a contemporary Medicaid program requires information technology strategies, as well as skills in contracts procurement and management, financial management, and data analytics. As with Medicaid Directors, these skills are in high demand in the private sector. Making salaries for senior Medicaid program leadership more competitive with the private sector requires fundamental changes in a state's personnel classification system. It is worth noting, however, these barriers are routinely surmounted for high-profile university athletic programs.

Finally, general Medicaid administrative expenses have not grown at the rate of program expenses. There are fewer people to administer more work. Faced with burdensome procurement and personnel systems, states have increasingly purchased rather than built administrative capacity. Seventy five percent of states now use Managed Care Organizations (MCOs) in Medicaid and over half of those that do have MCOs put at least 75% of their enrollees into managed care (Snyder & Rudowitz, 2016). Program development responsibilities are often outsourced to consultants—particularly in response to the surge of work created by the implementation of the ACA—as well as information system management functions.

Resources are a function of budgets, which are set by elected officials. Reflecting an electorate skeptical of the efficacy and efficiency of government services, state legislatures have been unwilling to enact the personnel and procurement reforms, and appropriate additional financial resources to build the administrative capacity complex Medicaid programs require.

## LESSONS LEARNED

In the face of these common challenges across states, Oregon's effort to develop CCOs offers five instructive lessons, which health policy leaders in other states can apply to their settings.

1. Managed Care in its current form may not be enough.

   While there are some traditional MCO responsibilities, which are retained in the CCO design—accepting state payments, administering benefits, contracting a provider network, and paying for services—CCOs change the structure and roles of MCOs in the state Medicaid program in two fundamental ways.

First, CCO governance is strictly prescribed to include provider and local community representation. In this sense, they are somewhat analogous to federal requirements for qualified community health centers, as a condition for federal grants and preferred rates. The theory is that unlike MCOs the newly formed CCOs will have a greater accountability to their local communities. No other states have similar requirements of their MCOs. The requirement also caused existing nonprofit MCOs in Oregon to either restructure with new governance, or become an administrator for newly formed CCOs.

Second, to emphasize this focus on geographic community, with the exception of two high-density geographies, CCOs have a geographic monopoly for all Medicaid enrollees in a given area: all enrollees in that area have their services administered by a single CCO. In determining that enrollees value choice of medical provider more than choice of benefits administrator, program designers are expecting that a CCO with a geographic monopoly will be better able to assess and influence the attributes of a community *outside* the health care system that affect population health—such as access to housing, food, and exercise-friendly environments.

Several state Medicaid programs, which were MCO pioneers, are now experimenting with some form of Accountable Care as envisioned by Medicare. Most of these—such as RI, NJ, MN, NY, and MA—are attempting to graft Accountable Care Organizations (ACOs) into their existing MCOs. With the exception of Vermont's state-wide ACO, no other state Medicaid program has fundamentally repurposed their MCOs like Oregon. The risk of creating a geographic monopoly is the loss of market or political power to that CCO—that the state Medicaid program will be unable to hold it accountable for service, network, performance, or costs in the way that choice and competition could. Program designers think that the state's contracting requirements and ability to create organizational alternatives as needed will be sufficient to address that threat.

**2.** Global budgeting can force choices and delivery system redesign.

Oregon's federal waiver is a critical element of its CCO strategy. Unlike most state Medicaid waivers, it is truly global in scope—covering all Medicaid services to all Medicaid enrollees and guaranteeing a maximum fixed rate of increase in per member costs for those services. This is in comparison to waivers negotiated in most other instances by states and the federal government, which are much narrower in scope—covering only certain populations and certain services, and rarely with per capita expense increases capped.

The CCO program design then aligns the incentives for each CCO with the state's—giving them a per-person capitation rate with very limited recourse for excess costs in the current year (or "downside risk") and already negotiated rates of increase for future years. In effect, the state has put the entire Medicaid program on a diet, making it clear to all who participate in it—CCOs and providers—that service prices and service utilization have to live within it, and state officials and legislators are powerless to provide any extra caloric appropriations.

With little capacity to negotiate for extra revenues, geographically focused delivery systems will then be forced to examine their expenses—what combination of benefits and service arrangements will be most effective in meeting the needs of the people who are their responsibility?

Geographic monopoly, consistent eligibility standards, and total cost responsibility are all intended to address the second challenge discussed in this chapter—the ability to act on evidence regarding the social determinants of health. Since CCOs will be responsible for all the population in a region for longer periods of time, they will be theoretically more inclined than MCOs to invest in the upstream social services that prevent medical costs, and more incented because of strict rates of growth imposed by the terms of the federal waiver.

Could a MCO strategy accommodate implementing such a broad Medicaid waiver? The Oregon design assumes no. In this view, ultimately, a fixed per capita budget for health care services involves community-wide discussions around trade-offs and priorities. A private corporation with no accountability to or responsibility for a geographically defined population does not have the practical or public authority for such a conversation.

In making health care budget limitations explicit and forcing cost-effectiveness decisions to the fore, Oregon is continuing its controversial and brave health policy work of the 1990s, when its Oregon Health Plan first implemented cost-effectiveness criteria explicitly into its benefit coverage decisions for its Medicaid program. The CCO effort now attempts to engage local communities in the work of matching resource demand to resource availability.

The program-wide focus on cost-effectiveness evaluation of potential benefits envisioned by the Oregon Health Plan is not a requirement for a state contemplating a broad Medicaid waiver. But such a waiver should not be considered lightly.

The chief risks of such efforts are both political and technical—the United States has systematically avoided the firm health care budgets that mark other political economies for these reasons. The trade-offs forced by fixed per-person budgets involve both patients and provider revenues, and affected parties will seek sympathetic ears for their concerns to change the offending decisions. Administering a fixed budget is complex—particularly when determining relative rate adequacy among categories of populations and geographic regions, and when adjudicating whether cost overruns are due to powers beyond the control of the accountable party. Such risk adjustment mechanisms—either prospectively for rates or retrospectively for enrollment—must be budget neutral and can encourage behavior by providers and CCOs to maximize revenue, not improve the cost effectiveness of care. Finally, the flexibility in benefit design offered by the waiver—flexibility intended to encourage the use of both nontraditional care such as community health workers and social services that could obviate the latter need for more expensive medical care—might never be utilized if it requires the diversion of already limited and heavily contested resources.

**3.** Invest in learning for improvement.

States implementing Medicaid managed care learn that new agency skills are required in areas such as contracting, rate setting, oversight, and assessment. Failure to acquire and develop these skills means that services to beneficiaries and accountability to legislatures and the public will both suffer.

Just as CCOs are intended to be something more than repurposed managed care, their oversight needs to be different. The CCO designers in Oregon, acknowledging the administrative capacity challenge discussed previously, developed two programs intended to increase the likelihood of CCO success.

First, its Quality Measurement Program has prioritized 33 CCO performance measures, which constitute success, then set forth measurement methodologies and performance standards, and allocated resources to compile and publish the results. Since CCOs are fundamentally delivery systems, it is provider performance, not insurance company performance, which is being measured. Of all the elements in the program, the commitment to public disclosure is perhaps most transformative. Although it depends upon a well-managed consensus development process, and adequate and effective measurement, only when provider performance is published reliably, in an objective, educational context, can it have the desired effect of improving the accountability of providers for improving their work.

Second, the Oregon Health Authority's Transformation Center, with its set of care transformation resources, technical assistance, and supports for entities engaged health care delivery system transformation and statewide Patient-Centered Primary Care Home program, is an ambitious attempt to facilitate both the goals of the CCO program and the process of delivery system transformation. It envisions a collaborative model of changed focused on participant learning facilitated through a state-wide utility. It is intended to work hand-in-hand with the quality measurement program, to help entities succeed at identified measures.

The design of these programs is consistent with awarding entire geographical markets to a single CCO. Improvement in performance will come from standards, measurement, technical assistance, and transparency, rather than a competitive model in which successful learning is incented through increased returns and market share. There is considerable evidence that while competition among providers has resulted in increased market share for the successful entities—particularly in tertiary medicine—it has not resulted in improved population health or lower overall cost trends. Given the lack of performance of a market-based model, it is not unreasonable to seek an alternative.

Both programs require ongoing investment in administrative infrastructure. While these amount to a fraction of the medical expenses in the state's Medicaid program, political will and foresight is nonetheless required to continue their funding. States attempting to learn from Oregon's experience in this area must make the commitment not only to the model for improvement, but also to providing adequate resources for it.

**4.** Moving beyond Medicaid is tough.

The CCO infrastructure was envisioned to accommodate other populations beyond Medicaid. With a geographically based, broadly inclusive, community-governed delivery system, and financing and population health management infrastructure built, CCOs could assume some or all health management responsibilities for others residing in the region. Under such an arrangement, population health efforts in geography would be strengthened and opportunities for cost shifting between payers reduced.

Although explicitly excluded from any legislation, presumably other populations for whom the Oregon Health Authority purchases medical benefits—state and local employees and school districts—would be the first to be considered for all or some CCO services. Other candidates would include enrollees in the state's Insurance Exchange and perhaps eventually privately insured and self-insured populations.

Such a march to integration has proven slow, however. The OHA has adopted the CCO quality measures for its other populations and uses similar allowable rates of expenditure growth in its insurer contracts for these populations. There has not been, however, a formal enrollment of other state-sponsored populations in CCOs.

The reasons for the slower-than-anticipated spread of the CCO model to other populations are both political and cultural in nature. A change in gubernatorial and agency leadership resulted in shifts in policy priorities and loss of administrative momentum. Providers—particularly large institutions—perceive benefit from having multiple sources of revenue rather than a large one with whom to negotiate. More fundamentally, public tolerance of shifts in provider networks and benefits administration is much greater for low-income populations receiving a perceived entitlement than for those receiving an employee benefits. Such arrangements must also be collectively bargained with public employee unions. States considering CCO-like arrangements would be wise to be modest in projecting their adoption beyond Medicaid populations.

**5.** Maintaining leadership and political support are crucial.

Any ambitious attempt to reshape the financing and delivery of care to Medicaid recipients while acknowledging finite state resources requires leadership and public support. The CCO program was strongly identified with Governor John Kitzhaber. It built on his work developing the Oregon Health Plan during his previous service as Governor. Kitzhaber was uniquely positioned to do this work. He was an emergency physician by training, passionate about the topic, and had 8 years of on-the-job training in his previous two terms as Governor in leading the development and implementation of state health policy.

He assumed office a second time with a vision for Medicaid, made it an administrative priority, hired a well-respected, highly skilled administrative team who shared that vision, and initiated negotiations with both the legislature and the Federal government. Once the waiver was negotiated with the Federal Government and passed by the state legislature, his team—many of them with previous public sector and managed care experience—set about implementing the ambitious agenda.

Even with such strong leadership and enshrinement in legislation and Federal Medicaid waivers, the CCO program has not been immune from wavering political support. Ironically, the passage and implementation of the ACA, although it shares similar broad goals with the CCO program, threatened its success by taxing the administrative and policy capacity of state government. OHA leadership was forced to focus on—and eventually lead—the state's beleaguered health insurance exchange. The premature departure of Governor Kitzhaber from office cast a shadow over the administration's various initiatives. Perhaps more significantly, it brought in a new Governor who has not made health policy a similar priority for her administration, and resulted in the departure of a cadre of competent, experienced administrative staff.

Now past its start-up phase, the CCO effort faces the challenges of continuing improvement in the face of political pressures and competing demands for leadership and resources. Most significantly, the program has created new institutions—the CCOs themselves—with new interests. Controlling all Medicaid revenue and an effective monopoly for particular geographies, the profitability and the administrative structure of the CCOs have become objects of public scrutiny (Lund-Muzikant, 2016).

Every public program creates its own set of institutions, bureaucracies, defenders, and detractors. A sweeping reorganization like the formation to CCOs has particularly large impacts. Building consensus on the policy goals of a CCO-like initiative and maintaining focus on their attainment require persistent skillful leadership from the Governor's office initially but similar skills from legislators who enshrine these policies and subsequent administrations who carry them forward.

## CONCLUSION

In terms of total public expenditures, Medicaid has now surpassed Medicare in size. Even if recent lower health care inflation persists, the sustainability of state Medicaid programs will be an increasing challenge for federal and state officials. It will continue to be a primary vehicle of insurance coverage for low-income populations. The increasing number of older Americans, particularly those with lower incomes and future assets, also means the program's demand on states—and the rest of their budget priorities—will increase.

In the face of these increasing budgetary demands, more states will continue to search for solutions that extend beyond reducing program eligibility and cutting provider rates—both of which merely shift expenses to other payers. Such solutions will involve changes in benefits and modes of service delivery, often coupled with a mechanism to force consideration and adoption of these measures, such as some form of global budgeting.

Early trials in these "Global Waivers" by states such as Rhode Island and Vermont offer lessons around the importance—and complexity—of negotiating the populations included, how to account for changes in economic conditions, the initial per-person rates and the targeted rates of increase as well as minimum standards for benefits, eligibility, provider network, and cost sharing. In developing its CCO model—including specifications for the governance of the risk-bearing entity, the granting in most cases of geographic exclusivity, and the adoption of a collaborative, transparent model for performance improvement—Oregon has deepened the scope of earlier global waiver proposals.

It is doubtful that other states could adopt the CCO model whole cloth—human nature and local politics dictate that a state develops health policy proposals that are customized to what is always cited as the "unique" or "special" nature of the state. However, the comprehensiveness of the CCO model is integral—geographic monopolies for health care delivery and financing are only conceivable with extensive accountability, transparency, and oversight.

The risks associated with such a model are significant. Will CCOs evolve into financial intermediaries and benefits administrators like MCOs? Will what is effectively a regional utility model—with public accountability and oversight—result in more social benefit than one predicated on individual choice? Will such a model function effectively between gubernatorial administrations and legislative assemblies whose leadership and priorities change? Who will cultivate and maintain political support crucial for such a program?

While such risks may seem daunting to a state leadership without the health policy expertise of Governor Kitzhaber and his administrative team, and the history of health policy innovation of Oregon, they are perhaps unavoidable given the size and scope of Medicaid in state government today and the demographic demands of tomorrow. Regardless of public sentiment about the scope of a state's Medicaid program, the herculean efforts required for the envisioning, negotiating, and implementation of fixed per capita health care financing and delivery organizations at the community level may be the minimum required for a sustainable, effective one.

## REFERENCES

Bradley, E. (2016). *Spending more and getting less*. Available from: https://es.slideshare.net/UHCF/e-bradley-quinnipiac42515.

Lund-Muzikant, D. (2016, May). *CCOS very profitable in 2015. The Lund report*. Available from: https://www.thelundreport.org/content/ccos-very-profitable-2015.

Rogan, E., & Bradley, E. (2016). *Investing in social services for states' health: Identifying and overcoming the barriers*. Milbank Memorial Fund.

Snyder, L., & Rudowitz, R. (2016). *Trends in state Medicaid programs: Looking back and looking ahead*. The Henry J. Jaiser Family Foundation. Available from: http://www.kff.org/report-section/trends-in-state-medicaid-programs-section-5-managed-care-and-delivery-system-reform/.

Chapter 17

# The Health Care Journey to Population Health: Guideposts From the Oregon Experience

**Somava S. Stout[1,2,3], Liz Powers[4], Rebecca Ramsey[5], Jennifer Richter[6], Kevin Ewanchyna[7] and Kristen Dillon[8]**

*[1]Institute for Healthcare Improvement, Cambridge, MA, United States, [2]Cambridge Health Alliance, Cambridge, MA, United States, [3]Harvard Medical School Center for Primary Care, Boston, MA, United States, [4]Winding Waters Medical Clinic, Enterprise, OR, United States, [5]CareOregon, Portland, OR, United States, [6]Yamhill Community Care Organization, McMinnville, OR, United States, [7]Samaritan Health Plans, Corvallis, OR, United States, [8]PacificSource Columbia Gorge CCO, Hood River, OR, United States*

> *I'm on the bus; in fact, I'm driving the bus. I'm "all in" for population health, for addressing the social determinants of health, for partnering with communities. I see that this is the only way we will achieve the Triple Aim. But I have no idea how to get from here to there. I'm in charge of an infrastructure of buildings and technology and staff designed to take care of people when they are sick. How do I transform my infrastructure and business model from a sick care system to a system designed [to] support people to thrive?*
>
> CEO, Mid-Sized Safety Net Health Care System

US health care delivery systems are being asked to undertake a monumental transformation on the journey from volume to value. As payment models shift from fee-for-service to alternate models based on global budgets or shared savings, health care organizations are feeling the shift from producing acute, largely reactive sick care services to prevention and proactive population health (Auerbach, 2016). It is difficult to think of another major industry that needs to so fundamentally change what it is about. There is no roadmap for this journey—it is an expedition into a new frontier of large-scale transformation. As health care leaders embark on the journey, they are building the plane as they are flying; the experiences of states like Oregon that are a decade into the journey provides critical insight on the routes that pioneers are discovering that are relevant nationally. This chapter offers

Health Reform Policy to Practice. DOI: http://dx.doi.org/10.1016/B978-0-12-809827-1.00017-3

a brief description of Oregon's transformation in a national context and a glimpse of stories at the leading edge of transformation. We use the stories of Oregon's transformation to delineate a potential framework that illustrates the interconnected portfolios of population health that health care systems learn they need to take on to achieve growth of health, well-being, and equity for people and places in the country.

The notion of the "Triple Aim" (Whittington, 2015) introduced by the Institute for Healthcare Improvement (IHI) in 2008, of improved population health, patient experience, and lower health care cost (Berwick, Nolan, & Whittington, 2008), has been adopted by hundreds of health care organizations and reform efforts in all parts of the country and at local, state, and national levels. As hundreds of health care systems have gone on the Triple Aim journey, in the context of the Affordable Care Act, several findings have become apparent: (1) health care systems have been less successful at forming partnerships with other sectors, in part because they do not have relationships with them. As a result, the vast majority of systems tend to work within their own walls, although the vast majority of what impacts health exists where people live, work, and play; (2) while the majority of high-risk, high-cost patients had social and behavioral drivers of poor health and cost outcomes, the vast majority of health care systems have not integrated this into their transformation strategies; and (3) while health care systems on the journey recognize these changes, there are few supports to help them make this transition; and the importance of shifting culture (Whittington, 2015). The level of collaboration across sectors that is needed to truly achieve population health, the integrated data systems that are needed to improve outcomes, and the financing mechanisms to sustain the journey simply are not in place. These collaborations need to be developed in order to support a critical mass of health care systems to go on this journey—along with a vast cultural shift that reorients everyone from the frontline to the C-suite about what they will need to do differently (Pollack, 2015).

Halfon et al. (2014) have described this shift as going from Era 1.0 to 3.0. In 1.0, health care operates as a sick care system. In Era 2.0, there is an effort to develop a coordinated health care system—with the patient-centered medical home and accountable care organizations serving as the primary building blocks for the journey. In Era 3.0, there is an effort to develop a community integrated health system that focuses on "population and community health outcomes; optimizing the health of populations over the life span and across generations" (Halfon et al., 2014). Major foundations and public health leaders have suggested that this transformation needs to be contextualized within a major cultural shift that is needed across sectors to prioritize health and well-being for both people and places (Lavizzo-Mourey, 2014).

The 100 Million Healthier Lives initiative (www.100mlives.org) was convened by IHI in 2014 to help accelerate the journey to population health, well-being, and equity with an emphasis on facilitating the shift in health care from Era 2.0 to Era 3.0 and supporting the development of a culture of health in people, organizations, and communities (Lavizzo-Mourey, 2014). Hundreds of partners joined the initiative. They recognized that an innovative and collaborative system was needed to realize transformative health improvements for 100 million people by 2020. To this end, the partners identified six core strategies:

1. Create thriving, equitable communities
2. Build bridges between health care, community, public health, and social service systems
3. Create a health care system that is good at health and good at care
4. Create peer to peer support systems
5. Create enabling conditions
6. Develop new mindsets.

As partners began to map out the different ways in which health care systems go on the journey to population health, we looked to places and organizations that were further along the journey—Vermont, Oregon, Kaiser Permanente, Southcentral Foundation (Alaska), Cambridge Health Alliance—to understand how health care systems were making progress (Koch, Stout, Landon, & Phillips, 2016). We learned, for instance, that transformation most often occurred in stages, in connected portfolios of work, often starting with a specific issue or population that was a common concern of partners across sectors. The work together on these first issues built the groundwork for collaboration—trust, an understanding of each other's assets, etc. In addition, the work often fell into related, balanced portfolios of work, which were previously separate and became more and more interconnected over time. They most often begin addressing issues related to their patient population, and as they change their financing and go deeper and deeper, begin to understand the interrelationship between the health and well-being of people and places and develop the trusting relationships that lead to better understand how real change is possible.

Over time, from studying these advanced systems and bringing our collective thought leadership together, the 100 Million Healthier Lives Health Systems Transformation Hub developed the following framework, which has served as a useful organizing system (Stout, 2017). It is important to note that in this initiative, health encompasses the World Health Organization domains of mental, physical, social, and spiritual well-being for people and places (Stiefel, Riley, Roy, Ramaswamy, & Stout, 2016). Rather than framing a linear journey toward population health, we have observed that health systems approach this journey based on an evolving understanding of their stewardship responsibility and impact. In Levels 1 and 2 of the framework, health care systems focus on patients and employees as the focus of the change effort. Levels 3 and 4 reflect an expanded sense of stewardship for place and a recognition that the well-being of people is directly related to the well-being of the place—the neighborhood and community—they live in. A health care system likely will have a portfolio of work across multiple levels—but rarely thinks strategically about these portfolios or balances them for impact. Health systems in the 100 Million Healthier Lives initiative are invited to strategically invest across these levels to maximize their impact on population health (Fig. 17.1).

***Level 1: Patient or Panel Health***—The focus is on proactively improving the mental and physical health of patients (or employees). Proactive population health management in the context of robust primary care transformation, integration of behavioral health services into primary care, extension of patient coaching and health education services—are all examples of activities at this level. Much of Oregon's transformation has focused on this level, with early evidence of significant results (Goldberg, 2016).

***Level 2: Patient Well-Being***—Health care organizations at this level screen for and address the social and spiritual drivers of health and well-being. Social drivers encompass both social connectedness and socioeconomic factors, such as food, housing, transportation, and income. Spiritual drivers describe factors that affect a sense of purpose, meaning, hope, and resilience. Examples of activities in this portfolio would include screening for social service needs for a segment of the population and making referrals to local social service agencies with tracking to completion. Many health systems start out doing these activities for high-risk, high-cost patients in their capitated portfolios, and expand the pool to include patients who are at rising risk or who are in the population as a whole because they recognize that these social factors drive future cost. The Pathways Community Hub model, which employs community health workers (CHWs) who screen people referred by any organization in a community comprehensively in

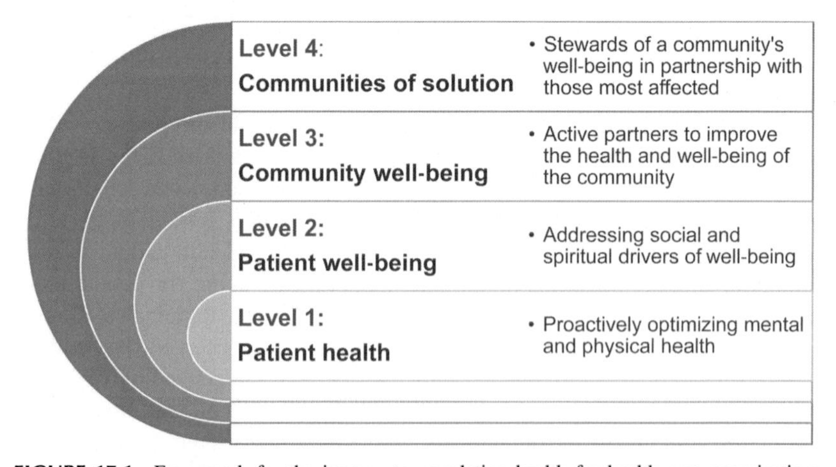


**FIGURE 17.1** Framework for the journey to population health for health care organizations. *From Stout, S., et al. Health System Transformation to Population Health–Emerging Framework, 2017, 100 Million Healthier Lives, Institute for Healthcare Improvement.*

terms of medical, social, and behavioral risk, places them into a pathway, tracks the pathway to completion, has linked incentives to pathway completion, and represents an advanced example of activities in this level (Redding et al., 2015). In more advanced systems, there may be interoperable integration between electronic medical records and the systems of social service agencies. Faith–health connections might be nurtured, as appropriate, and life coaching, education, and employment referrals might be made as a routine part of care. There is emerging evidence of the impact of addressing social and behavioral needs and ample evidence that these drivers are the predominant drivers of outcomes (McGinnis & Foege, 1993).

***Level 3: Community Well-Being***—The health care organization is committed to improving the well-being at the level of a community and is a partner at the table along with other leaders in public health and other sectors to support the health and well-being of a community as a whole. They might begin with a particular issue or a particular place that is of mutual concern across sectors and grow into full collaboration on a wide range of complex challenges. Early collective impact efforts fit into this level. They begin to discover what they can do together that they cannot do alone and begin to learn how to create system change together. Engagement may be by people in community benefits and occasionally staff engaged in an accountable care organization or patient-centered medical home. From childhood obesity to tobacco control to childhood asthma to efforts aimed at revitalization of a particular neighborhood, efforts at this level typically address concerns that are less complex and more amenable to change—the "low hanging fruit."

***Level 4: Community of Solutions***—At Level 4, health systems see themselves as stewards of a community's well-being, in partnership with leaders of diverse roles and levels in a community. People with lived experience (community members/supports who have direct experience with a challenge) are seen as crucial to the process and provide leadership and support in addressing the health and well-being of the community. Design and coproduction approaches are used to rapidly improve processes to meet the needs of people, and staff is allocated to support the process. Health care organizations adopt an anchor institution approach, such as the one described in the Democracy Collaborative monograph, *Can Hospitals Heal America's Communities?* (Norris & Howard, 2015). In this approach, health care leaders, along with leaders from other sectors, see themselves as long-term stewards of a community's well-being and leverage their assets in traditional and nontraditional ways to improve the health and well-being of a community, in accordance with the community's priorities. A health care system, applying this kind of anchor institution approach, might use its lever as an employer to support the cradle to career pipeline in a community of concentrated poverty or as a food purchaser, having an impact on the local economy for healthy fruit and vegetable availability within a food "desert." Together, they facilitate the removal of barriers, create sustainable policy change, and have the trust and relationship to move funds across sectors to meet the needs of the community in a dynamic, responsive, and proactive way.

These four levels are not linear; they represent nested portfolios of effort. At each level or portfolio, there are different levels of effort, different stakeholders to engage, and different financing, data, and policy systems to build. These levels are not discrete from one another. They are intended simply as an organizing system to help make a complex journey a little easier to navigate. The best way to understand what this might look like in real life is to examine the stories of real health care systems on the journey to population health. The Oregon transformation journey offers a glimpse of what this looks like against a policy backdrop of multipayer financing reform. The following stories, told by leaders at the frontline of change in their words, with a focus on describing both what has worked and what has been challenging, offer a rare glimpse of the journey.

## THE JOURNEY TO POPULATION HEALTH FRAMEWORK: PORTFOLIOS OF OREGON PRACTICE

### LEVEL 1: PATIENT OR PANEL HEALTH

#### Winding Waters Medical Clinic—Enterprise Oregon—Wallowa County—Northeast Oregon

**Liz Powers**
*Winding Waters Medical Clinic, Enterprise, OR, United States*

Wallowa County is a vast, rugged, mountainous frontier region in Northeast Oregon that is home to a ranching community of 7046 people, with only 2.2 persons per square mile. The remote location, sparse population, high unemployment rate, generational poverty, and limited services present significant barriers to health, historically producing high rates of poor physical health and disability.

Winding Waters Clinic was started 56 years ago by a pioneer doctor who practiced solo for 15 years until being joined by a second physician in 1975. In 2007, the clinic transitioned to a Rural Health Center. This brought additional stability to the practice, yet Winding Waters continued to struggle to meet community needs. Physicians in Wallowa County practice 66 miles from the nearest medical services and 154 miles from a hospital with interventional cardiac services. The clinic was faced with provider burn-out, reactive health care, excessive ER utilization, patient frustration, and poor health outcomes.

An elderly patient on a blood thinner was prescribed an antibiotic, and died a few days later of a hemorrhagic stroke, with his blood too thin. This was a tipping point for the clinic. When the clinic was recruited in 2008 to be a part of a demonstration project to help safety net primary care clinics become high-performing patient-centered medical homes, the Commonwealth Fund's Safety Net Medical Home initiative, Winding Waters seized the chance to begin its transformation journey.

Winding Waters focused on enhanced access as its first priority, expanding its regular hours from 30 hours per week to 64 hours per week (including evening and Saturday hours), moving to open access scheduling, providing walk-in/urgent care services for all comers, and providing 24/7 access to a physician via an on-call cell phone and a patient portal.

Given the pressures of provider shortages and clinical quality concerns, the transition from traditional care to team-based care was a natural next step. But it required significant cultural change: providers need to relinquish some control and staff needed to assume new levels of responsibility and accountability to patients, to their colleagues and to themselves. These early changes were too difficult for some staff. Change resistance and staff frustration resulted in the clinic losing a provider and suffering 50% staff turnover. Yet, despite the strain of losing staff, this proved to be instrumental in building a new culture focused around creativity, collaboration, and innovation to meet patient needs. Management leveraged defined transition points (adoption of an EHR, move to a new building) to change standard work, maximizing function, and efficiency of teams.

Other team changes followed. In 2012, Winding Waters partnered with the county mental health provider to co-locate a behavioral health coach within the clinic. Providers and patients alike immediately recognized the benefits of expanding the team's skill set and in 2013 a Health Resources and Services Administration (HRSA) small-practice quality improvement grant supported adding a full-time behaviorist and a full-time nurse care manager. Today there is a full-time nurse care manager as well as two full-time behavioral health coaches and a licensed clinical social worker together providing both health coaching around behavioral change and mental health counseling services within the clinic.

While the change was not easy, it has had dramatic results with Winding Waters' patients significantly reducing their need for acute care. For hospital visits, there has been an average of 7.25% reduction per year over a 6-year period (32.5% since 2008); for ER visits, an average of 6.4% reduction per year over a 6-year period (28.7% since 2008). Patients have also seen significant improvements in their health. Over a 3-year study period of the impact of adding behavioral health, 60% of obese patients lost weight, hypertension control (BP < 140/90) improved from 56% to 66%, diabetes control (HgbA1c < 8) improved from 64% to 74%, and cholesterol control (LDL < 100 for diabetic patients) improved from 54% to 61%.

Over the years, Winding Waters' definition of "team" has changed to also include community partners. In 2012, Winding Waters helped to establish a network of community providers (traditional and nontraditional care providers including the local Critical Access Hospital, primary care, alternative practitioners, pharmacies, public health, mental health, and community resource organizations) called the Wallowa Valley Network of Care. The Network meets monthly to identify needs and connects clients with needed services. In 2015 a collaboration between Winding Waters Medical Clinic (community health center), Wallowa Valley Center for Wellness (county mental health), and Building Healthy Families (early childhood education) created the Consortium for a Healthy Community. One result: In April 2016, Winding Waters and Building Healthy Families received an award for their partnership to improve developmental outcomes and kindergarten readiness in children 0–5 years of age.

Despite all the clinic changes and community partnerships, Winding Waters staff continued to face needs they could not meet: A 5-year-old who could not concentrate at school because of a tooth ache. A diabetic out of control because she could not afford insulin and did not have transportation to get to her medical visits. In order to truly impact health for the most vulnerable in Wallowa County, the clinic needed additional resources to expand the scope of services, and capacity to reach the most vulnerable people in the community. As a result of an extensive application process, Winding Waters Medical Clinic was recognized as a Federally Qualified Health Center (FQHC) in August 2015. This made additional resources available to expand both the clinic's internal team and staff's ability to connect with community partners.

With these additional resources, the clinic has grown from a staff of 27 to a staff of 48 during the first year of FQHC operations. This is making a positive difference to patients by improving access to care and services, and to the community by increasing the availability of family wage jobs. Health Center initiatives include the addition of dental care, expanded mental health services, acupuncture, transportation, child-care services, and telehealth. As a health center, Winding Waters is helping to overcome barriers to health (providing in some cases transportation, medication, housing, and food) and is working toward impacting factors that drive health outcomes (kindergarten readiness, overall education levels, employment status). The recent addition of a data analyst will allow the health center to highlight and address areas of ongoing need, and continue to drive change to improve health.

Patient and community response has been both validating and disheartening. Patients stop staff on the street to thank them for expanded access. Providers come to clinic with stories of patients stopping in front of them in the cross walk while they are driving to give a wide toothy grin, showing off their new dentures. Yet there is still unmet need and multiple challenges within a fragmented health system. It will take continued experimentation and innovation to achieve full integration and the most efficient use of resources. Winding Waters staff, patients, and community partners have traveled far on the journey of health care transformation, and look forward to what still lies ahead.

## LEVEL 2: PATIENT/PANEL WELL-BEING

### The Health Resilience Program: Addressing Mental, Physical, Social and Spiritual Needs

**Rebecca Ramsey**
*CareOregon, Portland, OR, United States*

Coordinated Care Organizations (CCOs) were created with the expectation that an integrated, holistic care model would be able to better meet the needs of Medicaid's most vulnerable enrollees, those with complex medical, behavioral, and social challenges, particularly those with high costs and acute care utilization as well as poor health and treatment outcomes. CareOregon's Health Resilience Program (Vartanian et al., 2016) now operates in 32 primary care and specialty clinics in four counties and three CCOs, including across the Portland metro region, partnering specially trained outreach workers with the clinic teams. These "Health Resilience Specialists" (HRSs) are Master's level behavioral health clinicians, either social workers or psychologists, with extensive experience in community mental health/addictions, and with interpersonal skills that foster engagement and relationship building. At any given time, the Health Resilience team is working with 700–800 individual Medicaid and Medicare members.

Soon after its launch in 2012, the program recognized that often-unrecognized psychosocial factors contribute to these patient's health challenges and poor outcomes:

- High prevalence of abuse, adversity, and trauma throughout their lives, resulting in difficulty problem solving and planning
- Distrust of traditional authority figures, including medical professionals
- Inadequate access to mental health assessment and services
- Cognitive and health literacy challenges
- Social isolation, lack of basic resources

Furthermore, 60% of the HRP population has substance use disorders; 23% has both substance use and mental health issues.

HRSs are embedded in clinical practices, where they are part of the care team. They spend the majority of time in the community meeting clients "where they are at," in their home, at a coffee shop, or wherever the client feels most at ease. HRSs work with clients over a period of 3–9 months, not only coordinating with their medical team, mental health team, substance abuse team, social service providers, and/or caregivers, but also by just "being there" for the client in a nonjudgmental, honest, and transparent "trauma-informed" manner. The program approach is to help members learn how to meet their own needs by "walking alongside" them and fostering their own resilience. This may mean accompanying them to their first mental health appointment, helping them choose an assisted living facility, purchasing for them a gym pass or a warm sleeping bag if they are living on the streets, or taking them grocery shopping while mentoring healthy food choices. One of the most important and easily underappreciated aspects of this program involves the HRSs modeling healthier methods to deal with the multitude of stresses that result from living in poverty; they may invite a client to take a walk in the Zoo, accompany them to a socialization center, or teach them mindfulness or relaxation techniques.

In many ways, the HRS acts as a "bridge" not only between parts of the health care delivery system, but also between the patient's lived experience—what has happened or is happening to them—and how the provider team perceives them. For many providers, this is an eye-opening experience.

CareOregon's Health Resilience Program also provides each HRS with weekly caseload supervision by a behavioral health supervisor, ensuring a more integrated approach. Staff self care is also a key component of the program given the impact of dealing with the daily struggles of the clients on those doing the work.

While the Health Resilience Program was developed with funding from a CMMI Challenge Grant (2012–15), it has been sustained and spread based on its strong support from clinic providers, high satisfaction from clients, and both acute care utilization and cost reductions. While both ED and hospital visits have decreased, primary care and behavioral health visits for the population have increased, with a net savings of over $3500 per program enrollee per year.

## LEVEL 3: COMMUNITY HEALTH AND WELL-BEING

### Yamhill Community Care's Early Learning Program

**Jennifer Richter**
*Yamhill Community Care Organization, McMinnville, OR, United States*

Yamhill County's nearly 100,000 residents focus on agriculture, forest products, manufacturing, and education and have an economy supported largely by its many unionized facilities such as Cascade Steel in McMinnville and White Birch Paper (Yamhill County, Oregon, 2016). Since its formation in 2012, Yamhill Community Care has devoted significant resources to upstream prevention efforts. In order to achieve the triple aim of providing better care to more people at lower costs, YCC recognizes the importance of preventative care and wrap-around supports that address the social determinants of health—including education and social support networks.

Before CCOs or Early Learning Hubs (https://oregonearlylearning.com/administration/what-are-hubs/) existed, partners from Head Start, Lutheran Community Services, and Public Health met regularly in a local breakfast café to discuss ways to partner to better serve the needs of families and children. Together they developed the Family CORE (Coordinated 0-5 Referral Exchange), which gave medical providers a simple way to refer families in need of additional supports to one central location. This partnership grew to include representation from Child Care Resources and Referral and Catholic Community Services to become the steering committee for the newly developing Early Learning Hub (www.yamhillcco.org/about-us/early-learning-hub).

Meanwhile, Yamhill Community Care appointed two of these steering committee members to its Board in order to actively address the intersection of health and education. When request for proposals (RFPs) were released for the development of Early Learning Hubs, the community felt that the CCO understood their mission to work collaboratively to provide coordinated systems that support healthy families and prepare children for success. Embedding the Early Learning Hub within the CCO has supported the work of the Family CORE and allowed it to grow. It has also facilitated the delivery of developmental screenings between early childhood and medical providers and has allowed for the pending creation of Maternal Medical homes, which support disenfranchised women before they give birth.

The development of the Early Learning Steering Committee into the Early Learning Council brought further partners to the table, including representatives from each of the county's seven school districts. This collaboration has allowed for several additional supports to the holistic wellness of Yamhill County families and children. These include cooking and nutrition classes, Reach Out and Read in select clinics, SNACK, and DASH (which address obesity and provide supports with nutrition, exercise, and other lifestyle choices). The implementation of Wellness to Learn integrated CHWs in each of McMinnville's six elementary schools to provide wrap-around supports for the families of children who might otherwise have fallen through the cracks.

During the 2016–17 school year, all kindergarten teachers, many preschool teachers, and child-care providers are being trained in Growing Early Mindsets, a curriculum that teaches children a growth mindset. Select schools are also implementing the Good Behavior Game and Family Check-up. Yamhill Community Care is also in the process of subcontracting with Lutheran Community Services to implement a Family Support Center in Sheridan School District.

Yamhill Community Care has also funded a community engagement coordinator who facilitates both parent education (in partnership with Polk County) as well as Service Integration Teams in each school district catchment area. These teams provide a platform for partners from social services, schools, the faith community, and more to address the needs of specific families.

Most recently, the collaboration in Yamhill County has facilitated the launch of YCC's Family Resiliency Community Conversations, a monthly gathering of cross sector providers to address factors that impact the development of strong children and healthy families, with topics from the societal value of quality child care to epigenetics to trauma-informed care in both clinical and educational settings.

Over the last couple of years, Yamhill Community Care has learned several key lessons. The intersection between Coordinated Care Organizations and Early Learning is becoming ever more apparent as we delve deeper into the social determinants of health. Having educators, social service providers, and health care leaders at the table, working on the same issues together, opens up ways to engage members in innovative ways that would not be possible within any one sector. While YCC has always sought feedback from all sectors, time has revealed that partner buy-in is even more critical. Stakeholders need to feel like they have a direct role in developing the system in order to retain a sense of ownership and investment in the outcomes. All partners whose contributions will be critical for eventual success should be invited to the table from the beginning in order to build one team, working together for the good of the community.

## HEALTH CARE CONTRIBUTING TO A STATEWIDE EFFORT

### Community Efforts to Reduce Tobacco Use

**Kevin Ewanchyna**
*Samaritan Health Plans, Corvallis, OR, United States*

The state of Oregon has made a concerted push to become tobacco free with multiple agencies and organizations working independently and together at every level, from state and local policy to taxes to behavior change efforts. Collectively these efforts across sectors have led to a 50% reduction in tobacco use statewide. The following story illustrates one health system's journey to contribute to these efforts in a coordinated way.

Samaritan Health Services (SHS), a 5-hospital integrated delivery system in the mid-valley and coastal region of Oregon, has long had a commitment to reducing the effects of tobacco on the communities in its three county service area. Years before CCOs, its own hospitals and clinics became tobacco free as did the associated clinics within its campus. They engaged multiple stakeholders within their system, solicited employee feedback and took particular care to ensure that some of their most vulnerable patients, those on the inpatient Psychiatry unit, received special support so that the policy did not complicate their recovery. But Samaritan still saw employees, patients, and community members using tobacco products close by the doors of the hospitals and clinics in the semiurban and rural communities that they serve.

Samaritan launched a broad taskforce to recruit other community organizations to take on tobacco use and prevention initiatives. Pharmacists, respiratory therapists, behavioral/mental health specialists, and independent primary care and specialty care providers all came together in the taskforce. For 8 months prior to their go-live date, they worked on branding and accessible messaging using a health literacy lens, with all key stakeholders adopting the same communication and marketing plan. As a result, four FQHCs joined the community campaign to become tobacco free and promote cessation, as did Oregon State University, with its nearby Corvallis campus of 25,000 students. The local Community College engaged as did the public health department. Campaigning to nearby homes and businesses also occurred ensuring that the community was engaged from the beginning.

The creation of the InterCommunity Health Network CCO (IHN) brought further opportunities to engage with new partners, particularly those from one of the northern coastal counties not affiliated with Samaritan Health. Community Health Centers and independent providers there subsequently took on the tobacco cessation campaign. The CCO also applied for and received a grant from an external funder to review and adopt tobacco cessation best practices and detailing to community practices. Success with tobacco cessation has led the CCO to take on other issues. The task force that formed around tobacco cessation has now become the CCO's Community Health Education HUB, responsible for convening and supporting initiatives focused on broad health issues. Its next objective is to improve outcomes for individuals with chronic pain as part of the statewide effort to deal with the State's prescription opioid crisis.

When the State Metrics Committee charged with setting the pay for performance measure and targets for the CCOs began discussing a tobacco cessation metric in 2014, the work of the IHN CCO helped demonstrate that CCOs could indeed effectively take on a broad community health issue. A tobacco cessation metric was adopted in 2016 and is now focusing the work of all CCOs on one of the State's leading causes of health morbidity.

The next steps in tobacco cessation for the CCO's Chief Medical Officer is a joint effort with the local Public Health Officer to convince their City Council and County Commissioners to pass legislation making it more difficult to sell tobacco products to those under 21.

In 2016, a new incentive metric regarding Tobacco Prevalence and Cessation methods was adopted by the Oregon Health Authority and the CCOs. Testimony by IHN CCO played a critical role in convincing State leaders that CCOs could and should take on addressing Oregon's leading cause of preventable death. Their work demonstrated the power of engaging not only providers within the health system, but also community leaders and organizations around issues with broad community health impact—one of the key competencies of the new Coordinated Care model.

## LEVEL 4: DEVELOPING COMMUNITIES OF SOLUTION

### Building Health in the Columbia Gorge

**Kristen Dillon**
*PacificSource Columbia Gorge CCO, Hood River, OR, United States*

The Columbia Gorge region is home to about 85,000 residents of Oregon and Washington. The region encompasses 10,000 square miles of rural and frontier land with only 2 cities of over 5000 people. Residents receive most of their health care from the four nonprofit hospitals and from affiliated and independent outpatient clinics. As is typical of rural areas in the United States, residents are generally older than average for their states. Twenty percent of the population identifies as Latino/Hispanic.

As part of Oregon's Medicaid reforms in 2012, the PacificSource Columbia Gorge CCO brought local leaders together in an unprecedented way, including the creation of an independent, nonprofit governing board for the CCO and other work, called the Columbia Gorge Health Council. With formation of the Health Council, the region took a crucial step forward in creating the relationships and accountability needed for collective decision making.

#### *Community Health Assessment*

The CCO was charged with conducting a Community Health Assessment (CHA), and leaders decided to scale up from all prior efforts. The 2013 CHA included communities in both states of Oregon and Washington, all four hospitals, six counties, three public health agencies, the region's Federally Qualified Health Center, and multiple nonprofit social service agencies as well as the nonprofit health plan administering the CCO. Bringing together quantitative inputs from mailed surveys with qualitative information from focus groups and users of services, the region's first Community Health Improvement Plan laid out shared priorities across the region's health organizations for the first time.

The CHA also met regulatory requirements for all participating institutions. For the hospitals, it created a plan against which to map community benefit spending and to support staff time to organize and resource the work. One of the region's hospitals, Providence Hood River Memorial, used community benefit dollars to fund a Collective Impact Health Specialist. He works at the region's United Way agency as a grant-writer and neutral convener of priority work. In the past 3 years, this program has brought in over $3 million in grant and federal funds to community organizations. The return in grant revenue has been 11 times the cost of the Health Specialist position.

As the program developed, the grant application process for individual hospital community benefit dollars has also been significantly simplified to reduce the burden on local applicant organizations. Starting in 2017, a

Regional Health Fund is being discussed to pool money contributed in equal amounts by the two Oregon hospitals and the CCO. This Fund structure would allow applicants to complete a single application and would increase alignment of the funders around shared priorities.

### *Collective Resources: Health Information Exchange, Community Engagement, and Community Health Workers*

Using shared savings dollars from the CCO, the Columbia Gorge Health Council has also been able to staff and fully fund deployment of a regional health information exchange. Following a full environmental scan, the decision was made to disband local efforts and merge with an organization that was already operating in another area of the state. Other technology investments have followed. The CCO health plan sponsor, PacificSource Health Plans, contributed staff time, and money to deploying the state's notification system for hospital and emergency department events, and the Health Council has licensed and deployed software specifically aimed at enrollment and care coordination. This technology aims to support the multiple social service and governmental agencies that do not have permission, nor the need, to access to the detailed medical records contained in the region's health information exchange.

The Community Advisory Council (CAC) of the Columbia Gorge Health Council is a legislatively required element of the CCO whose charge includes completion of the CHA. Legislation requires that a majority of participants are Medicaid members, and about 30 additional people attend meetings as nonvoting members, representing health systems and social service agencies. The Health Council has magnified the impact of this group through skilled facilitation that mediates power dynamics and continually elevates the voices of consumer members as experts in their own experience. The CAC's role has included writing a majority of the questions included in the mailed CHA surveys, approving the plan for listening sessions, and reviewing the final result to set the priorities in the Community Health Improvement Plan. The leaders of the health council and CAC aspire to have the group be a training ground for participation in leadership, whether in the CAC or other venues. The region's CAC has been recognized by the Oregon Health Authority as one of the most effective in the state.

Transformation of health care in the Columbia Gorge has included a priority on building and deploying an effective Community Health Worker (CHW) work force. A training center was established and has completed certification for over 80 CHWs, about half of whom had been working in the role prior to state certification standards. A community process was initiated to rethink how the community used CHWs given concern that the community's CHWs were spread across multiple social service agencies, were largely detached from health care institutions, and relied on grant funding.

CHWs were often restricted to work on a specific disease or intervention, when their clear preference was to be generalists who could serve the families and households they encountered in a holistic way. After multiple facilitated large group community meetings, a "Bridges to Health" care coordination hub model was agreed upon by 20 organizations to start in late 2016.

### *Preexisting Conditions and Local Factors*

The pace of transformation in the past 5 years rests on a foundation of organizational collaboration and personal relationships. The simple act of the CEOs of two competing hospitals initially agreeing to participate together in the CCO sent a signal to others in the region. Personal relationships among leaders in different sectors have allowed for aspirational conversations, risk-taking, and a willingness to continue to work together in the face of failure or disappointment. In a community where dual-allegiances are the norm, community leaders often serve on two or three boards and then as staff to another. In most cases, this arrangement has allowed for cross-fertilization and continued alignment between community organizations, as well as a careful process for understanding and disclosing conflicts of interest and possible divided allegiances.

While health systems have played a leadership role over the past 5 years, it has been as part of shared leadership with county government, social service agencies, independent health care providers, and community members. In addition, this work has gone on in parallel with clinical care changes, not as a step that followed optimizing panel management and population health within their own institutions. At least in this region, health systems have needed to work outside their walls to improve the resources in the community that are available to address food insecurity, unstable housing, or social isolation. As an experienced CHW said: "We're only as good as our resources." Health systems cannot address these issues within the patients they serve if the larger community's resources are deficient to meet the need.

The issues around deficient resources to meet basic needs brings up a painful reality for health systems. In many communities, the rising cost of health care has eaten into governmental budgets previously used for education, social services, and public programs like housing. Rising health care costs have squeezed household and governmental budgets in ways that make basic needs like quality child care and nutritious food at times unaffordable. Ironically, creating a community where residents' basic needs are met is likely a more effective strategy to build healthy communities over the long term, but a significant portion of the money that we need to do this work is going instead to expensive health care services.

## CONCLUSION

The challenge for health systems that truly want to partner in creating healthy communities is twofold. First, health systems need to right size their role in the work, to understand the nuance between doing the work, partnering in the work, and serving as an ally or funder to those doing the work. When health systems' leaders develop trust with other organizations and share power and resources, the work of creating health in a community accelerates.

Second, we can only move upstream if the resources to meet people's nonmedical needs are available. One of the drivers of scarcity in many communities is the money being spent on health care, both insurance and direct costs for care. Health care systems need to partner with payers and their communities to make themselves accountable for the cost and scale of their services in a transformed health care system.

The greater the development of trust and partnership, the more that payment systems can leverage the savings generated by creating a better system to grow and sustain the system. As flexible funds flow across sectors, more community members are part of leading the change, more of the work force is prepared in its culture and its competencies, and the more likely it is that health care organizations will be effective in making the transition from a reactive, sick care system to a health and well-being system. The stories of pioneering health care organizations in Oregon offer invaluable insight about how we might build transformation pathways for health care systems across the nation.

## ACKNOWLEDGMENT

The authors of this chapter would like to acknowledge the valuable support and contributions of David Labby, MD.

## REFERENCES

Auerbach, J. (2016). The 3 buckets of prevention. *Journal of Public Health Management & Practice*, *22*(3), 215–218.

Berwick, D., Nolan, T., & Whittington, J. (2008). The triple aim: Care, health, and cost. *Health Affairs*, *27*(3), 759–769.

Goldberg, B. (2016, December). *Lessons in safety net delivery system transformation: Oregon's health reforms*. Available from: http://caph.org/wp-content/uploads/2016/12/bruce-goldberg-caphsni-conference.pdf

Halfon, N., Long, P., Chang, D., Hester, J., Inkelas, M., & Rodgers, A. (2014). Applying a 3.0 transformation framework to guide large-scale health system reform. *Health Affairs*, *33*(11), 2003–2011.

Koch, U., Stout, S., Landon, B., & Phillips, R. (2016). From healthcare to health: A proposed pathway to population health. *Healthcare (Amst)*, Epub.

Lavizzo-Mourey, R. (2014). *Annual report*. Robert Wood Johnson Foundation. Available from: http://www.rwjf.org/en/library/annual-reports/presidents-message-2014.html

McGinnis, M., & Foege, W. (1993). Actual causes of death in the United States. *Journal of the American Medical Association*, *270*(18), 2207–2212.

Norris, T., & Howard, T. (2015). Can hospitals heal America's communities? Democracy Collaborative, Washington DC. Available from: http://democracycollaborative.org/healingcommunities

Pollack, S. (2015, April 27). *When health care transformation fails. Health Affairs Blog*. Available from: http://healthaffairs.org/blog/2015/04/27/when-health-care-transformation-fails/

Redding, S., Conrey, E., Porter, K., Paulson, J., Hughes, K., & Redding, M. (2015). Pathways community care coordination in low birth weight prevention. *Maternal and Child Health, 19* (3), 643–650.

Stiefel, M., Riley, C., Roy, B., Ramaswamy, R., & Stout, S. (2016). *100 Million Healthier Lives Measurement System: Progress to date*. Cambridge: Institute for Healthcare Improvement.

Stout, S. A. (2017). *Health system transformation to population health*. Cambridge: Institute for Healthcare Improvement.

Vartanian, K., Tran, S., Wright, B., Li, G., Holtorf, M., & Levinson, M. (2016). *The Health Resilience Program: A program assessment*. Portland: The Center for Outcomes Research & Education.

Whittington, J. N. (2015). Pursuing the triple aim: The first 7 years. *Milbank Quarterly*, *93*(2), 263–300.

Yamhill County, Oregon. (2016, December 16). Available from: https://en.wikipedia.org/wiki/Yamhill_County,_Oregon.

Chapter 18

# What Is Next for Oregon: Refining the CCO Model

Merwyn (Mitch) Greenlick
*Oregon Health & Science University, Portland, OR, United States*

## INTRODUCTION

Most social reform attempts either fail or require significant modification after a more or less successful start. Oregon's health care transformation experiment can be characterized by the latter. The transformation comprises the development of a set of Coordinated Care Organizations (CCOs), designed to provide or organize health care services for more than a million Medicaid-financed members of the Oregon Health Plan (OHP).

The original health care transformation bill was written during the 2011 session of the Oregon Legislative Assembly (House Bill 3650) when the House of Representatives, in which I served, was split 30–30 between Democrats and Republicans. There was a strong commitment by legislative leadership to craft a bill which, when signed into law, would bring nearly $2 billion from the Federal government, directly to Oregon's budget. But in order to reach political agreement on a final bill, many important conceptual differences were dealt with through obfuscation in the bill's final language, which passed each chamber with only a few dissenting votes.

The current group of 16 CCOs operates on a contract that is scheduled to expire during 2018. Negotiations or competition for new 5-year contracts will begin in 2018 for 2019–23 contracts. As such, there is now an opportunity to clarify a few critical issues that were left unsettled in the original legislation, but could have significance in the response to a call for the new contract.

This chapter reviews some of these areas of conceptual conflict and proposes my thoughts on changes in the design and conduct of the next generation of CCOs. It is clear that many other opinions will come forward as this debate emerges. I raise this specific set of issues because I believe that they are conceptually most critical. And I raise them in this forum in order to provide a conceptual basis for resolving them.

**Health Reform Policy to Practice. DOI: http://dx.doi.org/10.1016/B978-0-12-809827-1.00018-5**

Some of the changes proposed here could become effective at the beginning of 2019. The more dramatic changes could be designed to become effective at the start of the 2024 cycle, but would require plans for movement toward the new model during the upcoming contract period, adumbrating future design changes.

## THE CONCEPTUAL BASIS OF CCO DESIGN

In an article in *The Encyclopedia of Bioethics* (Garland and Greenlick, 1995), Michael Garland and I trace the development of health insurance models over the last 125 years and argue that for a variety of reasons two distinctive patterns have emerged. One pattern is the social insurance model. The second pattern is private (both nonprofit and for-profit) health insurance. For the past 125 years, these two patterns and their underlying ethical values have been the focus of political debate in the United States. The article points out that political liberals tend to frame the ethics of health insurance in terms of social solidarity, compassion, fairness, social justice, and rational government regulation (including government-owned and operated health systems, like those in the United Kingdom and the Scandinavian countries). The liberal value frame encompasses the idea of a right to health, both as a philosophical "human right" and as a legal goal to be established and enforced. The idea is embedded in the Universal Declaration of Human Rights adopted by the United Nations in 1948. Political conservatives, on the other hand, tend to frame the problems of health insurance in terms of personal responsibility to protect oneself from risk, behaving prudently, exercising freedom of choice, and relying on free-market dynamics to control costs and produce the highest aggregate social good.

The liberal view is expressed in America in social insurance models, such as Medicare, Medicaid, and the Veterans Administration health care systems. The conservative view has led to the development of private health insurance approaches currently available in the American marketplace. The former is designed to solve social problems such as the unavailability of health care for the seniors, the poor, and veterans with service-incurred injuries. The private health insurance market developed after World War II to facilitate the ability of private employers to offer health insurance as a benefit of employment, a fringe benefit that had emerged from wartime employment experiences.

The social insurance model produces health care systems, generally financed with governmental funds, providing comprehensive benefits, with the somehow vulnerable beneficiaries paying no or low premiums or out-of-pocket costs. The private insurance model features a wide range of benefit options, from very complete service benefit options to limited catastrophic indemnity products. Enrollees are generally expected to pay a relatively greater share of the premium costs and the plans are generally designed to cover a relatively smaller share of the medical care expenses. In this approach, employers have traditionally paid some of these expenses as a fringe benefit, and the government has played some role in the system through tax subsidies for employers providing fringe benefits.

In trying to accommodate these very conflicting views of the nature of health insurance in our attempt to transform Oregon's Medicaid system, we allowed structural flaws to emerge that could endanger the future stability of the transformation of the OHP, Oregon's unique Medicaid system. It is well to note that at the time we were putting the design of the OHP transformation in place we were also creating the Oregon Health Insurance Exchange. The Exchange, while facilitated by a government-organized marketplace, is heavily dominated by the private health insurance model. That seems appropriate to me, because the exchange is designed to make it feasible for individuals to buy private health insurance in an open marketplace, the ideal of the conservative approach to health insurance.

But that open marketplace concept is not an appropriate model for the OHP transformation. This task ultimately requires moving toward a purer social insurance model to achieve a sustainable system that can maximize the often cited "Triple-Aim"—to create a system that provides access to care for all in the population, is affordable, and improves the health of the population.

This chapter focuses on a set of CCO structural elements that need to be met to enable the transformation to grow into a stable and efficient health care system. The transformation was designed to test the hypothesis that it is possible to create a new production function for the delivery of health care services to the Medicaid population. The new way of delivering services would move us toward meeting the "Triple-Aim." As the transformation moves to the next phase, it is obviously necessary to ask some very basic questions, the most basic of which is "did some, or all, of the CCOs actually discover and implement a fundamentally new way to deliver health care services?" And there are many subquestions that derive from that question. For example, one of the principle characteristics of the model is that the CCOs were reimbursed on the basis of a global budget, comprising reimbursement for physical, behavioral, and oral health services. The underlying assumption was that an element of transformation included finding a way to integrate those services within the revolutionary new delivery model.

There is very little evidence that a revolution has occurred. While all of the CCOs have found some ways to improve the delivery of services to some of their members, most of the changes have been characterized as fairly traditional mechanisms at harvesting low-hanging fruit. For example, most CCOs have recognized that they could improve the service model by identifying members who tended to use the emergency department on a regular basis. They then put resources into coordinating the care needs of these members, including providing direct and easy access to primary care. This straightforward approach improved care for that small subgroup of the population and saved a great deal of money while providing improved care. A different story emerges with regard to integrating behavioral health services with physical health services. Some CCOs have embedded a mental professional into some primary care settings and there have even been examples of embedding a primary care medical provider into settings that provide services for those with severe, chronic mental health disorders. But there have been precious few examples of complete and seamless integration of physical and behavioral health services.

As we move toward phase 2 of the health care transformation, we need a systematic assessment of the extent to which the revolution has begun. And we need to reward the successful models and sanction the failures. But while we try to figure out how to accomplish that objective, there are some obvious matters that can be addressed that, if implemented, will improve the probability that the transformation will progress and thrive.

## CCO GOVERNANCE

How CCOs are governed is a central issue in looking toward the next phase of the OHP transformation. Two basic features in this system of care dictate how the system should be organized. The first is that the services are 100% financed by public dollars. Of equal importance is that the beneficiaries are, to some extent, medically, financially, and socially vulnerable. These factors impose a special set of safeguards that must be built into the design of the system to ensure that the public trust is safeguarded. Think about the difference between how our public schools are organized compared to how a supermarket chain is organized. Educating our children and delivering food to our families are both important functions in our society, but the functional responsibilities of governance are quite different in the two structures.

The tax-funded public school system is organized under quasi-governmental rules. Each school system is governed by a community-elected board, which operates under strict public-meeting rules. These rules require that meetings be held in public and require public disclosure of all financial transactions. School boards recognize that they serve an important social function, using taxpayer money to educate our future citizens. Consequently, it is clearly recognized that the boards are totally responsible to the citizens that elect them and recognize that they are operating under the umbrella of public trust.

By comparison supermarket corporations, while serving an essential social function, operate under an entirely different rubric. These corporations' main responsibility is to produce a profit for their shareholders. Their management strategy guides them to know that sustaining profits is dependent on efficiently and effectively serving their essential function. But returning profit to their shareholders is their main objective. These two types of organizations create very different organizational cultures, each culture designed specifically to facilitate achieving the organization's goals and objectives.

Governing the CCOs, and other social insurance models, should be much more analogous to the public school model than to the supermarket approach. Yet with regard to the 16 current CCOs, it is clearly a mixed bag, as you have seen in prior chapters. There are various governance approaches, including true community boards, investor-owned models, boards comprising a joint venture among various stakeholders, and CCOs owned by local physician organizations. Many are nonprofit, but most are some form of profit-making organizations.

*The first recommendation for the future governance is that each CCO be governed by a nonprofit, community-based organization.* The goal should be to have every CCO in the 2024 cycle organized this way. It can be argued that this is the only appropriate approach to maximize the long-range stability of the transformation. There is too much at stake, especially if CCOs are to provide a pathway to a truly transformed health care system for the population of Oregon generally. We must be able to absolutely trust that these organizations are operating solely in the public interest. That trust can only emerge if we can be confident that the sole objective of each organization is to deliver health care services in the most efficient and effective way possible. We cannot allow the organizational managers to become confused with the desire to return profits to organizational stakeholders.

We have already witnessed the purchase of one of the CCOs in Oregon by a Fortune 500 corporation. The original CCO owner (Agate Resources) was organized as a closed corporation in Eugene, with its stockholders being physicians in the community and officials of the CCO. The purchasing corporation, Centene, a national Medicaid contractor paid millions to purchase (merge with) Agate, resulting in individual Agate stockholders reaping millions as payoff for their agreement to sell their company. Centene's action is a cause for concern for many reasons, not the least of which is that Centene has recently abandoned the Medicaid market in two states, leaving the Medicaid program in those states holding the bag (The Lund Report, 2016).

There are two elements to that recommendation. The first is that the CCOs all become nonprofit, probably IRS 501-C-3 charitable organizations. The second is that the organizations become community-based. Free-market commentators have pointed out that the behavior of many nonprofit organization managers appears as entrepreneurial as managers of investor-owned businesses. And to a certain extent that is a correct observation.

But there is a significant difference between the social entrepreneurial behavior of nonprofit managers and that of managers of investor-owned companies. The nonprofit managers are not trying to maximize profits to be returned to stockholders, as are the managers of investor-owned companies. Richard Pratt, a former Chairman of the Federal Home Loan Bank, summed it up nicely with his statement "You can't expect private firms to do less in the pursuit of profit than law allows" (*Newsweek*, 1986). That is the heart of the matter. The CCOs must be expected to do everything possible in pursuit of the public, rather than the private, interest. [For a fuller discussion of the function of nonprofit organizations in the health care system, see Greenlick (1988)].

In aid of the second element of this first recommendation, that CCOs be community-based, come recommendations about the governance board of the CCO. House Bill 3650 gave far too much authority to the organizations that accepted risk in the CCO, leading to situations where the risk-bearing companies, usually hospitals, insurance companies, and/or physician groups, completely dominated the governance of CCOs. Further these privately dominated boards frequently make all critical decisions out of the glare of public scrutiny. That situation must change. By 2024 the governing board membership of each CCO should be constituted something like the following—50% local community members, openly selected in some way, 25% members representing risk-bearing entities in the community, and 25% members representing providers within the CCO.

In addition to the community-dominated board composition, it is absolutely critical that these boards operate under public-meeting laws and are liable under public information disclosure regulations. The situation that has emerged to this point is completely dysfunctional. Several of the CCOs operate completely in secret, coming to decisions behind closed doors, and hiding information from the public asserting that what is being hidden from public view is proprietary information. They make this argument even though all of this "proprietary" information completely relates to the expenditure of public funds.

It took a statute to force CCOs to hold *some* of the community advisory committee (CAC) meetings in public. In fact, at a hearing of the House Health Care Committee in 2013, a representative of a CCO told the committee that it objected to holding their CAC meetings in public because the membership of that committee was confidential and if the committee met in public its membership would be known. On the other hand, it is true that some CCOs have held all of their CAC meetings in public since the beginning.

The bill passed by the committee (House Bill 2960) forced CCOs to hold at least one CAC meeting a year in public. Hiding important information from the public threatens the sustainability of the transformation and must be changed as we move forward. Of course, it is difficult to change these governance matters instantly. Consequently, it would work to declare that these changes must be in place by 2024 and require the CCOs to outline their plans for moving as expeditiously as possible to this outcome during the coming 5-year contract period.

## PROTECTING CCO RESERVES

*The second recommendation is that the reserves of each community-based CCO be held in a community escrow fund in the State Treasury.* At the beginning of the program, each CCO had one or more organization that provided risk capital guaranteeing the stability of the operation. That was a significant public benefit contribution. But as the first contract period nears an end, each CCO has developed a reserve fund out of publically supplied operational funds. By the end of the first 5-year contract period, the reserves have grown substantially. OHA reported in the first CCO quarterly report that there were currently more than $900 million being held in CCO reserves.

As has emerged, these reserves have been provided, by public funds, and are held in the private account of each CCO. These funds belong to the community served by the CCO and must be protected for the benefit of that community. If, for example, Centene decides that continuing in the OHP market is no longer is its stockholders' best interest and pulls out of Oregon, the reserves they hold disappear with them. It is both possible and desirable to begin to move the reserves of CCOs into accounts in the Oregon Treasury to be held in trust for each community. I recommend that one-fifth of the amount required to be reserved be withheld, each year, from the capitation rate paid each CCO and that amount be deposited in a restricted Treasury account. The CCO would transfer an equivalent amount from its reserve account into its operating account to make up for its reduced capitation rate.

At the end of 5 years, each CCO's reserve would be tucked safely away in a Treasury account on behalf of the community served by each CCO. The Oregon Health Authority would establish, by rule, how funds could be withdrawn by the CCOs in order to protect the public interest. OHA could also determine by rule how, in the case of a change in CCO coverage in a community, a new CCO would or would not be able to access the accumulated reserves. But they would absolutely not be available for a CCO exiting the market to take with or to distribute to shareholders.

## NEW OUTCOME MEASURES

*The third recommendation is that a parsimonious set of outcome measures be created that comprise the minimum performance criteria for CCO continuation.* Currently, CCO performance is measured using a set of 17 "quality" measures that are used to distribute the portion of capitation held back to create a quality incentive payment pool. Each CCO has a performance benchmark on each of the 17 measures. Each CCO is judged annually on how well it did against performance targets on the 17 measures. The quality incentive pool is then distributed among the 16 CCOs on their performance, with CCOs meeting 13 of the 17 benchmarks getting a bigger payout than CCOs meeting many fewer benchmarks getting a smaller payout. The rules require all of the pool be distributed each year, regardless of the overall performance of the universe of CCOs and generally each CCO gets some quality payout, even if their performance was dismal. In addition, because of Federal regulations CCO performance is also reported on 16 additional outcome measures, creating a total of 33 performance criteria reported for each CCO.

That basically means none can be actually held accountable for their performance. If the Federal government has some reason to require reports on 33 measures, we probably have no choice but to send them that information. It is probably harmless if we choose to distribute a portion of the capitation rate on the basis of performance measured using 17 measures. That is, if the measures are valid measures of something. But that is not sufficient to actually produce a failsafe minimum performance standard for which a CCO can be held accountable. There must be a minimum performance required for a CCO to continue to contract to deliver services to the OHP population.

The Oregon Health Policy Board should be charged with this new task, to rethink the performance measurement process. This task asks the question "what defines a performance floor for any managed care organization to be allowed access to the market and to the public dollars that finance care under OHP?" As can be seen, this is bigger than just the CCO movement. But this exercise is about transforming the health care system, and the hope is that the CCO movement will shed light on how to transform all of health care. As argued, we have a special obligation when developing a social insurance model, but this is clearly a universal health care quest. It is the holy grail of health care measurement, defining what to expect from an organization that undertakes the delivery of health care services to a defined population.

The ideal would be the development of three measures—one measuring access to care, one measuring the efficiency of the system, and one measuring the care's effect on the health of the population served (or at least the quality of the care delivered). If we claim we are driven by the triple aim, we should be requiring that organizations with a critical responsibility in the system deliver against that aim. We have been seeking ideal measures for decades. Twenty years ago, it was noted "... existing population-based performance measures have minimal relevance to most consumers, yet their potential to shape the quality of care and hold health plans accountable is great" (Hanes & Greenlick, 1998). We have come a long way during these last 20 years, but we have not yet applied what we have learned to the transformation movement.

It is clear that this minimum set of measures will not comprise simple scales. It is likely that we will need to create complex indices measuring clear underlying concepts and then testing them in the real world to begin to explore pass/fail cut points on each of the developed indices. Since this has real-world implications, it will need to be developed and tested in the real world and understandable to critical stakeholders. There has been enough to criticize in the use of incomprehensible black boxes in the rate setting arena. While this process is clearly technical, it also needs to be understandable. But our technology is up to the task.

## ALTERNATIVE PAYMENT METHODS

*The fourth recommendation relates to alternative payment methodology.* While changing the way CCOs reimburse for care delivered within their system does not in its own right define a transformed system, the system cannot be reformed without dramatic changes away from the fee-for-service payment methodology. The State reimburses CCOs on the basis of risk-adjusted capitation rates. This comes about because of the basic nature of the Medicaid waiver that authorized the transformation. Oregon guaranteed that the annual inflation in the per capita expenditures for Medicaid services would be limited to 3.4%. At the time of the original waiver, the Federal government was predicting a Medicaid annual inflation rate of more than 5%. That guarantee provides a cap to the total Medicaid budget for the OHP. The problem then becomes how to equitably distribute the total budget among the CCOs providing OHP services. The risk-adjusted capitation rate is the answer.

But in the beginning when a CCO received its capitation rate, it had few or no restrictions on how it paid for services within its system. And since most organizations were operating within the existing community medical care system, it contracted to pay for services using the existing fee-for-service methodology or perhaps some form of discounted fee-for-service. This makes no sense on its face. There are many reasons why using fee-for-service methodology does not make sense for a capitated system. Not the least of which is that it provides all the wrong incentives for the service providers in the system.

When provider income is increased when the provider delivers more services, there is always the incentive to deliver more services, rather than delivering the right services. And there is no incentive for providers to use efficiencies, such as email and phone contacts when there is not a way to be paid for delivering these services. There is no incentive to bring personnel into the care model for which there is no fee-for-service payment available. Care coordinators, for example, cannot be reimbursed in the usual fee-for-service way, nor can other health care providers who increase the efficiency and effectiveness of the system.

Perhaps, the worst problem is that when ways are found to reduce demand for unnecessary or useless services, the income of providers is reduced. This leads to some very perverse incentives for providers paid on a fee-for-service basis.

There has been some effort among current CCOs to adopt "alternative payment methodologies." But most of the Alternative Payment Methods (APM) efforts have been to modify fee-for-service to include some pay-for-performance elements to the existing payment approach. That is not bad, but it still contains perverse incentives. Some CCOs have begun to explore capitation for primary care groups, or at least modified capitation approaches, adding capitation payments for some services to an underlying fee-for-service model. This has been a way to encourage the use of care coordinators into primary care. And there has been an effort to implement capitation payments for hospital services in some areas of the state, usually when a hospital is a key stakeholder in the CCO.

But as CCOs move forward into the next generation, the expectations for the level of APMs should be much greater and CCOs should be required to develop ways to move 75%–80% of their total reimbursements for services to APM approaches. This will be a stretch goal for some CCOs, but not impossible to meet. And moving in that direction will provide unexpected dividends for the CCOs in reducing the complexity of managing a delivery system.

To facilitate the movement away from the fee-for-service reimbursement models, it is incumbent on the Oregon Health Authority to reduce the demand for claims-form data, because it is inherently useless, even destructive, to demand claims data when providers are not being paid on the basis of submitted claims-forms. It is an administrative nuisance and produces data of questionable quality. Changing the payment approach obviously does not obviate the need for data, but it requires the development of imaginative new ways to gather data. As we get nearer to the time of universal use of electronic medical records, new data gathering methods should be relatively easy to develop. And these new methods should provide data that are much more likely to reflect the reality of care delivery.

## NEXT STEPS

First, it is necessary to say that the issues discussed here are not the only issues to be resolved as Oregon moves forward on the journey to transforming its health care system. But these issues seem to me to be the most critical and the most difficult to achieve. Implementing these recommendations is not going to be an easy task. This will take action by the State Legislature, the waiver of some regulations by the Center for Medicare and Medicaid Services, work by the Oregon Health Policy Board, changes in OHA's administration of the CCOs, and a great deal of change on the part of the CCOs themselves. And these things will not take place overnight. Careful thought should be put into phasing the steps in a way that makes them feasible and reduces unnecessary trauma to be faced by individual CCOs. It will help the extent to which the CCOs grow to believe that it is ultimately in their interest, and is definitely in the public interest, for these changes to be introduced over the next several years. The sustainability of the transformation will depend upon our ability to move the systems to the next phase. Or we face the threat of the demise of this great social experiment.

## REFERENCES

Greenlick, M. R. (1988). Profit and nonprofit organizations in health care: A sociological perspective. In J. D. Seay, & B. Vladeck (Eds.), *In sickness and in health: The mission of voluntary health institutions*. New York: McGraw-Hill.

Hanes, P., & Greenlick, M. (1998). *Grading health care: The science and art of developing consumer scorecards*. San Francisco, CA: Jossey-Bass.

Garland, M. J., & Greenlick, M. R. (1995). Health Insurance. In W. T. Reich (Ed.), *Encyclopedia of Bioethics*. London: Macmillan Publishing Company.

Pratt, R. (1986). *Newsweek*, November 10.

The Lund Report. (2016, June 22).

Chapter 19

# Leading by Example: Why Oregon Matters in the National Health Reform Discussion

**Senator T. Daschle**[1] **and Piper Nieters Su**[2]
[1]*The Bipartisan Policy Center, Washington, DC, United States,* [2]*McDermott + Consulting, Washington, DC, United States*

## THE AFFORDABLE CARE ACT EMERGES TO BUILD MOMENTUM FOR COVERAGE EXPANSION AND HEALTH SYSTEM TRANSFORMATION

On January 14, 2016, nearly 7 years into his presidency, President Obama was asked during a Twitter town hall to share his greatest memory of his time in office. Adhering to the medium's strict 140-character limit, he responded frankly, "the night the [Affordable Care Act] passed; standing on the [T]ruman balcony with all the staff who'd made it happen, knowing we'd helped millions." Some may have been surprised by his choice, given the number of historic events that have occurred during his tenure—from the rescue of the US financial system from near collapse to the capture of Osama Bin Laden. But students of history will agree that the passage of comprehensive health reform, following decades of failed attempts to overhaul our nation's health care system, may indeed be the most notable accomplishment of the last decade, both in terms of real-life impact and political considerations.

It would be impossible to underestimate the challenges faced by those who have tried to right-size a system that involves not only questions of illness, life, and death, but also millions of jobs and almost one-fifth of our nation's economy. And yet in 2010, after a year's worth of emotionally charged legislative debate, near death scenarios, and tortured congressional procedure, President Obama signed a sweeping health reform law that would reinvigorate and realign the engine that drives health care coverage and delivery across the country.

In the spring of 2008, most Americans focused their attention on the dramatic progression of the primary elections that would determine the

**Health Reform Policy to Practice. DOI: http://dx.doi.org/10.1016/B978-0-12-809827-1.00019-7**

candidates for the presidential election later that fall. John McCain emerged early on as the favorite for the Republican nomination, but a bruising battle was being waged on the Democratic side of the aisle. Political pundits first assumed that Senator Hillary Clinton, well known and supported by many in the party establishment, would emerge the victor despite a surprisingly strong challenge from a young, lesser known Senator from Illinois, Barack Obama.

As the 2008 race carried on, however, it became apparent that then Senator Obama's populist message of change and optimism was resonating with voters across the spectrum. And the need to improve health care for all sat at the heart of that message, as American families and businesses grappled with dramatically increasing costs in the system. Health care spending was growing at a rate double that of overall inflation, insurance premiums were up 33% over just a 5-year period, and the number of Americans without health insurance had grown to an alarming 46 million (over 15% of the US population) (DeNevas-Wait, 2009; Komissar, 2013; Schoen, 2016). And yet, for all of our spending, people's confidence in the system was declining, as anecdotes and academic studies both pointed to a subpar quality of care compared to international peers (Davis et al., 2007).

The emergence of health care as a central focus in the 2008 primary was not necessarily a factor that would play in Senator Clinton's favor, as many voters still associated the former first lady with the unsuccessful plan to overhaul the health care system in 1995, notably termed "Hillarycare." To the extent that Senator Obama was able to frame the 2008 primary around themes of "change," voters may have associated Senator Clinton's failed attempt to enact reform in the 1990s as evidence that she could not deliver on those themes. As such, while Senator Obama won the nomination and battled Senator McCain in the general election, policy experts in DC began to speculate about his legislative agenda should he win. What issue would he tackle first, and had the time finally come for policymakers inside the Beltway to address the daunting challenges of comprehensive health care reform?

Health care thought leaders quietly began developing possible approaches to reform should the issue take center stage in the new Administration, while Democratic leadership debated the wisdom of such a choice. Many argued that a battle over health care, which would inevitably focus on fundamental differences in opinion on the role of government more broadly, would be too risky to take on as one of the President's first initiatives. Every memorable attempt at wholesale reform in recent history had been met with failure, and the notion that a President would choose such an uphill battle as his first collaboration with Congress was viewed as politically precarious. And yet, soon after President Barack Obama stood in front of the United State Capitol and took the oath of office, it became clear that improvements in health care coverage and quality were the spoils that Democrats sought to secure with their newfound political capital.

For those who believed that this time the effort to reform health care could be successful, several considerations paved the pathway to enactment.

With the presidential election looming, the fall of 2008 brought the most severe economic downturn that the United States had seen since the Great Depression. "Too big to fail" proved to be an untrue mantra for the nation's banking system, as national unemployment rates skyrocketed to nearly 10% and families saw their savings and investments evaporate almost overnight (US Bureau of Labor Statistics, 2012). Federal and state governments struggled to respond to the unprecedented crisis as tax revenues rapidly declined and demand for public services increased. Policymakers were forced to consider legislative responses that were tremendous in scale and would have been politically impractical in any other environment, reinforcing the idea that government intervention was an essential response where more organic market dynamics were failing.

Addressing this crisis began even before the November election, as Congress was crafting a bailout package to shore up the nation's largest banks and experts began to contemplate a legislative package that could stimulate the economy. For many, this crisis mandated that the next President's early agenda focus on economic recovery and include only those policy initiatives that could directly support such recovery. Accordingly, one of the biggest questions for President Obama's team that was responsible for ensuring a smooth transition from one administration to the next was whether health care reform met that condition.

For those arguing in favor of tackling health reform, there were significant opportunities to highlight its connection to the country's broader economic woes. Health care spending represented over 16% of GDP in 2008, and businesses looking to cut costs during the downturn viewed skyrocketing premium increases as an unsustainable financial threat that would force them to reduce or withdraw employee coverage altogether (The Henry J. Kaiser Family Foundation, 2011). American families struggling to meet even their most basic expenses found the cost of insurance and services untenable, often dropping coverage and postponing medical care indefinitely. Notably, health care debt was also cited as the number one cause of personal bankruptcy filings around this time (Himmelstein et al., 2009). Amidst this turmoil, costs were rising, access was diminishing, and a unique opportunity was emerging to drive change by launching a policy initiative that many considered long overdue. But the ultimate shape of the health reform law would undeniably be impacted by the concerns and context of the time, and very few would have predicted what a political flashpoint it would remain for years to come.

The development of the Affordable Care Act was no small feat, and the law aimed to accomplish three primary goals: (1) ensure access to affordable health coverage for all Americans; (2) improve the quality of health care; and (3) curtail spending growth in health care. It was an ambitious goal for a system overrun with complexity and relatively little accountability. Moreover, Democrats leading the postelection strategy for both Congress and the White House quickly concluded that the push for reform must have bipartisan support and could not add to the federal deficit.

As a result of these combined considerations, the substantive debate on health care coverage began to shift toward a private market solution that would rely on regulated insurance market competition to provide coverage and contain costs rather than establishing a universal, public plan operated by the federal government (sometimes called "Medicare for all"). In fact, this approach was modeled on the reform initiative in Massachusetts that was led by Mitt Romney during his tenure as governor and had successfully bridged the ideologies of both Republicans and Democrats in the state. The assumption that using this model would appeal to Republicans in Congress by potentially limiting the government's financial obligation would ultimately prove to be somewhat misguided but the die was already cast—coverage reform was to be built upon commercial insurance exchanges that are operated by individual states as well as expansion of Medicaid to all Americans living below certain income levels.

The decision to utilize state-based exchanges and Medicaid as vehicles for coverage expansion had a number of ramifications that are still being felt today. By design, health reform's success would rely on the authority and activity of individual states to a degree that was unprecedented in decades. Up until that point, the state–federal partnership to operate Medicaid was their primary collaboration on health care, and both parties would admit that it was sometimes strained by tensions over funding and oversight. But passage of the Affordable Care Act (ACA) meant that federal policymakers would be dependent on the ability of states to realize its vision. This offered states greater leverage in negotiations around future initiatives, a dynamic that was only reinforced by the Supreme Court's later decision to interpret Medicaid expansion as a voluntary program where states would have to opt-in to participate.

On March 23, 2010, President Obama signed the most comprehensive health reform bill in recent history surrounded by dozens of Democratic lawmakers. The legislative debate over the law had taken twice as long as expected and faced near-certain death at multiple points, and the attempt to bring Republican support to the initiative ultimately failed. There were no Republican lawmakers standing behind the President at the signing ceremony, and "Obamacare," as the law is now known, survived multiple attempts to fatally wound the law at the Supreme Court level. It was the beginning of what would become a promising but unpredictable pathway to transformation that has yet to reveal its ultimate outcome.

## STATE GOVERNORS EMERGE AS POWERFUL FORCES FOR REFORM UNDER THE ACA

While the jury is still out on the long-term success of the ACA's high-profile initiatives—public insurances exchanges, Medicaid expansion, and delivery reform—there is no doubt that it changed the health care landscape almost immediately. The law sparked a period of innovation and market investment that continues today, and state governors quickly emerged as one of the stakeholder groups most eager to build upon this momentum. Armed with newfound bargaining power and a serious need to address their own health system challenges, governors began developing their own models for reform in the hopes that this new wave of innovation would allow them to utilize greater federal flexibility and funding to help achieve their goals.

The politics surrounding the ACA drove a number of policy decisions that were intended to mitigate the concerns of lawmakers and the public, including fears that the effort represented the federal government's takeover of health care payment and delivery. As a result, the law included design features aimed at preserving the authority and autonomy of stakeholders beyond federal policymakers, such as state governors. While the law did include a strong role for federal regulators in the newly created insurance markets, it also offered the states the opportunity to build and operate their own exchanges within certain guidelines. The fact that so few states exercised this option and that 27 states ended up using the platform facilitated by the federal government was, in fact, a surprise to many of those involved in developing the law. But while governors ultimately shied away from the overwhelming demands of standing up new exchange markets, they did quickly come to recognize the possible benefits presented by the momentum of the national reform initiative.

Prior to the ACA, the perception among many federal policymakers was that states were expected to participate as a partner in the Medicaid program but were less instrumental in charting the path on broader health care coverage and cost issues. Less appreciated was the role states play in impacting care delivery and market incentives through their regulatory authorities, like those related to commercial insurance or provider licensing. The ACA arguably turned that notion on its head, affording states authority over the operation of the newly created insurance exchanges and expanding coverage to middle class families through state-operated Medicaid programs. The law's creation of the Center for Medicare and Medicaid Innovation (CMMI) also offered new options for granting flexibility as part of reform, giving governors a robust platform to pursue creative solutions that were previously discouraged by the federal regulatory framework.

Although not anticipated, the Supreme Court's decision in National Federation of Independent Business v. Sebelius further empowered governors by shifting states' ACA-mandated participation in expansion of the Medicaid program to a voluntary option (132 S. Ct. 2566, 2012). In that case, the Court reasoned that the penalties imposed in the ACA for states that did not comply with expansion of the Medicaid program were so coercive that they failed to offer states a genuine choice in the receipt of federal funds. As such, the decision interpreted expansion as a voluntary rather than mandatory program, essentially giving states a new seat at the federal negotiating table overnight. This development, combined with the other program flexibilities and authority granted to states in the ACA, set the stage for a new dynamic in the relationship between state and federal policymakers at a time when governors needed it most.

Just as the national health reform effort in 2009 was driven by concerns about cost and access in the American health care system, states were increasingly forced to address these issues in their own budgets and communities. In fact, their pressure to control spending growth was even more acute at the state level because while the federal government can spend at a deficit to meet program needs, state budgets are limited by requirements to spend only what they accrue in revenue on an annual basis. As such, governors have faced tremendously difficult policy trade-offs as they try to keep pace with the rapidly growing cost of Medicaid and other state health programs. This tension was further exacerbated by the economic downturn in 2008, forcing state leaders to consider unprecedented program and provider rate cuts in Medicaid as well as cutbacks in other policy areas like education, transportation, and infrastructure. Some states, however, seized the opportunity presented by the aligning forces of financial crisis, increased federal flexibility and surging momentum for change.

The last 5 years have been one of the most active periods of state reform and innovation seen since the creation of the Medicaid program in 1965. In many states, the appeal of a generous federal investment in Medicaid expansion and looming financial pressure brought many stakeholders to the negotiating table to consider payment and delivery approaches that otherwise would have been political impossibilities. Moreover, the interest in care innovation and rapid rollout of transformation programs at the national level led many to conclude that failing to engage in more progressive models would ultimately leave them behind in rapidly changing health care markets. And so a wide range of new state-driven reform models are beginning to emerge—from the inpatient all-payer model in Maryland to the statewide transformation plan in New York to the episodic payment initiative in Arkansas. But in 2011, one model in particular was gaining the attention of policymakers across the country, and the comprehensive coordinated care model being built in Oregon quickly emerged as a leader among its peers.

## OREGON HARNESSES OPPORTUNITY TO ADOPT AN AMBITIOUS PAYMENT AND DELIVERY TRANSFORMATION PLAN

While 2010 is often remembered as the year of national health reform, it would be a mistake to overlook an equally ambitious and noteworthy effort launched that same year in Oregon by Governor John Kitzhaber. The state is no stranger to novel efforts to offer health care coverage to those in need—previous coverage expansion initiatives included the development of a prioritized treatment list for covered health conditions and a lottery to determine who would receive insurance when the state had limited resources to meet demand (http://www.oregon.gov/oha/HPA/CSI-HERC/Documents/Brief-History-Health-Services-Prioritization-Oregon.pdf). And as a physician in his third term as governor, Kitzhaber was uniquely qualified to lead a discussion on comprehensive health reform. He was popular among constituents, experienced as an emergency care physician and voiced an unwavering commitment to getting a bipartisan initiative across the finish line.

Perhaps most importantly, Oregon was also facing a serious state crisis with an estimated funding shortfall of $3.5 billion for its 2009–11 budgets. With over 16% of the state's general fund committed to health care, it was clear that closing the gap would require tough choices within state health programs, and stakeholders across the state were bracing for inevitable rate and service reductions. But rather than proceeding down the unattractive path of traditional cost containment, Oregon's policymakers began to unite behind a plan to address the state's needs through an innovative reform plan that would realign the state's payment and delivery system and push accountability for quality, access, and spending to local communities. And with the support of an early infusion of federal support through Medicaid expansion and a new waiver agreement, Oregon was able to avoid the negative impact of its short-term budget crisis while gaining stakeholder commitments to address market issues through longer term initiatives.

While the Oregon process for reaching agreement around a reform plan was no less challenging than the drama seen at the federal level, the outcome was more promising. Just 2 years after the enactment of the ACA, Governor Kitzhaber signed Oregon's own sweeping health reform law, surrounded by bipartisan supporters and stakeholders eager to see the state improve its health care system. Even the agreement itself is noteworthy, as the state reform envisioned in Oregon's plan went even further than the ambition of the ACA, seeking to guarantee significant reductions in health care spending as well as fundamentally transform the delivery of care in communities across the state. Oregon's health transformation plan, now in its fifth year of operation, is one of the most closely watched developments for federal and state policymakers who have a genuine interest in discovering the elements for success in correcting a health care market that is failing to meet expectations in terms of quality and value.

There is no doubt that the Oregon model has significantly impacted the national reform discussion, even considering that it is too early to fully understand the effort's long-term outcome. Initial quality and cost metrics indicate that the model is driving the state's health care system in the right direction—controlling cost growth at a reasonable level and demonstrating incremental improvements in selected quality and access measures (Oregon Health Authority, 2016). But some experts remain skeptical that these trends will continue, and local observers report that transformation has come more slowly than they initially hoped it would.

In some ways, however, the long-term outcomes may be less relevant in understanding why the effort is so meaningful now to broader reform efforts and the models being developed by Oregon's state peers. This is because the effort's value lies not only in its measured outcomes, but also in its ability to significantly change the policy community's view of what is politically and operationally possible in state-driven transformation. The Oregon model included many elements that were first of their kind in one comprehensive package and, as such, established a new paradigm for what federal and state partners could pursue in their efforts.

In order to appreciate how Oregon's initiative has shaped the outlook for state-driven market reform, it is necessary to understand how it came to pass as well as the fundamental underpinnings of its operation. The plan's use of an innovative federal waiver framework, its uniquely comprehensive approach to its payment and delivery models, and its focus on local ownership and autonomy all offer lessons in new approaches to reform that could yield promising results as other policymakers and stakeholders continue to invest in health care transformation.

The first unique aspect of Oregon's reform initiative is found in the legal foundation it negotiated with the Centers for Medicare and Medicaid Services (CMS). As a state–federal partnership program, Medicaid program design is largely regulated by federal requirements but there are available mechanisms that a state can use to customize its individual program by developing a formal agreement with the federal government to waive those requirements. There are several different waivers that states may pursue to obtain greater program flexibility, and complex reform proposals will typically utilize multiple waiver authorities to build a broad blueprint that accommodates a wide range of policy needs. Over the years, states have been creative in developing proposals under the Medicaid waiver framework, and while CMS has shown a willingness to accommodate those proposals, there remain barriers to innovation due to mandatory waiver conditions, such as budget neutrality.

Under the budget neutrality requirement, CMS is only allowed to grant waiver requests if the state can demonstrate that its alternative program approach will cost the federal government the same amount of money as the traditional approach. Typically, this is one of the most challenging aspects of a waiver negotiation because in calculating the anticipated financial impact of a proposal, both the state and CMS are making market assumptions in their respective modeling. In many cases, those assumptions are different and thus result in different financial projections that can derail an otherwise reasonable waiver request.

Oregon has historically been creative in its use of the Medicaid waiver process as it sought expanded coverage for populations that would not typically be eligible for Medicaid coverage. But its strategy for constructing the waiver proposal for its 2012 reform initiative was ambitious even compared to those previous efforts, pushing the boundaries of CMS' previous budget neutrality assumptions and other widely understood program norms.

In most waiver negotiations, CMS reviews the immediate services that it anticipates a state will provide under a waiver and then estimates what the cost of those services will be relative to the traditional program approach on an annual basis. If the alternative approach costs more, the waiver would fail to meet the budget neutrality requirement. This approach, however, posed significant challenges as state governors looked to tackle large-scale reform in an economic downturn. They simply did not have the capital needed to shore up their current program spending needs and finance the infrastructure needed to support widespread delivery innovation.

In order to combat this problem, Oregon pitched a different way of thinking about budget neutrality in the discussion of its reform plan waiver proposal. Understanding that it could not meet the traditional test in an environment where the upfront costs of reform would inevitably exceed immediate savings, Oregon proposed that CMS looks at the waiver request in terms of its return on investment. Utilizing this approach, the state proposed that CMS invests a substantial amount of money at the beginning of the initiative in exchange for the state's guarantee that it would reduce federal spending enough across the life of the waiver term that it would cover the cost of that upfront investment. If the state failed to reduce spending growth at a level that generated the expected return, it would have to repay CMS out of its general funds.

As discussed in earlier chapters, the waiver negotiation between Oregon and CMS was at times contentious with each party trying to protect the essential conditions it needed while sharing a desire to develop innovative solutions to traditional program barriers. The end result, however, was a progressive waiver agreement that set an important new precedent for calculating budget neutrality. The federal government agreed to provide Oregon with $1.9 billion for its use in transformation efforts at the start of the 5-year waiver agreement and, in exchange, the state guaranteed that it would reduce the expected growth in Medicaid spending by 2% over the life of the waiver (US Centers for Medicare and Medicaid Services, 2012). This structure represented a more expansive understanding of the types of arrangements a state could pursue to satisfy this condition, which was quickly adopted by several other states seeking to realign the fundamental payment and delivery models in their Medicaid programs and beyond.

The ability to use an early federal investment to fund reforms that promised longer term savings changed the dynamics of state-driven health reform for several reasons. First, it offered a feasible financial structure for large-scale changes within the constraints of state budgets that must be balanced on annual basis. Second, it offered a new incentive for market stakeholders to join the negotiation table and ensure the success of enacted reforms. The immediate federal funding could help avoid program spending cuts that would negatively impact providers and patients, but the long-term savings guarantee meant that they had to remain constructive partners in reform if they wanted to avoid repayment of unrealized savings. And finally, it signaled a willingness on the part of CMS to entertain new approaches to meeting waiver conditions. For many outside observers, the fact that CMS was able to think differently about something as fundamental as budget neutrality sparked energized discussion about what other types of flexibility would now be possible.

In addition to demonstrating flexibility on how the federal government could fund reform, the Oregon waiver agreement also challenged traditional limitations on the types of interventions that a state could purchase with those dollars. Under the traditional Medicaid program, federal funds can only be used for the delivery of medical services with a few very limited exceptions. Thus, if a state wanted to cover the cost of a nonclinical intervention that it believed would improve the health of a Medicaid beneficiary, it could only do so if it paid for that service with nonfederal funds. But many experts argue that such a narrow definition of health care is shortsighted and fails to recognize the environmental factors that may help or hinder a beneficiary's medical condition. In fact, Governor Kitzhaber became one of the most vocal supporters of this position, often citing the clinical benefits of purchasing an air conditioner for use in the home of a patient with chronic cardiopulmonary problems. If the Medicaid program would cover the cost of the environmental intervention for this patient, he argued, it could likely prevent much more expensive medical services like avoidable emergency room visits and hospitalizations.

Although there are a few previous occasions where the Medicaid program historically agreed to cover the costs of nonmedical services, the language that Oregon was able to negotiate with CMS in this regard represents another noteworthy precedent for purposes of health reform and beyond. Under this agreement, Oregon can utilize its Medicaid funding to purchase a broader range of goods and services that it deems to be beneficial to a patient's health. By allowing the state the flexibility to address beneficiary needs in nontraditional ways, such as offering the services of a community health worker to help a patient manage issues related to a disease or chronic condition, CMS opened up new opportunities for effective, low-cost interventions that were essential to Oregon's ability to meet its spending growth targets. Moreover, it finally moved the Medicaid program closer toward the much-touted goal of addressing true population health by allowing local communities to offer practical benefits that could prevent, as well as treat, expensive medical interventions.

Beyond the infrastructure of its Medicaid waiver design, Oregon's health plan also broke new ground in terms of its ambition and scope. First and foremost, it incorporated a financial model that empowered local communities with greater authority to make spending decisions but also held those communities to a higher level accountability by introducing a capitated payment model. Similar to the pay for performance and accountable care concepts that were beginning to emerge in the post-ACA discussion at the national level, Oregon's reform plan recognized the need to design its financial incentives to drive the program behaviors it desired at both the community and provider level. By creating regional Coordinated Care Organizations (CCOs) that would be allocated a finite amount of funding to serve the state's health plan beneficiaries, the state was implementing a degree of financial accountability that was unique among its peers. While many states were considering introducing elements of accountable payment models into their Medicaid programs, Oregon was pushing ahead with an ambitious methodology that held communities accountable to total cost of care thresholds almost immediately upon enactment of the new reform initiative. These regional organizations, known as CCOs, became the central point of accountability within Oregon's plan and currently take responsibility for the financial and clinical performance of their respective communities. Under the plan, the state pays CCOs a risk-adjusted global payment amount for each beneficiary in their region, minus a 4% withhold amount that is allocated to an incentives pool that rewards providers for quality improvement.

The introduction of greater financial risk into provider payment methods has been a topic of much discussion throughout the health care system. The Medicare program is moving more aggressively toward shifting financial accountability to providers through payment models like Accountable Care Organizations (ACOs) and bundled payments, and commercial insurance plans are following closely behind as they introduce similar concepts into the individual and employer-sponsored coverage markets. Likewise, other states are adopting payment models that require providers to focus on more efficient care delivery, such as Arkansas' shift toward episodic payment for a growing list of medical services. But even among these other efforts, Oregon's CCO payment model continues to stand apart as one of the most comprehensive, pushing full financial risk down to the community level and requiring the CCOs to distribute that risk further through the use of alternative payment models.

Given the lingering questions around the impact of these financial risk models on care quality and access at the national and local level, there is no doubt that all eyes will remain focused on Oregon's experience as its reform effort moves forward. The outcomes that the state documents—both positive and negative—are inevitably influencing the decisions of other policymakers at the state and national levels as they calculate how much financial risk they can responsibly introduce into their health care programs and at what pace such changes should be implemented.

Another noteworthy financial feature of the Oregon model is its commitment to "braiding" the payment streams from previously discrete insurance programs. America's health care market and delivery system has long been challenged by its siloed nature, with multiple funding mechanisms supporting, but not coordinating, our clinical infrastructure and frontline delivery of care. Historically, this created a system where duplication and inefficiency prevail as stakeholders juggle misaligned incentives and disparate streams of communication. As a result, the American health care "system" is hardly a system at all, but rather a patchwork of programs with little ability to drive consistent quality or cost effectiveness.

At the outset, one of the primary goals of Oregon's reform initiative was to integrate the financial as well as the clinical aspects of its delivery system. The model incorporated a platform that pulled together the payment mechanisms from as many programs as possible so that local communities could focus on performance incentives across their market, rather than constantly trying to reconcile benefits and reimbursement from different payers. The focus in aligning multiple funding mechanisms not only provides simplification within the health care system but also allows communities and providers to implement delivery initiatives at scale, which in turn offers an opportunity to improve efficiency and communication. It also sets clear performance pathways so that stakeholders respond to incentives that are consistent across patient populations and remove the constant shift of dollars from one program to another as providers are forced to do in the traditional fee-for-services environment.

In fact, one of the unique features of this most recent wave of state innovation is the comprehensive nature of the efforts, with a number of states moving beyond the boundaries of "Medicaid reform." States have historically tried to influence their local health care markets by leveraging their Medicaid authority, similar to the federal government's effort to use its Medicare purchasing power to shape the national payment and delivery system. While this approach can meet discrete objectives, it frequently yields unintended consequences like cost shifting and patient steering. Thus, both state and federal policymakers now recognize the need to develop reform initiatives with a much broader lens, creating alignment between payers and programs to the greatest extent possible. Oregon's transformation plan is one of the most ambitious in its scope, incorporating not just its Medicaid population but expanding to include state employees, potentially teachers unions and even its prison population. And while the breadth of the effort may have increased its political complexity at the outset, it offers the state the ability to drive incentives across a much greater portion of the market and increases the likelihood that the new payment model can fundamentally change behavior in a positive way.

The importance of the multipayer alignment in payment and delivery reform simply cannot be overstated, and it is an objective where current efforts have been less successful than many had hoped. For instance, CMS initiated a number of efforts specifically targeted at improving alignment across payers, including harmonizing quality measurement, coordinating benefit design or utilizing consistent payment methodologies between programs. But it is fair to say that those efforts have been more difficult and progressed more slowly than originally anticipated, and policy experts are continually working to identify better approaches to finding such alignment.

In this respect, the Oregon model could prove a highly valuable experience because it not only aims to enhance consistency across payers but also aggressively combines the payment and delivery systems of significant programs from the outset of the effort. And, perhaps most importantly, Oregon is moving forward with its intention to expand to the Medicare population in it's next phase of implementation. Indeed, all eyes will remain on the state as it negotiates with CMS for its new waiver, which is slated to begin in 2018. If that agreement allows the Oregon CCO model to cover beneficiaries across Medicare, Medicaid, and the state employee health program, it could represent the most comprehensive state transformation initiative undertaken in decades.

Another area where the Oregon initiative included a comprehensive approach was in the scope of the included benefits themselves. A number of the state's peers are pursuing large-scale delivery and payment models that represent vastly different approaches to care. But most of those models are approaching care transformation on a benefit by benefit basis—often excluding expanded service categories like behavioral health and dental care. While there is still much merit in those models, Oregon's comprehensive plan to integrate traditional medical care with mental health and dental care has the potential to demonstrate a greater range of lessons and outcomes. Few would argue that the current approach of separating those services is the best model for the system or the patient. But incorporating all of those elements in a payment reform model requires disruption of multiple care delivery traditions and introduces additional complications in terms of stakeholder politics, which can be significant barriers to enacting reform.

The Oregon model is especially ambitious because it includes a more comprehensive "bundle" of services that fall under the risk-based payment model. This commitment stemmed from a recognition that the greatest potential for success—both in terms of clinical improvement and cost efficiency—could be found in the ability to coordinate care across as many factors impacting a patient's health as possible. And anecdotal evidence, as well as early program results, indicates that this approach is indeed generating results. For instance, one might attribute a portion of the reduction in emergency room visits to the fact that fewer patients were being seen for dental-related pain. Rather, those patients were now being seen in a dentist's office for routine procedures instead of seeking care in a more expensive acute care environment.

In fairness, much of the care delivery integration in Oregon is still under development and has yet to occur because of the complexity of the task. While many communities are pursing models to connect patients more effectively to interventions like behavioral health services, most are struggling to fully integrate those services into the primary and acute care environment. On the other hand, efforts to engage patients in an environment where such coordination is possible are showing promising results, with an 86% increase in medical home enrollment among health plan beneficiaries. Longer term experience is needed to assess whether Oregon's more comprehensive model will yield its intended results, but simply pursuing this approach, the state may be raising the bar for what is achievable under both state and federal transformation initiatives.

Last but not least, the Oregon health reform plan is clearly influencing other program models across the country even in terms of its basic structure. One principle that is evident throughout the plan is a dedicated focus to building transformation from the ground up. Clearly designed around the concept that change must be driven at the community and regional level, Oregon's initiative places resources and accountability in the hands of local stakeholders. Even the basic premise of developing reform through regional hubs of accountability—the CCOs—signaled that this was an effort that would rise and fall on the efforts of individual communities rather than predetermined visions of centralized state authorities. This regional approach is now seen in significant state reform efforts throughout the country—such as Alabama, North Carolina, and New York—and represents a break with more traditional notions that statewide implementation was the most effective means of reform.

The value of local leadership and autonomy is seen in Oregon's inclusion of community stakeholders not just in development of its reform plan but in its ongoing operation. Strong engagement at the community level was apparent from the outset of the reform effort in 2010 when Governor Kitzhaber convened a series of committees and workgroups comprised of local stakeholders to inform the development of the Oregon Action Plan for Health. Local leaders also made up the majority of the 45 members of the Health System Transformation Team that was responsible for developing the specific proposal for the CCO-centered reform approach. And throughout the plan development process, state officials actively incorporated the work and relationships already underway in communities across the state. In fact, the flexibility offered to communities in organizing their local CCO—whether it is provider-led, plan-led, or a collaborative of community organizations—is a direct result of the state's effort to embrace rather than disrupt the approaches that were already working at the local and regional level.

In addition to the role local stakeholders played in the planning process, many of those same leaders have now taken ownership for the performance of their regions even beyond the formal CCO activities. The Oregon plan was explicit in its desire to incorporate community input on an ongoing basis through formal organization requirements as well as stakeholder engagement in activities like quality measurement and collaborative learning. For example, each CCO is required to solicit input from its Community Advisory Council comprised of local representatives from a wide range of interested communities, including patients, providers, local government officials, and other health-related organizations. This process ensures that decision making is transparent and reflects the concerns and advice of the broad community beyond just the entities that formed the CCO.

Likewise, many CCOs are utilizing feedback from Clinical Advisory Panels that develop recommendations on care processes and other needs with the help of local providers and health professionals. This focus on developing relationships and care protocols based on local needs and practices, rather than following generalized state guidelines and prescriptive models, has been embraced to a greater degree in Oregon than in most other reform initiatives. And by many accounts, much of the early success seen in the outcomes of the Oregon model is attributable to this community-based approach. By delegating ownership and accountability to local leaders, the state is seeking to drive changes that will more effectively address the circumstances and needs of patients in diverse regions.

It is also noteworthy that even in circumstances where the state is seeking to enable transformation at scale, much of its effort is designed to identify successful local practices and translate them statewide. Similar to the creation of CMMI in the ACA, Oregon also included a Transformation Center as part of its reform initiative. Initially created through funding from a CMMI Statewide Innovation Model grant, the Oregon Transformation Center's primary task is to assist communities in successfully building the care processes and programs that allow them to successfully transition to a more accountable environment. But here again, the approach of the Center's team is to support collaboration between regions so that they can learn from one another's experiences as well as identify best practices at the local level that can then be taught at scale to other communities.

At the outset, it was clear that the Center was not going to operate as a central regulator dictating general practices across the state but rather serving as a facilitator to help steer local leaders through issues that arise as they transform their systems. And by most accounts, the support that the Transformation Center has offered thus far has been a lifeline for many of those providers and leaders seeking to improve care on the frontlines. Community organizations can use supplemental grants and expertise to build new programs and processes to address unique needs, and they can lean on the Center's staff, many of whom are clinically trained, to guide them through regulatory challenges and other barriers that emerge as they pursue delivery changes.

While much of the health care transformation discussion across the country has focused on the concept of community-based care, there are relatively few programs that have demonstrated that they have mastered this approach. One of the most challenging aspects of changing the care delivery system to focus more on the whole health and wellness of a population is bringing together a diverse set of service providers who have not historically worked together as a team. Aside from building effective communication between medical care providers, which can be challenging on its own, communities now have to build an infrastructure to support collaboration across an even broader spectrum of patient needs—from health care to housing to transportation to nutrition and beyond. And while the Oregon model provides one of the most robust platforms to facilitate coordination across local resources, many regions are still working to find the right solutions to effectively support this collaboration. The state's ability to keep local stakeholders engaged and committed to this process will be key to the longer term success of its transformation plan, and experts across the country will be watching to determine whether the multipronged approach included in the CCO model is able to bridge the gaps that currently plague other efforts to realize the goals of "population health."

Most experts will agree that the Oregon transformation plan incorporates many of the key elements that the health care community identifies as important to expanding access, improving quality, and controlling cost growth. But implementation of the model is in its early stages, as are many of the delivery and payment reforms seen in other parts of the country, so it is difficult to draw concrete conclusions about its success. The first 3 years of outcomes data released by the state are certainly promising—15 out of 16 CCOs met their performance targets on at least 12 of the 17 quality measures included in their payment incentive program, and performance is consistently improving year to year on most of the over 50 measures tracked within the program. In addition, 95% of the state's population has insurance coverage, and each one of the 16 CCOs is meeting their financial stability requirements under the capitated payment model. All of these are positive indicators that the initiative, while ambitious, is on the right pathway to meet its goal of better health, better care, and lower costs. Indeed, even having the ability to track the state's performance in such a comprehensive fashion is yet another unique feature of the model, which includes a level of transparency and detail on access and outcomes measurement that is atypical of other ongoing payment and delivery reform programs.

Although many indicators in Oregon are trending positive, there are still a number of outstanding questions that should be monitored before rendering a verdict in on its long-term impact. Expert observers will be watching as the model evolves and communities move beyond the "low hanging fruit" of early interventions. Can the CCOs maintain the current rates of performance improvement, and will these communities be able to maintain financial stability in the wake of unexpected challenges or developments? How will CCO and community relationships develop as current leadership transitions and new interests emerge? Will the model expand to include other large populations, such as Medicare beneficiaries, and how will such expansion impact its overall operation? These are just a few of the issues that will arise over the next few years that are critically important to understanding the strengths and weaknesses of its various approaches. Regardless of these longer term questions, there is no question that the effort is valuable to the broader transformation discussion in its current form. By pursuing such a comprehensive and truly disruptive solution to its health care challenges, Oregon is raising the bar for other states and federal programs developing their own pathways for payment and delivery transformation.

## UNDERSTANDING THE OUTLOOK FOR FUTURE REFORM

Seven years into the ACA and 5 years into the Oregon transformation plan, both efforts are showing signs of success as well as identifying opportunities for improvement. Coverage expansion under the ACA is providing over 18 million individuals with health insurance, and the Medicare and Medicaid programs have launched dozens of payment and delivery reform efforts across the country. But implementation of the ACA has not been seamless and even Democratic leaders agree that the law could be improved to address a number of issues, especially those related to the affordability of and availability of plans on the public exchange market. With the election of Donald Trump and Republican majorities in both the House and Senate in November 2016, the question moving forward is not simply what about the ACA should be fixed but rather whether the major components of the ACA will survive into the new Administration. A firm pledge to immediately repeal major portions of the health reform law was one of the primary promises made by Republicans during the campaign season, and the incoming leadership has also indicated a desire to pursue significant reforms in both the Medicare and Medicaid programs more broadly. This dynamic will inevitably result in more change within the health care system and poses new opportunities as well as risks for state innovation efforts currently underway.

While the fate of public exchanges has largely dominated the public dialogue about the ACA, equally important is the fate of the many payment and delivery reform programs that have emerged under the umbrella goals of transformation. Many of these initiatives are programs developed under the pilot authority and budget of CMMI, although others are fully integrated into the traditional Medicare and Medicaid programs. Alternative payment models like ACOs, bundled payments and risk-based payment methodologies are rapidly emerging in the public program landscape, and providers and other health care stakeholders are in the midst of significant changes as they adapt to the incentives of these new programs. But these initiatives are still in their infancy and there is much debate on whether they can and will achieve the articulated goals of quality improvement and cost containment. As such, the next 3–5 years will be critically important in evaluating their impact and adjusting approaches as needed.

The next Administration and Congress will be tasked with the very important decision of whether to continue the current structure and investment in accountable care models. During his campaign, President-elect Trump voiced support for the general concepts of value and transparency but offered few details on his health care policy priorities. And a number of the existing CMMI models are the subject of criticism from congressional Republicans and some health policy experts in terms of their ability to rapidly reduce program spending and deliver meaningful quality improvement. To date, these criticisms are somewhat speculative because much of the activity in accountable payment has been focused on developing models and introducing them to providers and patients. The time is coming, however, when policymakers will need to evaluate whether those efforts are meeting their goals and if they represent effective approaches to creating value within our health care system. New leaders in Congress and the Administration will be responsible for defining how we judge the performance of various "value-based" care models in addition to whether they are successful in fulfilling those criteria.

Much of the health policy discussion in recent years has occurred in the context of partisan disagreement, with Republicans unwilling to consider improvements to the ACA and Democrats fiercely defending their policy choices across the board. But outside of the politics of the ACA, there has actually been a significant amount of bipartisan agreement in terms of the broad vision for America's future health care system. The passage of the Medicare Access and CHIP Reauthorization Act of 2015 (MACRA) included a new payment methodology for physician services within Medicare Part B that moves aggressively toward holding providers accountable for their clinical and financial performance and was passed with support from an overwhelming majority of lawmakers on both sides of the aisle. And while Republicans have previously expressed negative views of CMMI specifically, they frequently endorse concepts like risk-based payment, quality measurement, and pay for performance. Thus, while policymakers may debate the value of individual programs, there will likely remain a clear consensus that the general direction of these efforts should be focused on increased provider and plan accountability for access, quality, and cost across the health care system.

The significant investment in recent years in state-based transformation will also become ripe for evaluation and adjustment in the next few years. While Republicans and Democrats both voice support for the ability of states to lead innovation, their respective opinions on the broader role and structure of the Medicaid program do not share such agreement. In fact, both the Trump Administration and the Republican majority in Congress have put forward proposals to significantly alter the fundamental financing structure of the Medicaid program, transitioning the current "matching" methodology to a per capita capitation model that would limit the amount of federal funding states could draw down for their program based upon the number of people enrolled in their program. While these proposals also allow states additional flexibility in their program design and coverage, the funding reductions projected as a result of the changes rise to a level that has caused concern for governors in both political parties. In fact, the Congressional Budget Office estimates that the per capita capitation methodology included in the American Health Care Act of 2017, the ACA repeal bill passed by House Republicans in April, would reduce federal spending for the Medicaid program by over $800 billion in the next decade (Congressional Budget Office, 2017). A change of this magnitude, combined with the accompanying rollback of the ACA's expanded coverage option and the spillover effect on other health care programs, could most certainly pose a chilling effect for some states considering innovative reforms across their health care markets. While efforts to provide regulatory flexibility at the administrative level could help mitigate some impacts of the funding reductions, it remains to be seen whether that flexibility could remedy harmful effects on state programs and their beneficiaries. As such, policymakers will need to consider whether federal reforms could have the unintended effect of reversing the reform gains already realized by states and introducing a barrier to future gains (Congressional Budget Office, 2017).

It goes without saying that the state innovation initiatives in operation today hold tremendous promise but their long-term success cannot yet be quantified. Their success, and the willingness of governors to continue these efforts, will carry great weight in the coming federal debate on the future of health care reform. Many of these programs will not only be evaluated but also need to be reauthorized or expanded under the terms of their current agreements with CMS. For instance, Oregon's transformative waiver will need to be renewed in 2017, and Maryland's all-payer model will also need to be renewed in 2018. Both models offer opportunity for significant expansion to new populations and service categories if their perceived outcomes warrant such action. Moreover, the Section 1332 of the ACA could offer states even greater tools for exploring new approaches to coverage and delivery beginning in 2018 if it remains intact in light of the anticipated ACA repeal effort. While these new authorities could offer greater opportunities to innovate, they must be paired with resources that are adequate to meet the demands of such innovation, and reforms that significantly reduce funding for state programs could pose barriers to future efforts. Thus, it is hardly an overstatement to assume that the ongoing debate over Medicaid and the ACA could impact patients for many years to come.

Given the tremendous amount of state innovation activity over the past 5 years and the bipartisan support that activity initially received at the federal level, the potential for additional transformation efforts is great in the new Congress and Administration. But realizing that potential depends heavily upon the yet to be seen results of the ambitious initiatives blazing the trail for future state-based efforts. Likewise, a continued commitment to shared responsibility between federal and state governments is critically important to the sustainability of existing efforts as well as new ones. The best evidence for success is already being developed in thoughtful state models across the country, and all eyes will remain on ambitious state leaders like Oregon as the health care community plots the next steps on the pathway to progress on reform.

## REFERENCES

Congressional Budget Office (2017). H.R. 1628, American Health Care Act of 2017: Cost Estimate (May 24, 2017). Available from: https://www.cbo.gov/publication/52752

Davis, K., Schoen, C., Schoenbaum, S. C., et al. (2007). *Mirror mirror on the wall: An international update on the comparative performance of American Health Care*. New York: The Commonwealth Fund.

DeNevas-Wait, C. (2009). *Income, poverty and health insurance coverage in the United States: 2008*. Available from: http://www.census.gov/content/dam/Census/library/publications/2009/demo/p60-236.pdf

Himmelstein, D. U., et al. (2009). Medical bankruptcy in the United States 2007: The results of a national study. *The American Journal of Medicine*, *122*(8), 741–746.

Komissar, H. (2013). *The effect of rising health care costs on middle-class economic security*. Washington, DC: AARP Public Policy Institute.

Oregon Health Authority. (2016). *Oregon's Health System Transformation: CCO Metrics 2015 Final Report*. Available from: http://www.oregon.gov/oha/HPA/ANALYTICS-MTX/Documents/2015_Performance_Report.pdf

Schoen, C. (2016). *The Affordable Care Act and the U.S. economy: A five year perspective*. New York: The Commonwealth Fund.

The Henry J. Kaiser Family Foundation. (2011). *Snapshots: Health care spending in the United States and selected OECD countries*. Available from: http://kff.org/health-costs/issue-brief/snapshots-health-care-spending-in-the-united-states-selected-oecd-countries/

US Bureau of Labor Statistics. (2012). *Spotlight: The recession of 2007–2009*. Available from: https://www.bls.gov/spotlight/2012/recession/pdf/recession_bls_spotlight.pdf

US Centers for Medicare and Medicaid Services. (2012). *Oregon Health Plan Waiver 21-W-00013/10 and 11-W-00160/10*. Available from: http://www.oregon.gov/oha/HPA/HP-Medicaid-1115-Waiver/Pages/2012-2017-Demonstration.aspx

# Index

*Note*: Page numbers followed by "*f*," "*t*," and "*b*" refer to figures, tables, and boxes, respectively.

## I



## J



## K



## L

## M



## N



## O

# P

# Q

## R



## S



## T

Printed in the United States
By Bookmasters